Applied Swarm Intelligence

Editor
Yaniv Altshuler
Massachusetts Institute of Technology
Cambridge, Massachusetts, USA

CRC Press
Taylor & Francis Group
Boca Raton London New York

CRC Press is an imprint of the
Taylor & Francis Group, an **informa** business
A SCIENCE PUBLISHERS BOOK

Cover credit: **By Yaniv Altshuler**

First edition published 2025
by CRC Press
2385 NW Executive Center Drive, Suite 320, Boca Raton FL 33431

and by CRC Press
4 Park Square, Milton Park, Abingdon, Oxon, OX14 4RN

Library of Congress Cataloging-in-Publication Data (applied for)

ISBN: 978-0-367-22687-9 (hbk)
ISBN: 978-1-032-93224-8 (pbk)
ISBN: 978-0-429-27637-8 (ebk)

DOI: 10.1201/9780429276378

Typeset in Times New Roman
by Prime Publishing Services

Preface

In the ever-evolving field of Artificial Intelligence, swarm intelligence stands as a pivotal paradigm that emphasizes the power of collective behavior and decentralized decision-making. This book, "Applied Swarm Intelligence," aims to bridge the gap between theoretical research and real-world applications, offering a comprehensive exploration of how swarm-based approaches can address complex and dynamic problems across various domains.

The inception of this work was driven by a fascination with the natural world—how simple creatures, following simple rules, can achieve complex and adaptive behaviors that far exceed the capabilities of any single individual. This book endeavors to translate these natural phenomena into technological advancements that can enhance human capabilities and efficiencies in unprecedented ways.

The scope of this book spans several cutting-edge applications, from optimizing ride-sharing dynamics to deploying autonomous drone swarms for surveillance and infrastructure defence. Each chapter not only delves into the current state of research but also pushes the boundaries by presenting new findings and methodologies that advance the field.

The target audience for this work includes academic researchers, industry professionals, and enthusiasts in the fields of computer science, robotics, and artificial intelligence. It is structured to provide both depth and breadth in understanding swarm intelligence, making it accessible to newcomers while still offering valuable insights for seasoned experts.

This book is more than just an academic text; it is a call to rethink how we approach problem-solving and innovation in the age of AI. It invites you, the reader, to explore how the principles of swarm intelligence can be applied to your own work, potentially transforming challenges into opportunities.

Enjoy your journey through the intriguing world of AI and swarm intelligence.

Yaniv Altshuler
USA

Contents

Chapter 1

Recent Developments in the Theory and Applicability of Swarm Research

Yaniv Altshuler
Massachusetts Institute of Technology, yanival@mit.edu

1.1 Overview

Swarm Intelligence (SI) is a collective behavior exhibited by groups of simple agents, such as ants, bees, and birds, which can achieve complex tasks that would be difficult or impossible for a single individual. The collective behavior of these organisms is characterized by decentralized decision-making, self-organization, adaptive responses to environmental changes, and emergent properties that are not present in individual organisms. SI algorithms emulate these features to solve complex optimization, control, classification, clustering, routing, and prediction problems in diverse domains, such as engineering, robotics, biology, economics, social sciences, and humanities [64].

SI algorithms can be classified into two main categories: swarm-based algorithms and swarm-inspired algorithms [40]. Swarm-based algorithms involve the simulation of a population of individuals (agents) that interact with each other and their environment to achieve a collective goal. Examples of swarm-based algorithms include Ant Colony Optimization (ACO) [145], Particle Swarm Optimization (PSO) [69], Artificial Bee Colony (ABC) [92], and Firefly Algorithm (FA) [163]. Swarm-inspired algorithms, on the other hand, extract specific mechanisms or principles from natural swarms and incorporate them into conventional optimization or machine learning algorithms. Examples of swarm-inspired algorithms include Artificial Immune Systems (AIS) [59], Bacterial Foraging Optimization (BFO) [121], and Grey Wolf Optimizer (GWO) [113].

The success of SI algorithms is attributed to their ability to explore a large search space efficiently, converge to optimal or near-optimal solutions, and handle multiple objectives or constraints simultaneously. The collective intelligence of the swarm enables the sharing and exchange of information, the exploitation of promising regions, and the avoidance of suboptimal regions. Furthermore, the decentralized and distributed nature of the swarm allows for scalability, robustness, fault-tolerance, and adaptivity to dynamic or uncertain environments [29].

Despite their advantages, SI algorithms face several challenges and limitations, such as premature convergence, scalability issues, sensitivity to parameter settings, lack of theoretical guarantees, and difficulty in interpreting or explaining the obtained results. Researchers have proposed various approaches to overcome these challenges, such as hybridization with other optimization or machine learning techniques, dynamic adaptation of parameters, incorporation of domain knowledge, and rigorous analysis of convergence properties. The content of this introductory chapter is based in part on a work initially published in [20].

1.2 Applications

The advancement of technology has spurred a growing demand for multi-agent and swarm robotics solutions to address an ever-expanding range of complex and diverse challenges. With the emergence of distributed systems, it has become increasingly clear that relying solely on a single robot may not be the optimal approach for many application domains. Instead, teams of robots are being called upon to work in a coordinated and intelligent fashion, leveraging the power of redundancy to achieve greater efficiency and reliability.

The benefits of multi-agent systems stem from their ability to harness the collective intelligence of multiple entities, allowing them to tackle complex tasks that would be beyond the capability of a single robot. This approach provides the flexibility to scale up or down the number of robots based on the task at hand, while also providing redundancy to ensure mission success even in the face of individual robot failures. Moreover, multi-agent systems can leverage complementary skills and diverse perspectives, leading to improved problem-solving capabilities and more robust decision-making.

Swarm robotics takes the concept of multi-agent systems a step further by drawing inspiration from the collective behavior of natural swarms, such as ants, bees, and birds. Swarm robotics seeks to emulate the self-organizing and adaptive behavior of swarms in order to create distributed systems that can operate autonomously and efficiently. By leveraging simple local interactions between agents, swarm robotics can achieve complex global behaviors, such as exploration, foraging, or assembly, without the need for centralized control or explicit communication. The emergence of swarm robotics opens up exciting new possibilities for applications in fields such as search and rescue, environmental monitoring, and precision agriculture.

In [63] a detailed description of swarm-robotics application domains is presented, demonstrating how large-scale decentralized systems of autonomous robotic agents can be significantly more effective than a single robot in many areas. However, when designing such systems it should be noticed that simply increasing the number of robots assigned to a task does not necessarily improve the system's performance — multiple robots must intelligently cooperate to avoid disturbing each other's activity and achieve efficiency.

In nature, "simple minded" animals such as ants, bees or birds cooperate to achieve common goals and exhibit amazing feats of collaborative work. It seems that these animals are "programmed" to interact locally in such a way that the desired global behavior is likely to emerge even if some individuals of the colony die or fail to carry out their task for other reasons. A similar approach may be considered for coordinating a group of robots without a central supervisor, by using only local interactions between the robots. When this decentralized approach is used, much of the communication overhead (typical of centralized systems) is saved, the hardware of the robots can be fairly simple, and better modularity is achieved. A properly designed system should be readily scalable, achieving reliability through redundancy.

There are several key advantages to the use of such *intelligent swarm robotics*. First, such systems inherently enjoy the benefit of parallelism. In task-decomposable application domains, robot teams can accomplish a given task more quickly than a single robot, by dividing the task into sub-tasks and executing them concurrently. In certain cases, a single robot may simply be unable to accomplish the task on its own (e.g. to carry a large and heavy object).

Second, decentralized systems tend to be, by their very nature, more robust than centralized systems (or systems comprised of a single complex unit). Generally speaking, a team of robots may provide a more robust solution by introducing redundancy, and by eliminating any single point of failure. While considering the alternative use of a single sophisticated robot, we should note that even the most complex and reliable robot may suffer an unexpected malfunction, which will prevent it from completing its task. When using a multi agents system, on the other hand, even if a large number of the agents stop working for some reason, the entire group will often still be able to complete its task, although perhaps slower. For example, for exploring a hazardous region (such as a minefield or the surface of Mars), the benefit of redundancy and robustness offered by a multi agents system is quite obvious, and it is in this context that Rodney Brooks wrote his famous "Fast, cheap and out of control" report [45].

Another advantage of the decentralized swarm approach is the ability of dynamically reallocating sub-tasks between the swarm's units, thus adapting to unexpected changes in the environment. Furthermore, since the system is decentralized, it can respond relatively quickly to such changes, due to the benefit of locality, meaning — the ability to swiftly respond to changes without the need of notifying a hierarchical "chain of command". Note that as the swarm becomes larger, this advantage becomes increasingly important.

In addition to the ability to respond quickly to changes, the decentralized nature of such systems also improves their scalability. The scalability of multi-agent systems is derived from relying on the "emergence" of task completion by inherently low communication and computation overhead protocol implemented by the agents. As the tasks assigned nowadays to multi-agent based systems become increasingly complex, so does the importance of the high scalability of the systems.

Finally, by using heterogeneous swarms, even more efficient systems could be designed, thanks to the utilization of different types of agents whose physical properties enable them to perform more efficiently in certain special tasks.

Significant research effort has been invested during the last few years in design and simulation of multi-agent robotics and intelligent swarm systems (see e.g. [43,53,61,87,88,107,153,154]).

Such designs are often inspired by biology (see [74, 95] for evolutionary algorithms, [35] or [44, 164] for behavior based control models, [62,67,108,130,146] for flocking and dispersing models, [37,86,157] for predator-prey approaches and [56] for models inspired by the behavior of cats, by physics [54, 84, 94], sociology [97, 134, 149], network theory [23, 26, 31, 126] or by economic applications [6,30,32,33,80,103,112,118,133,158,168].

A swarm based robotics system can generally be defined as a highly decentralized group of extremely simple robotic agents, with limited communication, computation and sensing abilities, designed and deployed to accomplish various tasks. Tasks that have been of particular interest to researchers in recent years include synergetic mission planning [7, 73], emergency detection using decentralized sensing capabilities [25], patrolling [3, 4, 16], fault tolerance cooperation [98, 119, 160], network security [128], adversarial learning modeling [114], financial systems modeling [139], crowds modeling [11], swarm control [57, 109], human design of mission plans [105, 106], role assignment [50, 52, 116, 144, 167], multi-robot path planning [4, 76, 104, 135, 147, 161], traffic control [5, 8, 125], formation generation [34, 41, 79, 83], formation keeping [36, 38, 155], exploration and mapping [124, 129, 133], target tracking [85, 120], collaborative cleaning [10, 15, 19, 150], control architecture for autonomous drones [17, 18] and target search [89, 100]. In the realm of transportation and ride-sharing, the application of swarm intelligence offers innovative solutions for modeling and optimizing traffic flow and vehicle routing [137], mirroring the efficiency and adaptability seen in natural swarms to enhance the coordination and efficiency of shared transportation systems [8].

Unfortunately, the mathematical and geometrical theory of such multi-agents systems is far from being satisfactory, as pointed out in [39,42,70,115] and many other papers.

Our interest is focused on developing mathematical tools necessary for the design and analysis of such systems. For example, in [151] it was shown that a number of agents can arrange themselves equidistantly in a row via a sequence of linear adjustments, based on a simple "local" interaction. The convergence of the configuration to the desired one is exponentially fast. A different way of cooperation between agents, inspired by the behavior of ant-colonies, is described in [46]. There it was proved that a sequence of ants engaged in deterministic chain pursuit will find the shortest (i.e. straight) path from the ant-hill to the food source, using only local interactions. In [47] the behavior of a group of agents on $\mathbf{Z}^2$ is investigated, where each ant-like agent is pursuing his predecessor, according to a discrete biased-random-walk pursuit model on the integer grid. The average paths of such a sequence of agents engaged in a chain of probabilistic pursuit was shown to converge to the "straight line" between the origin and destination, and this too happens at an exponential speed.

An in-depth analysis of the effect of certain geometric properties on the search efficiency of a collaborative swarm of autonomous drones appears in [13, 14]. Meanwhile, an example of an analytic complexity bounds for this problem can be found in [21, 22]. A work that analyzed the effect of a stochastic framework for the same problem is presented in [127].

1.3 Decentralized Intelligence Architectures and The Swarm Paradigm

A key principle in the notion of swarms, or multi agent robotics, is the simplicity of the individual agent. The notion of "simplicity" here means that the agents should be *significantly simpler* than a "single sophisticated system", which can be constructed for the same purpose. As a result, the capabilities and the resources of such simple agents are assumed to be very limited, with respect to the following aspects:

- Memory resources — basic agents should be assumed to contain only $O(1)$ memory resources (i.e. the size of the memory is independent of the size of the problem or the number of agents). This usually imposes many interesting limitations on the agents. For example, agents can remember only a limited history of their activities so far. Thus, protocols designed for agents with such limited memory resources are usually very simple and attempt to solve a given problem by relying on some (necessarily local) basic patterns arising in the environment. The task is completed by a repetition of these patterns by a large number of agents.
- Sensing capabilities — defined according to the specific nature of the problem. For example, for agents moving along a 100×100 grid, a reasonable sensing radius may be 3 or 4, but certainly not 40.
- Computational resources — although agents are assumed to employ only limited computational resources, a formal definition of this constraint is hard to define. In general, most of the time-polynomial algorithms may be used, provided that the amount of memory the agents have is sufficient.
- Communication is very limited too — the issue of communication in multi agents systems has been extensively studied in recent years. Distinctions between implicit and explicit communication are usually made. Implicit communication occurs as a side effect of other actions, or "through the world" (see, for example [117]), whereas explicit communication is a specific act intended solely to convey information to other robots on the team. Explicit communication can be performed in several ways, such as a short range point to point communication, a global broadcast, or by using some sort of distributed shared memory. Such memory is often referred to as a *pheromone*, used to convey small amounts of information between the agents [74, 152, 165, 166]. This approach is inspired from the coordination and communication methods used by many social insects — studies on ants (e.g. [2,82]) show that the pheromone based search strategies used by ants in foraging for food in unknown terrains tend to be very efficient. Additional information can be found in the relevant NASA survey, focusing on "intelligent swarms" comprised of multiple "stupid satellites" [131, 132] or the following survey conducted by the US Naval Research Center [136]. The lack of explicit communication poses a challenge for various special configuration sets, such as symmetric environments [14].

In the spirit of designing a system which uses as simple agents as possible, we aspire that the agents will have as little communication capabilities as possible. With respect to the taxonomy of multi-agents discussed in [68], we would be interested in using agents of the types *COM-NONE* or if necessary of type *COM-NEAR* with respect to their communication distances, and of types *BAND-MOTION*, *BAND-LOW* or even *BAND-NONE* (if possible) with respect to their communication bandwidth. Therefore, although a certain amount of implicit communication can hardly be avoided (due to the simple fact that by changing the environment, the agents are constantly generating some kind of implicit information), explicit communication should be strongly limited or avoided altogether, in order to fit our paradigm (note that in many works in this field, this is not the case, and communication, as well as memory, resources, are often being used in order to create complex cooperative systems).

In summary, when designing intelligent swarm systems, we must assume (and often even *aspire for*) having individual agents that are myopic, mute, senile and rather stupid.

1.4 Limitations

While swarm intelligence has been applied successfully in many fields, including optimization, robotics, and networking, it also has limitations that need to be taken into account. One of the main limitations of swarm intelligence is its sensitivity to initial conditions and parameter settings. Small changes in the initial configuration or the parameters of the swarm can have a significant impact on its behavior and performance, leading to suboptimal solutions or even failure to converge. This problem is exacerbated in large-scale systems, where the number of variables and interactions increases exponentially [29].

Another limitation of swarm intelligence is its vulnerability to perturbations and disturbances. Swarms are designed to be robust and resilient to individual failures or disruptions, but they can be vulnerable to systemic disturbances [140], such as environmental changes, resource depletion, or external attacks. These disturbances can destabilize the swarm beyond its self-emergent macroscopic regularities [141], leading to disintegration, divergence, or oscillations.

Real-world examples of these limitations include the behavior of ant colonies in changing environments. Ants use swarm intelligence to forage for food and build nests, but they are also susceptible to disturbances such as climate change or human intervention. In some cases, ant colonies can collapse or become maladapted to their environment due to the loss of critical resources or the disruption of communication channels.

Another limitation of swarm intelligence is related to the trade-off between exploration and exploitation. Swarms can achieve impressive results by exploring a large search space and exploiting the best solutions found. However, there is a risk of getting stuck in local optima or suboptimal regions of the search space, especially if the swarm lacks diversity or adaptability [101]. In some cases, the swarm may require a balance between exploration and exploitation to achieve the best results, which can be challenging to achieve in practice [30].

A related limitation is the scalability of swarm intelligence [22]. While swarms can scale up to thousands or millions of agents, the computational and communication overheads can become prohibitive in large-scale systems. The swarm may require efficient algorithms for coordination, decision-making, and resource allocation, which can be difficult to design and optimize. Such limitations may manifest, for example, when swarm intelligence is employed in traffic management systems. Swarms of autonomous vehicles or drones can optimize traffic flow and reduce congestion by coordinating their movements and avoiding collisions [8]. However, these systems require efficient algorithms for path planning, decision-making, and communication, as well as robust mechanisms for handling uncertainties and unexpected events.

Another example is the application of swarm intelligence in social networks. Swarms of agents can learn and adapt to social dynamics by interacting with each other and with the environment [24]. However, these systems are also susceptible to biases, echo chambers, and polarization, which can affect their ability to explore new ideas and perspectives [142].

1.5 Swarm Search with Communication

While decentralized swarms have been the main focus of swarm-based search algorithms due to their scalability and simplicity, there are also several promising works that utilize synchronization or communication among the agents. These parallel swarms often employ communication to enhance the efficiency of the search process, such as parallel ant colony optimization, parallel particle swarm optimization, and other parallel metaheuristic approaches.

Parallel ant colony optimization (PACO) [122] is an example of a parallel swarm algorithm that utilizes communication among agents. PACO algorithms allow multiple agents to cooperate by sharing pheromone information, which helps in quickly identifying the optimal solution. For instance, PACO has been used in multi-robot coverage problems, where a group of robots are required to explore an unknown environment while avoiding collisions with each other. By sharing pheromone information, the robots can quickly converge to a solution, even in complex and large environments [74, 75].

Parallel particle swarm optimization (PPSO) [99] is another example of a parallel swarm algorithm that uses communication among agents. PPSO is a variant of particle swarm optimization (PSO) that allows multiple agents to communicate with each other to improve the search process. For instance, PPSO has been used to optimize complex systems such as power grids, where the agents need to communicate to efficiently manage the distributed resources [81].

In cases where decentralized swarms may not be sufficient, parallel swarms can be beneficial. For example, in situations where the problem space is complex, the search space is vast, and the search process is time-critical, parallel swarm algorithms can offer a significant advantage over decentralized swarms. In such scenarios, communication among agents can help to identify the optimal solution more quickly and efficiently.

However, one of the main drawbacks of parallel swarm algorithms is the increased complexity of the communication mechanisms, which may require significant computational resources [55]. Additionally, communication can also lead to increased synchronization overhead, which may impact the scalability of the algorithm. Thus, in cases where the problem space is relatively simple, decentralized swarm algorithms may still be a better choice.

While decentralized swarms remain the main focus of swarm-based search algorithms, parallel swarm algorithms that utilize communication among agents have shown significant promise in enhancing the efficiency of the search process. These algorithms have been used in various applications, such as multi-robot coverage problems and power grid optimization. However, the increased complexity of communication mechanisms and synchronization overhead should also be considered when deciding on an appropriate approach for a given problem.

1.6 The Convergence of Swarm Intelligence and Modern AI Systems

The fusion of swarm intelligence (SI) with modern artificial intelligence (AI) systems can potentially mark a significant leap in the evolution of AI technologies. This convergence brings together the decentralized, self-organizing principles of swarm intelligence with advanced AI methodologies, leading to robust, scalable, and efficient AI solutions.

Theoretical Foundations

The integration of SI and AI is fundamentally grounded in the concept of harnessing the collective behaviors observed in natural biological systems. These systems, exemplified by phenomena such as the coordinated movements of bird flocks or the intricate colony organization of ants, embody a form of intelligence that is distributed across many individuals. Each individual in these systems follows simple rules, and their interactions lead to the emergence of complex, adaptive behaviors. This attribute of biological systems provides a rich source of inspiration for enhancing AI algorithms.

SI models are particularly characterized by their distributed intelligence and lack of centralized control, making them inherently robust and scalable [51]. In contrast to traditional AI systems that rely on central decision-making algorithms, SI models operate based on the collective behavior of numerous agents, each making simple, local decisions that contribute to the achievement of a global goal. This decentralized approach is not only more resilient to individual failures but also allows for the parallel processing of information, leading to increased efficiency and adaptability. This is particularly essential in stochastic environments where numerous traditional AI algorithms encounter limitations [27].

When these SI principles are integrated into AI systems, such as neural networks, genetic algorithms, or deep learning models, they bring a distinct set of advantages. For instance, in neural networks, incorporating swarm-based approaches can lead to more efficient methods of optimizing network weights and structures [114]. This can result in faster convergence during training and improved network performance. Genetic algorithms, inspired by the principles of natural selection and genetics, benefit from swarm intelligence by enhancing their ability to explore and exploit the solution space, thereby finding optimal or near-optimal solutions more effectively [48].

Deep learning models, which have gained significant traction for their ability to process and learn from large amounts of data, can also benefit from the incorporation of SI principles. By leveraging the distributed processing capability of SI, deep learning models can handle more complex, multi-dimensional data sets with increased efficiency. This integration can lead to the development of more robust models that are better equipped to adapt to changing data patterns and environments, thus enhancing their problem-solving capabilities. The convergence of SI with AI therefore represents a symbiotic amalgamation of two potent computational paradigms. This fusion can not only enhance the adaptability and efficiency of problem-solving in AI systems but also opens up new avenues for the development of sophisticated, resilient, and scalable AI solutions. As we continue to explore this convergence, it is likely to yield transformative advancements in the field of AI, offering novel approaches to tackle increasingly complex challenges.

Application Areas

The amalgamation of SI and AI has led to notable advancements in various fields, among others:

- **Robotics**: Swarm robotics, drawing inspiration from SI, utilizes AI for executing complex tasks such as coordinated movement and obstacle avoidance. This field mimics the collective behavior of natural swarms, translating it into the coordinated functioning of robotic units. In search and rescue missions, swarm robots can cover vast areas, swiftly adapting to new obstacles and efficiently dividing tasks among themselves, thereby accelerating search operations and enhancing the likelihood of successful rescues [28]. Similarly, in exploring unknown terrains, such as in planetary exploration or underwater surveys, these robots can autonomously navigate challenging environments, collect data, and even make decisions based on collective intelligence, all while maintaining robustness against individual robot failures.

- **Data Mining and Analysis**: The integration of AI algorithms with SI techniques significantly bolsters data mining capabilities. In clustering and predictive analysis, for example, swarm-based approaches like particle swarm optimization [58] can be used to fine-tune the parameters of machine learning models, leading to more accurate predictions and better data categorization. This integration proves particularly beneficial in handling large-scale, complex datasets, where traditional methods might struggle with the dimensionality and intricacy of the data. The collective decision-making process inherent in SI can navigate through vast data spaces more effectively, uncovering patterns and insights that might be missed by conventional algorithms [77].
- **Optimization Problems**: In the realm of logistics and supply chain management, SI algorithms, when merged with AI techniques, offer significant improvements in optimizing routes and reducing operational costs [1]. The nature of these problems often involves complex variables and constraints, where traditional optimization methods may not be efficient or feasible. SI algorithms, such as ant colony optimization, can mimic the pheromone trail-laying behavior of ants to find the most efficient routes, adapting to dynamic changes like traffic conditions or delivery urgencies. This not only improves the efficiency of logistics operations but also contributes to substantial cost savings and enhanced service quality [110].
- **Financial Engineering**: In the field of financial engineering, the integration of SI and AI is revolutionizing the way financial markets are analyzed and traded. For instance, swarm-based algorithms can be employed to optimize trading strategies, where multiple AI agents simulate different trading scenarios to identify the most profitable approaches under varying market conditions [102]. These agents can adapt to real-time market changes, learning and evolving to enhance predictive accuracy and risk assessment. Additionally, SI principles are being applied in portfolio optimization, where a distributed approach to asset allocation and risk management can lead to more resilient and

diverse investment portfolios, capable of withstanding market volatility and achieving balanced returns [139].

- **Homeland Security**: In homeland security, the convergence of SI and AI is being utilized to enhance surveillance, threat detection, and emergency response [26]. Swarm robotics, for example, can be deployed for reconnaissance missions, providing real-time data and analysis in sensitive or hazardous environments [49, 72]. These robotic swarms, equipped with AI-driven sensors and decision-making capabilities, can autonomously navigate and monitor large areas, detecting anomalies or potential threats efficiently. Additionally, SI algorithms are being used in cyber-security to detect and counteract sophisticated cyber threats through distributed, self-organizing defense mechanisms [123]. This approach mimics the collective defense behaviors found in natural systems, offering a robust and adaptive strategy to safeguard critical information infrastructure.
- **Agriculture**: The application of swarm intelligence in agriculture is revolutionizing traditional farming practices. Utilizing SI principles, agricultural robots can perform tasks such as planting, weeding, and harvesting more efficiently. These robots, operating as a coordinated swarm, can adapt to various field conditions and crop statuses, optimizing their tasks in real-time [143]. This approach not only improves the precision and efficiency of agricultural operations but also minimizes waste and reduces the environmental impact. Additionally, swarm-based systems can be used for detailed monitoring of crop health, soil quality, and pest control, employing AI-driven analysis to make informed decisions about irrigation, fertilization, and pesticide distribution, thus enhancing yield and sustainability in farming practices [48].

Challenges and Limitations

Despite the promising potential of integrating swarm intelligence (SI) with artificial intelligence (AI), this fusion is met with several challenges and limitations that need to be addressed for optimal implementation.

- **Algorithm Design Complexity**: One of the foremost challenges lies in the complexity inherent in designing algorithms that effectively merge SI and AI principles. This complexity stems from the need to balance the decentralized, self-organizing nature of SI with the often centralized, computation-heavy processes of AI. Ensuring that these fundamentally different approaches work in harmony without compromising the strengths of each other requires intricate algorithm design and rigorous testing.

- **Stability and Predictability**: Another significant challenge is maintaining stability and predictability in these integrated systems. SI models, by their nature, can exhibit emergent behaviors that are difficult to predict or control, particularly when scaled up. When combined with AI systems, especially those that involve learning and adaptation, the unpredictability can be amplified, making it challenging to ensure consistent and reliable outcomes, especially in dynamically changing environments.
- **Scalability and Resource Management**: As SI systems often involve a large number of agents, scaling these systems while integrating with resource-intensive AI algorithms presents a challenge. Effective resource allocation, including computational power and memory, becomes crucial as the system scales. Balancing the scalability of SI with the resource demands of AI is a key hurdle to overcome in practical applications.
- **Adaptation to Real-World Conditions**: Translating the theoretical models of SI and AI into real-world applications is another challenge. Real-world environments are often more complex and unpredictable than controlled settings. Designing SI-AI systems that can adapt to these varying and unforeseen conditions, while still performing optimally, is a significant challenge.
- **Ethical and Societal Implications**: The integration of SI with AI also raises ethical and societal questions. Issues like privacy, security, and the potential for misuse need to be carefully considered. Ensuring that these systems are designed with ethical considerations in mind and do not inadvertently harm or disadvantage individuals or groups is crucial.
- **Interdisciplinary Collaboration**: Finally, the convergence of SI and AI requires an interdisciplinary approach, combining expertise in biology, computer science, engineering, and more. This necessitates collaboration across different fields, which can be challenging due to differing terminologies, methodologies, and research cultures.

The convergence of swarm intelligence with modern AI systems represents a frontier in AI research and development. By harnessing the collective and decentralized attributes of SI alongside the advanced computational capabilities of AI, this integration holds the promise of transforming various sectors, from robotics to data analysis. As this field evolves, it will undoubtedly unveil new possibilities and solutions to complex problems, underscoring the transformative potential of AI technologies.

1.7 Ethical Considerations and Societal Impact

The integration of swarm intelligence (SI) with artificial intelligence (AI) extends beyond technical development, encompassing a range of ethical considerations and societal impacts that necessitate thorough examination.

Privacy and Data Security

Privacy and data security are paramount concerns in SI-AI systems, especially when handling large datasets that may include personal information. For instance, in the context of smart city implementations, where SI-AI systems monitor traffic patterns, utility usage, or public safety, there is a potential risk of infringing on individual privacy [9, 12]. Ensuring data anonymization and robust protection against breaches is critical [91]. Similarly, in healthcare applications, where SI-AI could analyze patient data for better diagnosis and treatment plans, safeguarding sensitive health information is a prime ethical obligation.

Accountability and Decision-Making

Decision-making processes in SI-AI systems raise questions of accountability, particularly in high-stakes scenarios [90]. In autonomous vehicles, for instance, where decisions made by SI-AI systems can have life-or-death consequences, establishing who is responsible—the manufacturer, the software developer, or the AI system itself—becomes complex. This necessitates a legal framework to assign and manage liability [111].

Impact on Employment

The impact of SI-AI systems on employment is a significant concern. The automation of tasks in sectors like manufacturing, where SI-AI-driven robots could replace human assembly line workers, may lead to job displacement. Transitioning the workforce to more AI-complementary roles and emphasizing education and training for new technological skills are vital steps to mitigate this impact [148].

Bias and Fairness

Bias in SI-AI systems is a critical issue. For example, in financial services, if a SI-AI system is used for credit scoring and the training data has historical biases against certain demographics, the system might unfairly deny credit to those groups. Rigorous testing for biases and ensuring diversity in training datasets are necessary to promote fairness [96].

Societal Impact and Public Perception

The societal impact and public perception of SI-AI systems are significant. Deployment of swarm drones for surveillance might lead to concerns about privacy invasion and the emergence of a surveillance state [93]. Ensuring ethical deployment and maintaining transparent communication with the public are essential to address these concerns [71].

Use in Military and Defense

The use of SI-AI in military applications, such as autonomous drone swarms for reconnaissance or combat [78], presents ethical challenges. These include concerns about the escalation of unmanned warfare and the ethical implications of delegating life-and-death decisions to AI systems [156]. International regulations and ethical standards are necessary to govern the use of such technology.

In summary, while SI-AI convergence offers significant advancements, it is crucial to address its ethical challenges and societal implications proactively. This requires technical safeguards, robust policy frameworks, and continuous dialogue with stakeholders to ensure responsible and beneficial use of these technologies.

1.8 Opportunities and Future Research

Swarm intelligence and swarm systems have received considerable attention in recent years due to their potential for solving complex problems in various fields, such as robotics, optimization, and network design. As a result, there are numerous opportunities for future research in this area.

One promising avenue for future research is the development of more sophisticated algorithms and models for swarm intelligence. While current approaches have shown promise, there is still much to be done in terms of improving the efficiency and adaptability of swarm systems [66]. Researchers may explore new ways to optimize the communication and coordination of swarm agents, or develop new approaches for dealing with the inherent uncertainty and complexity of real-world environments [65].

Another important area for future research is the application of swarm intelligence to real-world problems. While there have been many successful demonstrations of swarm systems in laboratory settings, there is a need for more research on how to apply these systems to real-world problems. This may involve working with industry partners to develop practical solutions that can be deployed in the field, or collaborating with government agencies to address societal challenges such as disaster response or urban planning [159, 162].

In addition to these technical challenges, there are also important ethical and social considerations to be addressed. As swarm systems become more advanced and pervasive, there may be concerns around issues such as privacy, security, and control. Researchers may need to explore new ways to address these concerns, such as developing transparent and accountable algorithms, or working with policymakers to establish appropriate regulations and standards [60].

Overall, there are numerous opportunities for future research in swarm intelligence and swarm systems. By continuing to explore these systems and their potential applications, researchers can help to unlock new solutions to complex problems and contribute to the advancement of science and technology.

Conclusion

The study of swarm intelligence has revealed that even seemingly simple organisms, such as ants, can exhibit complex and sophisticated collective behaviors when allowed to work together in a synergistic manner. This insight has led researchers to investigate the potential for applying this approach to artificial intelligence and robotics, with promising results.

In this special issue, a number of research studies have been presented that demonstrate the power of swarm intelligence in producing complex and adaptive behaviors. By studying the ways in which ants and other social insects cooperate and communicate with one another, researchers have been able to develop algorithms and models that can be applied to a wide range of problems.

One of the key insights from these studies is that individual agents within a swarm do not necessarily need to be highly intelligent or even aware of the larger goals of the group. Rather, by following simple rules and responding to local cues, they can collectively produce intelligent and adaptive behaviors that emerge at the swarm level.

This approach has numerous potential applications, from optimizing traffic flow to coordinating the movements of swarms of robots in search and rescue operations. By harnessing the power of swarm intelligence, researchers are exploring new ways to tackle complex problems that would be difficult or impossible for any individual agent to solve alone.

Overall, the research presented in this special issue provides compelling evidence that even the simplest organisms can exhibit remarkable intelligence and adaptability when working together in a synergistic manner. By taking inspiration from nature, researchers are opening up exciting new avenues for developing advanced technologies that can benefit society in countless ways.

In summary, let us cite a statement made by a scientist after watching an ant making its laborious way across a wind-and-wave-molded beach [138]:

> An ant, viewed as a behaving system, is quite simple. The apparent complexity of its behavior over time is largely a reflection of the environment in which it finds itself.

Such a point of view, as well as the results of the research presented in this special issue, lead us to believe that even simple, ant-like beings, when allowed to synergistically collaborate, can yield a complicated, adaptive and quite efficient macroscopic behavior, in the intelligent swarm-level scope.

References

[1] Mohd Nadhir Ab Wahab, Samia Nefti-Meziani, and Adham Atyabi. A comprehensive review of swarm optimization algorithms. *PloS one*, 10(5):e0122827, 2015.

[2] F.R. Adler and D.M. Gordon. Information collection and spread by networks of partolling agents. *The American Naturalist*, 140(3): 373–400, 1992.

[3] N. Agmon, S. Kraus, and G.A. Kaminka. Multi-robot perimeter patrol in adversarial settings. In *IEEE International Conference on Robotics and Automation (ICRA 2008)*, 2339–2345, 2008.

[4] N. Agmon, V. Sadov, G.A. Kaminka, and S. Kraus. The impact of adversarial knowledge on adversarial planning in perimeter patrol. In *AAMAS '08: Proceedings of the 7th international joint conference on Autonomous agents and multiagent systems*, pages 55–62, Richland, SC, 2008. International Foundation for Autonomous Agents and Multiagent Systems.

[5] A. Agogino and K. Tumer. Regulating air traffic flow with coupled agents. In *AAMAS '08: Proceedings of the 7th international joint conference on Autonomous agents and multiagent systems*, pages 535–542, Richland, SC, 2008. International Foundation for Autonomous Agents and Multiagent Systems.

[6] S. Aknine and O. Shehory. A feasible and practical coalition formation mechanism leveraging compromise and task relationships. In *Proceedings of the IEEE/WIC/ACM international conference on Intelligent Agent Technology*, pages 436–439, 2006.

[7] R. Alami, S. Fleury, M. Herrb, F. Ingrand, and F. Robert. Multi-robot cooperation in the martha project. *IEEE Robotics and Automation Magazine*, 5(1):36–47, March 1998.

[8] Tal Altshuler, Yaniv Altshuler, Rachel Katoshevski, and Yoram Shiftan. Modeling and prediction of ride-sharing utilization dynamics. *Journal of Advanced Transportation*, 2019, 2019.

[9] Y. Altshuler, N. Aharony, A. Pentland, Y. Elovici, and M. Cebrian. Stealing reality: When criminals become data scientists (or vice versa). *Intelligent Systems, IEEE*, 26(6):22–30, nov.-dec. 2011.

[10] Y. Altshuler, A.M. Bruckstein, and I.A. Wagner. Swarm robotics for a dynamic cleaning problem. In *IEEE Swarm Intelligence Symposium*, pages 209–216, 2005.

[11] Y. Altshuler, M. Fire, N. Aharony, Y. Elovici, and A Pentland. How many makes a crowd? on the correlation between groups' size and the accuracy of modeling. In *International Conference on Social Computing, Behavioral-Cultural Modeling and Prediction*, pages 43–52. Springer, 2012.

[12] Y. Altshuler, W. Pan, and A Pentland. Trends prediction using social diffusion models. In *International Conference on Social Computing, Behavioral-Cultural Modeling and Prediction*, pages 97–104. Springer, 2012.

[13] Y. Altshuler, I.A. Wagner, and A.M. Bruckstein. Shape factor's effect on a dynamic cleaners swarm. In *Third International Conference on Informatics in Control, Automation and Robotics (ICINCO), the Second International Workshop on Multi-Agent Robotic Systems (MARS)*, pages 13–21, 2006.

[14] Y. Altshuler, I.A. Wagner, and A.M. Bruckstein. On swarm optimality in dynamic and symmetric environments. volume 7, page 11, 2008.

[15] Y. Altshuler, I.A. Wagner, and A.M. Bruckstein. Collaborative exploration in grid domains. In *Sixth International Conference on Informatics in Control, Automation and Robotics (ICINCO)*, 2009.

[16] Y. Altshuler, I.A. Wagner, V. Yanovski, and A.M. Bruckstein. Multi-agent cooperative cleaning of expanding domains. *International Journal of Robotics Research*, 30:1037–1071, 2010.

[17] Y. Altshuler, V. Yanovski, I.A. Wagner, and A.M. Bruckstein. The cooperative hunters - efficient cooperative search for smart targets using uav swarms. In *Second International Conference on Informatics in Control, Automation and Robotics (ICINCO), the First International Workshop on Multi-Agent Robotic Systems (MARS)*, pages 165–170, 2005.

[18] Y. Altshuler, V. Yanovsky, A.M. Bruckstein, and I.A. Wagner. Efficient cooperative search of smart targets using uav swarms. *ROBOTICA*, 26:551–557, 2008.

[19] Y. Altshuler, V. Yanovsky, I. Wagner, and A. Bruckstein. Swarm intelligence—searchers, cleaners and hunters. *Swarm Intelligent Systems*, pages 93–132, 2006.

[20] Yaniv Altshuler. Recent developments in the theory and applicability of swarm search. *Entropy*, 25(5):710, 2023.

[21] Yaniv Altshuler and Alfred M. Bruckstein. The complexity of grid coverage by swarm robotics. In *ANTS 2010*, pages 536–543. LNCS, 2010.

[22] Yaniv Altshuler and Alfred M. Bruckstein. Static and expanding grid coverage with ant robots: Complexity results. *Theoretical Computer Science*, 412(35):4661–4674, 2011.

[23] Yaniv Altshuler, Shlomi Dolev, Yuval Elovici, and Nadav Aharony. Ttled random walks for collaborative monitoring. In *NetSciCom 2010 (Second International Workshop on Network Science for Communication Networks)*, San Diego, CA, USA, 3 2010.

[24] Yaniv Altshuler, Yuval Elovici, Armin B Cremers, Nadav Aharony, and Alex Pentland. *Security and privacy in social networks*. Springer Science & Business Media, 2012.

[25] Yaniv Altshuler, Michael Fire, Erez Shmueli, Yuval Elovici, Alfred Bruckstein, Alex Sandy Pentland, and David Lazer. The social amplifier—reaction of human communities to emergencies. *Journal of Statistical Physics*, 152(3):399–418, 2013.

[26] Yaniv Altshuler, Alex Pentland, Shlomo Bekhor, Yoram Shiftan, and Alfred Bruckstein. Optimal dynamic coverage infrastructure for large-scale fleets of reconnaissance uavs. *arXiv preprint arXiv:1611.05735*, 2016.

[27] Yaniv Altshuler, Alex Pentland, and Alfred M Bruckstein. Collaborative patrolling swarms in stochastically expanding environments. In *Swarms and Network Intelligence in Search*, pages 155–185. Springer, 2018.

[28] Yaniv Altshuler, Alex Pentland, and Alfred M Bruckstein. Introduction to swarm search. In *Swarms and Network Intelligence in Search*, pages 1–14. Springer, 2018.

[29] Yaniv Altshuler, Alex Pentland, and Alfred M Bruckstein. *Swarms and Network Intelligence in Search*. Springer, 2018.

[30] Yaniv Altshuler, Alex Sandy Pentland, and Goren Gordon. Social behavior bias and knowledge management optimization. In *Social Computing, Behavioral-Cultural Modeling, and Prediction*, pages 258–263. Springer, 2015.

[31] Yaniv Altshuler, Rami Puzis, Yuval Elovici, Shlomo Bekhor, and Alex Sandy Pentland. On the rationality and optimality of transportation networks defense: a network centrality approach. *Securing Transportation Systems*, pages 35–63.

[32] Yaniv Altshuler, Erez Shmueli, Guy Zyskind, Oren Lederman, Nuria Oliver, and Alex Pentland. Campaign optimization through behavioral modeling and mobile network analysis. *Computational Social Systems, IEEE Transactions on*, 1(2):121–134, 2014.

[33] Yaniv Altshuler, Erez Shmueli, Guy Zyskind, Oren Lederman, Nuria Oliver, and AS Pentland. Campaign optimization through mobility network analysis. *Geo-intelligence and visualization through big data trends*, pages 33–74, 2015.

[34] T. Arai, H. Ogata, and T. Suzuki. Collision avoidance among multiple robots using virtual impedance. In *Proceedings of the IEEE/RSJ International Conference on Intelligent Robots and Systems*, pages 479–485, 1989.

[35] R.C. Arkin and T. Balch. Aura: Principles and practice in review. *Journal of Experimental and Theoretical Artificial Intelligence*, 9(2/3):175–188, 1997.

[36] T. Balch and R. Arkin. Behavior-based formation control for multi-robot teams. *IEEE Transactions on Robotics and Automation*, 14(6):926–939, 1998.

[37] M. Benda, V. Jagannathan, and R. Dodhiawalla. On optimal cooperation of knowledge sources. *Technical Report BCS-G2010-28, Boeing AI Center*, 1985.

[38] K. Bendjilali, F. Belkhouche, and B. Belkhouche. Robot formation modelling and control based on the relative kinematics equations. *The International Journal of Robotics and Automation*, 24(1):79–88, 2009.

[39] G. Beni and J. Wang. Theoretical problems for the realization of distributed robotic systems. In *IEEE Internal Conference on Robotics and Automation*, pages 1914–1920, 1991.

[40] Gerardo Beni. From swarm intelligence to swarm robotics. In *International Workshop on Swarm Robotics*, pages 1–9. Springer, 2004.

[41] R.M. Bhatt, C.P. Tang, and V.N. Krovi. Formation optimization for a fleet of wheeled mobile robots — a geometric approach. *Robotics and Autonomous Systems*, 57(1):102–120, 2009.

[42] E. Bonabeau, M. Dorigo, and G. Theraulaz. *Swarm Intelligence: From Natural to Artificial Systems*. Oxford University Press, US, 1999.

[43] Manuele Brambilla, Eliseo Ferrante, Mauro Birattari, and Marco Dorigo. Swarm robotics: a review from the swarm engineering perspective. *Swarm Intelligence*, 7(1):1–41, 2013.

[44] R.A. Brooks. A robust layered control system for a mobile robot. *IEEE Journal of Robotics and Automation*, RA-2(1):14–23, 1986.

[45] R.A. Brooks and A.M. Flynn. Fast, cheap and out of control, a robot invasion of the solar system. *Journal of the British interplanetary Society*, 42:478–485, 1989.

[46] A.M. Bruckstein. Why the ant trails look so straight and nice. *The Mathematical Intelligencer*, 15(2):59–62, 1993.

[47] A.M. Bruckstein, C.L. Mallows, and I.A. Wagner. Probabilistic pursuits on the integer grid. *American Mathematical Monthly*, 104(4):323–343, 1997.

[48] Quang-Thanh Bui, Quoc-Huy Nguyen, Xuan Linh Nguyen, Vu Dong Pham, Huu Duy Nguyen, and Van-Manh Pham. Verification of novel integrations of swarm intelligence algorithms into deep learning neural network for flood susceptibility mapping. *Journal of Hydrology*, 581:124379, 2020.

[49] David J Bursac. Autonomous robotic weapons: Us army innovation for ground combat in the twenty-first century. Technical report, US Army School for Advanced Military Studies Fort Leavenworth United States, 2015.

[50] C. Candea, H. Hu, L. Iocchi, D. Nardi, and M. Piaggio. Coordinating in multi-agent robocup teams. *Robotics and Autonomous Systems*, 36(2–3):67–86, 2001.

[51] Amrita Chakraborty and Arpan Kumar Kar. Swarm intelligence: A review of algorithms. *Nature-inspired computing and optimization: Theory and applications*, pages 475–494, 2017.

[52] G. Chalkiadakis and C. Boutilier. Sequential decision making in repeated coalition formation under uncertainty. In *Proceedings of the 7th international joint conference on Autonomous agents and multiagent systems*, pages 347–354, 2008.

[53] G. Chalkiadakis, E. Markakis, and C. Boutilier. Coalition formation under uncertainty: bargaining equilibria and the bayesian core stability concept. In *AAMAS '07: Proceedings of the 6th international joint conference on Autonomous agents and multiagent systems*, pages 1–8, New York, NY, USA, 2007. ACM.

[54] D. Chevallier and S. Payandeh. On kinematic geometry of multi-agent manipulating system based on the contact force information. In *The Sixth International Conference on Intelligent Autonomous Systems (IAS-6)*, pages 188–195, 2000.

[55] Shu-Chuan Chu, John F Roddick, and Jeng-Shyang Pan. A parallel particle swarm optimization algorithm with communication strategies. *J. Inf. Sci. Eng*, 21(4):809–818, 2005.

[56] Shu-Chuan Chu, Pei-Wei Tsai, and Jeng-Shyang Pan. Cat swarm optimization. In *PRICAI 2006: Trends in Artificial Intelligence: 9th Pacific Rim International Conference on Artificial Intelligence Guilin, China, August 7-11, 2006 Proceedings 9*, pages 854–858. Springer, 2006.

[57] R. Connaughton, P. Schermerhorn, and M. Scheutz. Physical parameter optimization in swarms of ultra-low complexity agents. In *AAMAS '08: Proceedings of the 7th international joint conference on Autonomous agents and multiagent systems*, pages 1631–1634, Richland, SC,

2008. International Foundation for Autonomous Agents and Multiagent Systems.

[58] Micael Couceiro, Pedram Ghamisi, Micael Couceiro, and Pedram Ghamisi. *Particle swarm optimization*. Springer, 2016.

[59] Dipankar Dasgupta. An overview of artificial immune systems and their applications. *Artificial immune systems and their applications*, pages 3–21, 1999.

[60] Michael Decker. Caregiving robots and ethical reflection: the perspective of interdisciplinary technology assessment. *AI & society*, 22:315–330, 2008.

[61] S.A. DeLoach and M. Kumar. *Intelligence Integration in Distributed Knowledge Management*, chapter Multi-Agent Systems Engineering: An Overview and Case Study, pages 207—224. Idea Group Inc (IGI), 2008.

[62] J. Deneubourg, S. Goss, G. Sandini, F. Ferrari, and P. Dario. Self-organizing collection and transport of objects in unpredictable environments. In *Japan-U.S.A. Symposium on Flexible Automation*, pages 1093–1098, 1990.

[63] M.B. Dias and A. Stentz. A market approach to multirobot coordination. *Technical Report, CMU-RI - TR-01-26, Robotics Institute, Carnegie Mellon University*, 2001.

[64] Marco Dorigo and Thomas Stützle. *Ant colony optimization: overview and recent advances*. Springer, 2019.

[65] Marco Dorigo, Guy Theraulaz, and Vito Trianni. Reflections on the future of swarm robotics. *Science Robotics*, 5(49):eabe4385, 2020.

[66] Marco Dorigo, Guy Theraulaz, and Vito Trianni. Swarm robotics: Past, present, and future [point of view]. *Proceedings of the IEEE*, 109(7):1152–1165, 2021.

[67] A. Drogoul and J. Ferber. From tom thumb to the dockers: Some experiments with foraging robots. In *Proceedings of the Second International Conference on Simulation of Adaptive Behavior*, pages 451–459, 1992.

[68] G. Dudek, M.R.M. Jenkin, E. Milios, and D. Wilkes. A taxonomy for multi-agent robotics. *Autonomous Robots Journal*, 3(4):375–397, 1996.

[69] Russell Eberhart and James Kennedy. Particle swarm optimization. In *Proceedings of the IEEE international conference on neural networks*, volume 4, pages 1942–1948. Citeseer, 1995.

[70] A. Efraim and D.Peleg. Distributed algorithms for partitioning a swarm of autonomous mobile robots. *Structural Information and Communication Complexity, Lecture Notes in Computer Science*, 4474:180–194, 2007.

[71] Ray Eitel-Porter. Beyond the promise: implementing ethical ai. *AI and Ethics*, 1:73–80, 2021.

[72] Erhan Erkut and Armann Ingolfsson. Catastrophe avoidance models for hazardous materials route planning. *Transportation Science*, 34:165–179, May 2000.

[73] U. Visser et. al. *RoboCup 2007: Robot Soccer World Cup XI*, volume 5001 of *Lecture Notes in Computer Science*. Springer Berlin, 2008.

[74] A. Felner, Y. Shoshani, Y. Altshuler, and A.M. Bruckstein. Multi-agent physical a* with large pheromones. *Journal of Autonomous Agents and Multi-Agent Systems*, 12(1):3–34, 2006.

[75] A. Felner, R. Stern, A. Ben-Yair, S. Kraus, and N. Netanyahu. Pha*: Finding the shortest path with a* in unknown physical environments. *Journal of Artificial Intelligence Research*, 21:631–679, 2004.

[76] C. Ferrari, E. Pagello, J. Ota, and T. Arai. Multirobot motion coordination in space and time. *Robotics and Autonomous Systems*, 25:219–229, 1998.

[77] Elliackin Figueiredo, Mariana Macedo, Hugo Valadares Siqueira, Clodomir J Santana Jr, Anu Gokhale, and Carmelo JA Bastos-Filho. Swarm intelligence for clustering—a systematic review with new perspectives on data mining. *Engineering Applications of Artificial Intelligence*, 82:313–329, 2019.

[78] Ann Finkbeiner. Military technology: Death by remote control. *Nature*, 534(7609):618, 2016.

[79] J. Fredslund and M.J. Mataric. Robot formations using only local sensing and control. In *Proceedings of the International Symposium on Computational Intelligence in Robotics and Automation (IEEE CIRA 2001)*, pages 308–313, 2001.

[80] B.P. Gerkey and M.J. Mataric. Sold! market methods for multi-robot control. *IEEE Transactions on Robotics and Automation, Special Issue on Multi-robot Systems*, 2002.

[81] Dennis Gies and Yahya Rahmat-Samii. Reconfigurable array design using parallel particle swarm optimization. In *IEEE Antennas and Propagation Society International Symposium. Digest. Held in conjunction with: USNC/CNC/URSI North American Radio Sci. Meeting (Cat. No. 03CH37450)*, volume 1, pages 177–180. IEEE, 2003.

[82] D.M. Gordon. The expandable network of ant exploration. *Animal Behaviour*, 50:372–378, 1995.

[83] N. Gordon, I.A. Wagner, and A.M. Bruckstein. Discrete bee dance algorithms for pattern formation on a grid. In *IEEE International Conference on Intelligent Agent Technology (IAT03)*, pages 545–549, October 2003.

[84] J. Hagelbäck and S.J. Johansson. Demonstration of multi-agent potential fields in real-time strategy games. In *AAMAS '08: Proceedings of the 7th international joint conference on Autonomous agents and multiagent*

systems, pages 1687–1688, Richland, SC, 2008. International Foundation for Autonomous Agents and Multiagent Systems.

[85] I. Harmatia and K. Skrzypczykb. Robot team coordination for target tracking using fuzzy logic controller in game theoretic framework. *Robotics and Autonomous Systems*, 57(1):75–86, 2009.

[86] T. Haynes and S. Sen. *Evolving Behavioral Strategies in Predators and Prey*, volume 1042 of *Lecture Notes in Computer Science*, chapter Adaptation and Learning in Multi-Agent Systems, pages 113—126. Springer, Berlin, 1986.

[87] S. Hettiarachchi and W. Spears. Moving swarm formations through obstacle fields. In *International Conference on Artificial Intelligence*, 2005.

[88] Michael G Hinchey, Roy Sterritt, and Chris Rouff. Swarms and swarm intelligence. *Computer*, 40(4), 2007.

[89] G. Hollinger, S. Singh, J. Djugash, and A. Kehagias. Efficient multi-robot search for a moving target. *The International Journal of Robotics Research*, 28(2):201–219, 2009.

[90] Andreas Holzinger, Georg Langs, Helmut Denk, Kurt Zatloukal, and Heimo Müller. Causability and explainability of artificial intelligence in medicine. *Wiley Interdisciplinary Reviews: Data Mining and Knowledge Discovery*, 9(4):e1312, 2019.

[91] Carter Jernigan and Behram F.T. Mistree. Gaydar: Facebook friendships expose sexual orientation. 14(10), 2009.

[92] Dervis Karaboga et al. An idea based on honey bee swarm for numerical optimization. Technical report, Technical report-tr06, Erciyes university, engineering faculty, computer ..., 2005.

[93] Hisham O Khogali and Samir Mekid. The blended future of automation and ai: Examining some long-term societal and ethical impact features. *Technology in Society*, 73:102232, 2023.

[94] S. Kirkpatrick and J.J. Schneider. How smart does an agent need to be? *International Journal of Modern Physics*, C 16:139–155, 2005.

[95] T. Klos and G.J. van Ahee. Evolutionary dynamics for designing multi-period auctions. In *AAMAS '08: Proceedings of the 7th international joint conference on Autonomous agents and multiagent systems*, pages 1589–1592, Richland, SC, 2008. International Foundation for Autonomous Agents and Multiagent Systems.

[96] Akhil Alfons Kodiyan. An overview of ethical issues in using ai systems in hiring with a case study of amazon's ai based hiring tool. *Researchgate Preprint*, pages 1–19, 2019.

[97] Peter M Krafft, Julia Zheng, Wei Pan, Nicolás Della Penna, Yaniv Altshuler, Erez Shmueli, Joshua B Tenenbaum, and Alex Pentland.

Human collective intelligence as distributed bayesian inference. *arXiv preprint arXiv:1608.01987*, 2016.

[98] S. Kraus, O. Shehory, and G. Taase. Coalition formation with uncertain heterogeneous information. In *Proceedings of the second international joint conference on Autonomous agents and multiagent systems*, pages 1–8, 2003.

[99] Soniya Lalwani, Harish Sharma, Suresh Chandra Satapathy, Kusum Deep, and Jagdish Chand Bansal. A survey on parallel particle swarm optimization algorithms. *Arabian Journal for Science and Engineering*, 44:2899–2923, 2019.

[100] S.M. LaValle, D. Lin, L.J. Guibas, J.C. Latombe, and R. Motwani. Finding an unpredictable target in a workspace with obstacles. In *Proceedings of the 1997 IEEE International Conference on Robotics and Automation (ICRA-97)*, pages 737–742, 1997.

[101] David Lazer and Allan Friedman. The network structure of exploration and exploitation. *Administrative Science Quarterly*, 52(4):667–694, 2007.

[102] Alexander Lipton and MARCOS LÓPEZ DE PRADO. A closed-form solution for optimal ornstein–uhlenbeck driven trading strategies. *International Journal of Theoretical and Applied Finance*, 23(08):2050056, 2020.

[103] Yang-Yu Liu, Jose C Nacher, Tomoshiro Ochiai, Mauro Martino, and Yaniv Altshuler. Prospect theory for online financial trading. *PloS one*, 9(10):e109458, 2014.

[104] V.J. Lumelsky and K.R. Harinarayan. Decentralized motion planning for multiple mobile robots: The cocktail party model. *Autonomous Robots*, 4(1):121–136, 1997.

[105] D. MacKenzie, R. Arkin, and J. Cameron. Multiagent mission specification and execution. *Autonomous Robots*, 4(1):29–52, 1997.

[106] E. Manisterski, R. Lin, and S. Kraus. Understanding how people design trading agents over time. In *AAMAS '08: Proceedings of the 7th international joint conference on Autonomous agents and multiagent systems*, pages 1593–1596, Richland, SC, 2008. International Foundation for Autonomous Agents and Multiagent Systems.

[107] S. Mastellone, D.M. Stipanovi, C.R. Graunke, K.A. Intlekofer, and M.W. Spong. Formation control and collision avoidance for multi-agent non-holonomic systems: Theory and experiments. *The International Journal of Robotics Research*, 27(1):107–126, 2008.

[108] M.J. Mataric. Designing emergent behaviors: From local interactions to collective intelligence. In J.Meyer, H.Roitblat, and S.Wilson, editors, *Proceedings of the Second International Conference on Simulation of Adaptive Behavior*, pages 432–441. MIT Press, 1992.

[109] M.J. Mataric. *Interaction and Intelligent Behavior.* PhD thesis, Massachusetts Institute of Technology, 1994.

[110] Michalis Mavrovouniotis, Changhe Li, and Shengxiang Yang. A survey of swarm intelligence for dynamic optimization: Algorithms and applications. *Swarm and Evolutionary Computation*, 33:1–17, 2017.

[111] John A McDermid, Yan Jia, Zoe Porter, and Ibrahim Habli. Artificial intelligence explainability: the technical and ethical dimensions. *Philosophical Transactions of the Royal Society A*, 379(2207):20200363, 2021.

[112] N. Michael, M.M. Zavlanos, V. Kumar, and G.J. Pappas. Distributed multi-robot task assignment and formation control. In *IEEE International Conference on Robotics and Automation, 2008 (ICRA 2008)*, pages 128–133, 2008.

[113] Seyedali Mirjalili, Seyed Mohammad Mirjalili, and Andrew Lewis. Grey wolf optimizer. *Advances in engineering software*, 69:46–61, 2014.

[114] Itay Mosafi, Eli David, Yaniv Altshuler, and Nathan S Netanyahu. Dnn intellectual property extraction using composite data. *Entropy*, 24(3):349, 2022.

[115] R. Olfati-Saber. Flocking for multi-agent dynamic systems: Algorithms and theory. *IEEE Transactions on Automatic Control*, 51(3):401–420, 2006.

[116] E. Pagello, A. D'Angelo, C. Ferrari, R. Polesel, R. Rosati, and A. Speranzon. Emergent behaviors of a robot team performing cooperative tasks. *Advanced Robotics*, 2002.

[117] E. Pagello, A. D'Angelo, F. Montesello, F. Garelli, and C. Ferrari. Cooperative behaviors in multi-robot systems through implicit communication. *Robotics and Autonomous Systems*, 29(1):65–77, 1999.

[118] Wei Pan, Yaniv Altshuler, and Alex Pentland. Decoding social influence and the wisdom of the crowd in financial trading network. In *Privacy, Security, Risk and Trust (PASSAT), 2012 International Conference on and 2012 International Confernece on Social Computing (SocialCom)*, pages 203–209. IEEE, 2012.

[119] L.E. Parker. Alliance: An architecture for fault-tolerant multi-robot cooperation. *IEEE Transactions on Robotics and Automation*, 14(2):220–240, 1998.

[120] L.E. Parker and C. Touzet. Multi-robot learning in a cooperative observation task. *Distributed Autonomous Robotic Systems*, 4:391–401, 2000.

[121] Kevin M Passino. Biomimicry of bacterial foraging for distributed optimization and control. *IEEE control systems magazine*, 22(3):52–67, 2002.

[122] Martín Pedemonte, Sergio Nesmachnow, and Héctor Cancela. A survey on parallel ant colony optimization. *Applied Soft Computing*, 11(8):5181–5197, 2011.

[123] A. Pentland and Y. Altshuler. *Social Physics and Cybercrime*, chapter New Solutions for Cybersecurity, pages 351–364. MIT Press, 2018.

[124] M. Pfingsthorn, B. Slamet, and A. Visser. *RoboCup 2007: Robot Soccer World Cup XI*, volume 5001 of *Lecture Notes in Computer Science*, chapter A Scalable Hybrid Multi-robot SLAM Method for Highly Detailed Maps, pages 457–464. Springer Berlin, 2008.

[125] S. Premvuti and S. Yuta. Consideration on the cooperation of multiple autonomous mobile robots. In *Proceedings of the IEEE International Workshop of Intelligent Robots and Systems*, pages 59–63, 1990.

[126] R. Puzis, Y. Altshuler, Y. Elovici, S. Bekhor, Y. Shiftan, and A.S. Pentland. Augmented betweenness centrality for environmentally-aware traffic monitoring in transportation networks. Journal of Intelligent Transportation Systems. 17(1): 91–105, 2013.

[127] Eyal Regev, Yaniv Altshuler, and Alfred M Bruckstein. The cooperative cleaners problem in stochastic dynamic environments. *arXiv preprint arXiv:1201.6322*, 2012.

[128] M. Rehak, M. Pechoucek, P. Celeda, V. Krmicek, M. Grill, and K. Bartos. Multi-agent approach to network intrusion detection. In *AAMAS '08: Proceedings of the 7th international joint conference on Autonomous agents and multiagent systems*, pages 1695–1696, Richland, SC, 2008. International Foundation for Autonomous Agents and Multiagent Systems.

[129] I.M. Rekleitis, G. Dudek, and E. Milios. Experiments in free-space triangulation using cooperative localization. In *IEEE/RSJ/GI International Conference on Intelligent Robots and Systems (IROS)*, 2003.

[130] W. Ren and N. Sorensena. Distributed coordination architecture for multi-robot formation control. *Robotics and Autonomous Systems*, 56(4):324–333, 2008.

[131] Christopher Rouff, Walter Truszkowski, James Rash, and Mike Hinchey. Formal approaches to intelligent swarms. In *Software Engineering Workshop, 2003. Proceedings. 28th Annual NASA Goddard*, pages 51–57. IEEE, 2003.

[132] Christopher A Rouff, Walter F Truszkowski, JL Rash, and MG Hinchey. A survey of formal methods for intelligent swarms. *Greenbelt, MD: NASA Goddard Space Flight Center*, 2005.

[133] S. Sariel and T. Balch. Real time auction based allocation of tasks for multi-robot exploration problem in dynamic environments. In *In Proceedings of the AAAI-05 Workshop on Integrating Planning into Scheduling*, pages 27–33, 2005.

[134] S. Savarimuthu, M. Purvis, and M. Purvis. Tag based model for knowledge sharing in agent society. discussion paper 2009/01. Discussion paper, Department of Information Science, University of Otago, Dunedin, New Zealand, 2009.

[135] R. Sawhney, K.M. Krishna, K. Srinathan, and M. Mohan. On reduced time fault tolerant paths for multiple uavs covering a hostile terrain. In *AAMAS '08: Proceedings of the 7th international joint conference on Autonomous agents and multiagent systems*, pages 1171–1174, Richland, SC, 2008. International Foundation for Autonomous Agents and Multiagent Systems.

[136] Alan C Schultz and Lynne E Parker. *Multi-robot Systems: From Swarms to Intelligent Automata: Proceedings from the 2002 NRL Workshop on Multi-robot Systems*. Springer Science & Business Media, 2013.

[137] Erez Shmueli, Itzik Mazeh, Laura Radaelli, Alex Sandy Pentland, and Yaniv Altshuler. Ride sharing: A network perspective. In *Social Computing, Behavioral-Cultural Modeling, and Prediction*, pages 434–439. Springer, 2015.

[138] H.A. Simon. *The Sciences of the Artificial, Second Edition*. MIT Press, 1981.

[139] Shahar Somin, Yaniv Altshuler, Goren Gordon, Alex Pentland, and Erez Shmueli. network dynamics of a financial ecosystem. *Scientific reports*, 10(1):1–10, 2020.

[140] Shahar Somin, Yaniv Altshuler, Alex 'Sandy'Pentland, and Erez Shmueli. Beyond preferential attachment: falling of stars and survival of superstars. *Royal Society Open Science*, 9(8):220899, 2022.

[141] Shahar Somin, Yaniv Altshuler, Erez Shmueli, et al. Remaining popular: power-law regularities in network dynamics. *EPJ Data Science*, 11(1):61, 2022.

[142] Shahar Somin, Goren Gordon, and Yaniv Altshuler. Social signals in the ethereum trading network. *arXiv preprint arXiv:1805.12097*, 2018.

[143] Konstantina Spanaki, Erisa Karafili, Uthayasankar Sivarajah, Stella Despoudi, and Zahir Irani. Artificial intelligence and food security: swarm intelligence of agritech drones for smart agrifood operations. *Production Planning & Control*, 33(16):1498–1516, 2022.

[144] P. Stone and M. Veloso. Task decomposition, dynamic role assignment, and low-bandwidth communication for real-time strategic teamwork. *Artificial Intelligence*, 110(2):241–273, 1999.

[145] Thomas Stützle, Marco Dorigo, et al. Aco algorithms for the traveling salesman problem. *Evolutionary algorithms in engineering and computer science*, 4:163–183, 1999.

[146] H. Su, X. Wang, and Z. Lin. Flocking of multi-agents with a virtual leader. *IEEE Transactions on Automatic Control*, 54(2):293–307, 2009.

[147] P. Svestka and M.H. Overmars. Coordinated path planning for multiple robots. *Robotics and Autonomous Systems*, 23(3):125–152, 1998.

[148] Jim Torresen. A review of future and ethical perspectives of robotics and ai. *Frontiers in Robotics and AI*, 4:75, 2018.

[149] G. Trajkovski and S.G. Collins. *Handbook of Research on Agent-Based Societies: Social and Cultural Interactions*. Idea Group Inc (IGI), 2009.

[150] I.A. Wagner, Y. Altshuler, V. Yanovski, and A.M. Bruckstein. Cooperative cleaners: A study in ant robotics. *The International Journal of Robotics Research (IJRR)*, 27(1):127–151, 2008.

[151] I.A. Wagner and A.M. Bruckstein. Row straightening via local interactions. In *Technical report CIS-9406, Center for Intelligent Systems, Technion, Haifa*, 1994.

[152] I.A. Wagner and A.M. Bruckstein. Ants: agents, networks, trees and subgraphs. *Future Generation Computer Systems Journal*, 16(8):915–926, 2000.

[153] I.A. Wagner and A.M. Bruckstein. From ants to a(ge)nts: A special issue on ant—robotics. *Annals of Mathematics and Artificial Intelligence, Special Issue on Ant Robotics*, 31(1–4):1–6, 2001.

[154] I.A. Wagner, M. Lindenbaum, and A.M. Bruckstein. Efficiently searching a graph by a smell-oriented vertex process. *Annals of Mathematics and Artificial Intelligence*, 24:211–223, 1998.

[155] P.K.C. Wang. Navigation strategies for multiple autonomous mobile robots. In *Proceedings of the IEEE/RSJ International Conference on Intelligent Robots and Systems (IROS)*, pages 486–493, 1989.

[156] Sherry Wasilow and Joelle B Thorpe. Artificial intelligence, robotics, ethics, and the military: A canadian perspective. *AI Magazine*, 40(1):37–48, 2019.

[157] A. Weitzenfeld. A prey catching and predator avoidance neural-schema architecture for single and multiple robots. *Journal of Intelligent and Robotic Systems*, 51(2):203–233, 2008.

[158] M.P. Wellman and P.R. Wurman. Market-aware agents for a multiagent world. *Robotics and Autonomous Systems*, 24:115–125, 1998.

[159] Alan FT Winfield, Christian Blum, and Wenguo Liu. Towards an ethical robot: internal models, consequences and ethical action selection. In *Advances in Autonomous Robotics Systems: 15th Annual Conference, TAROS 2014, Birmingham, UK, September 1-3, 2014. Proceedings 15*, pages 85–96. Springer, 2014.

[160] H. Work, E. Chown, T. Hermans, and J. Butterfield. Robust team-play in highly uncertain environments. In *AAMAS '08: Proceedings of the 7th international joint conference on Autonomous agents and multiagent systems*, pages 1199–1202, Richland, SC, 2008. International Foundation for Autonomous Agents and Multiagent Systems.

[161] A. Yamashita, M. Fukuchi, J. Ota, T. Arai, and H. Asama. Motion planning for cooperative transportation of a large object by multiple mobile robots in a 3d environment. In *In Proceedings of IEEE International Conference on Robotics and Automation*, pages 3144–3151, 2000.

[162] Guang-Zhong Yang, Jim Bellingham, Pierre E Dupont, Peer Fischer, Luciano Floridi, Robert Full, Neil Jacobstein, Vijay Kumar, Marcia McNutt, Robert Merrifield, et al. The grand challenges of science robotics. *Science robotics*, 3(14):eaar7650, 2018.

[163] Xin-She Yang. Firefly algorithm, stochastic test functions and design optimisation. *International journal of bio-inspired computation*, 2(2):78–84, 2010.

[164] Xin-She Yang, Zhihua Cui, Renbin Xiao, Amir Hossein Gandomi, and Mehmet Karamanoglu. *Swarm intelligence and bio-inspired computation: theory and applications*. Newnes, 2013.

[165] V. Yanovski, I.A. Wagner, and A.M. Bruckstein. Vertex-ants-walk: a robust method for efficient exploration of faulty graphs. *Annals of Mathematics and Artificial Intelligence*, 31(1–4):99–112, 2001.

[166] V. Yanovski, I.A. Wagner, and A.M. Bruckstein. A distributed ant algorithm for efficiently patrolling a network. *Algorithmica*, 37:165–186, 2003.

[167] X. Zheng and S. Koenig. Reaction functions for task allocation to cooperative agents. In *Proceedings of the 7th international joint conference on Autonomous agents and multiagent systems*, pages 559–566, 2008.

[168] R. Zlot, A. Stentz, M.B. Dias, and S. Thayer. Multi-robot exploration controlled by a market economy. In *Proceedings of the IEEE International Conference on Robotics and Automation*, volume 3, pages 3016–3023, 2002.

Chapter 2

Modeling and Prediction of Ride Sharing Utilization Dynamics

Tal Altshuler[1], Yaniv Altshuler[2], Rachel Katoshevski[1], and Yoram Shiftan[1]

[1]Department of Civil and Environmental Engineering, Israel Institute of Technology

[2]Massachusetts Institute of Technology, Media Lab

2.1 Introduction

The increasing availability of portable technologies gives new fuel to studies on metropolitan transportation optimization, pushing urban design one step closer towards the long sought concept of "smart cities" [1, 2]. Mobile devices and ubiquitous connectivity make it easier than ever to collect data on the way people live in cities and big-data analytic methods facilitate the extraction of actionable insights from it. City administrators and policy makers can in turn act upon such results to enhance city management, channeling current advancements in data analysis for the immediate improvement of urban quality of life.

Many of the fundamental problems in big cities nowadays relate to cars. The high number of vehicles congests the streets, vehicles standing in traffic jams increase air pollution while also increasing traveling times, significantly increasing passengers' stress levels. Availability of large-scale datasets accompanied with recent advancements in the analysis of big-data and the development of novel models of human mobility, give rise to new possibilities to study urban mobility.

Such new models include for example the work of [3] in which large-scale mobile phone data was analyzed in order to characterize individual mobility, showing that human travel patterns are far from random, and are efficiently describable by a single spatial probability distribution. Similarly, [4] show that mobile phone data can be used as a proxy to examine urban mobility and [5] analyzes social network data of different cities to find that mobility highly correlates with the distribution of urban points of interest. Mobile technologies are also the enablers of many successful consumer applications, such as Waze [6], that provide traffic-aware city navigation by using data provided by the community. Alternative ways of moving in the city, such as autonomous mobility-on-demand and short-term car rental have been identified among the possible solutions to the ever-growing transport challenge [7].

Ride sharing has the potential of improving traffic conditions by reducing the number of vehicles on the roads, reducing the emission of CO_2 and the fuel consumption per person and giving the riders the opportunity to socialize with people (that otherwise would have been fierce "road competitors"). A recent study [8] shows that traffic in the city of Madrid can be reduced by 59% if people are willing to share their home-work commute ride with neighbors. Even if they are not willing to ride with strangers, but only with friends of friends (for safety issues), the potential reduction is still up to 31%. Another recent study [9] had shown that on-demand route-free public transportation based on mobile phones outperforms standard fix-route assignment methods when comparing traveling times. These results encourage the deployment and policies supporting ride sharing in urban settings.

However, despite such evidences and others, ride-sharing adoption rate in cities worldwide is slower than what can be expected given the clear benefits of ride-sharing [10, 11]. One important reason, as suggested by [12–14] and others is the uneven, and often unstable, potential benefits associated with ride-sharing. When the value that can be extracted from using a service such as *Lyft* [15], *Uber* [16] or *Sidecar* [17] is high at one part of the city, but significantly lower at another neighborhood, or worse – suddenly decreases for a period of two days – potential users of the service are much likely to opt for a private car usage [18].

In this work we propose a data-driven framework to dynamically predict the impact, or *potential utilization*, of ride sharing in a city, at different times, and in different regions. Specifically, the technique we propose provides both policy makers as well as ride-charing operators tools for assessing the future benefit of ride-sharing, encapsulated through the percent of rides saveable through merging of nearby departures and destinations. Simply put, a shared taxi service can use this proposed technique in order to know ahead of time what the ride-sharing demand is going to be (at various places in the city), whereas municipal services can dynamically change tolls and service fees in order to incentivize the use of ride-sharing in "low hours" that are predicted in advance.

Our method is based on analyzing the network features of the dynamic O-D matrix as represented by data collected by various sources, such as mobile phone call records, or sensors mounted on the taxis themselves. In our research we show a clear correlation between such properties and the portion of "merge-able rides". We have analyzed the efficacy of our proposed network-oriented method using a dataset of over 14 Million taxi trips taken in New York City during January 2013 [19].

This work is structured as follows: Section 2.2 presents an overview of the relevant related research in the field. In Section 2.3 we discuss the data and analytic methodologies that were used for this work: starting with the calculation of the average ride-sharing potential as a function of the maximum delay a taxi-user would be willing to sustain, we demonstrate that more than 70% of the rides can be shared when users are willing to undertake up to 5 minutes delay. We then demonstrate that urban ridesharing potential is not only highly dynamic, but that it can also be predicted using the analysis of the rides that took place in the city a few hour beforehand. We present a method for comprising a dynamically changing network using the taxi-rides, and analyzing the topological properties of this network (Section 2.4). We analyze the dynamics of these properties over time, and demonstrate our ability to accurately predict changes in the utilization of ride-sharing several hours in advance. Concluding remarks and suggestions for future works are contained in Section 2.5.

This work was first published in [20].

2.2 Related Work

Network features are often used to predict events or properties that are external to the network, but still influence it [21–23]. A network can often be built on easily available data and serve as an important source for predictions regarding various (seemingly unrelated) events and large-scale decision-making processes [24–29]. Features of a phone call network can signal the occurrence of an emergency situation [30] or predict trust among individuals [31], and specific behaviors in a Twitter account can identify a spamming account [32], whereas similar techniques can be used to distinguish human-controlled financial activity from automatic "trading bots" [33]. Such discoveries had sparked the interest of researchers in different research fields, who could benefit from this new ability to model large-scale human behavioral patterns and dynamics. One of the fields most influenced by this evolving research thrust was the data-driven study of human mobility and its potential application for Intelligent Transportation Systems [11,34–37].

It has been recently shown that in trying to detect semantic network events (such as an accident or a traffic jam) it is crucial to understand the underlying structure of the network these events are taking place at [38–40], the role of the link weights [41], as well as the response of the network to node and link removal [42]. Past research [43] had pointed out the existence of powerful patterns in the placement of links, or that clusters of strongly tied together individuals tend to be connected by weak ties [41]. It was also shown that this finding provides insight into the robustness of the network to particular patterns of link and node removal, as well as into the spreading processes that take place in the network [44, 45]. In addition, recent work had demonstrated the trade-off between the number of individuals (the width of the data) and the amount of information available from each one (the depth of the data), with respect to the ability to accurately model crowd behavior [46–50]. An analytical approach to this problem discussing the (surprisingly large) amount of personal information that can be deduced by an "attacker" who has access to one's personal interactions' meta-data can be found in [51–54].

One of the first works that examined the statistical distribution of event appearance in mobility and communication networks have found that these follow a power law principle [55], and that such distributions are significantly affected by anomalous events that are external to the networks [56–58].

A method for filtering mobile phones Call Data Records (CDRs) in space and time using an agglomerative clustering algorithm in order to reconstruct the origin-destination urban travel patterns was recently suggested in [59]. Other works had examined the evolution of node groups in the network, aiming for the development of algorithms capable of identifying "new clusters" – a certain kind of anomalous network pattern [60].

Recent works that have been analyzing data collected by the pervasive use of mobile phones have broadly supported the notion that most of human mobility patterns are affected by a relatively small number of factors, easily modeled, and very predictable [4, 61–64]. A comprehensive survey of ride-sharing literature can also be found in [65] and another relevant study that developed a spatial, temporal, and hierarchical decomposition solution strategy for ride-sharing as presented in [66].

To date, much of the research related to ride sharing has focused on understanding the characteristics of ride sharing trips and users. In a recent survey of app-based, on-demand rideshare users in San Francisco, researchers found that 45% of ridesharers stated they would have used a taxi or driven their own car had ridesharing not been available, while 43% would have taken transit, walked, or cycled [67].

A recent work by Santi et al. [68] introduces a way of quantifying the benefits of sharing. The study applies to a GPS dataset of taxi rides in New York City and uses the notion of *shareability network* to quantify the impact and the feasibility of taxi-sharing. When passengers have a 5 minutes flexibility on the arrival time, and they are willing to wait up to 1 minute after calling the cab, over 90% of the sharing opportunities can be exploited and 32% of travel time can be saved. The authors have also shown that the problem is computationally tractable when we look for sharing a taxi among two people with the option of in-route picking up. Furthermore, sharing solutions involving more people are not tractable, but do not provide a significant improvement with respect to solutions involving only two people. Similar results have been demonstrated using a theoretical model analyzing *Autonomous Mobility On Demand* system, demonstrating that a combined predictive positioning and ridesharing approach is capable of reducing customer service times by up to 29% [69].

An extensive simulation infrastructure for ride sharing analysis is suggested in [70], allowing the initialization and tracking of a wide variety of realistic scenarios, monitoring the performance of the ride sharing system from different angles, considering different stakeholders' interests and constraints. The simulative infrastructure is claimed to use an optimization algorithm that is linear in the number of trips and makes use of an efficient and fully parallelized indexing scheme.

In another study by Cici et al. [8] mobile phone data and social network data were used to estimate the benefits of ride sharing on the daily home-work commute. Mobile phone data are easier to collect than GPS traces, and have a higher penetration, providing a good sample of a city's mobility. Social network data are used to study the effect of friendship on the potential of ride sharing, showing that if people want to travel only with friends then expected ride sharing benefits are negligible. On the other hand, when people are willing to ride with friends of friends the achieved efficiency resembles that of the variant that also allows riding with strangers (implying that safety issues may have a significant effect on the actual success of a ride sharing solution).

A similar study has been presented by [71] calculating shareability curves using millions of taxi trips in New York City, San Francisco, Singapore, and Vienna, showing that a natural re-scaling collapses them onto a single, universal curve. The authors presented a model that predicts the potential for ride sharing in any city, using a few basic urban quantities and no adjustable parameters. A related study [72] analyzed the assumption that seamless integration of ride-sharing and public transit may offer fast, reliable, and affordable transfers to and from transit stations in suburban areas thereby enhancing mobility of the residents.

The issue of pricing policies in ride sharing services has gained significant attention recently. with the booming expansion of commercial ride sharing services such as *Uber*, *Lyft* and others. The work of [73] studies dynamic pricing policies for ride sharing platforms. As such platforms are two-sided this requires economic models that capture the incentives of both drivers and passengers. In addition, such platforms support high temporal-resolution for data collection and pricing. The combination of the latter requires stochastic models that capture the dynamics of drivers and passengers in the system.

In [74] the authors highlight the impact of the demand pattern of the underlying network on the platform's optimal profits and aggregate consumer surplus. In particular, the authors establish that both profits and consumer surplus are maximized when the demand pattern is balanced across the network's locations. In addition, the authors show that profits and consumer surplus are monotonic with the "balancedness" of the demand pattern (as formalized by the pattern's structural properties).

The work of [75] proposes a recommendation framework to predict and recommend whether and where should ride sharing users wait in order to maximize their chances of getting a ride. In the framework, a large-scale GPS data set generated by over 7,000 taxis in a period of one month in Nanjing, China was utilized to model the arrival patterns of occupied taxis from different sources.

The recent work of Alexander and Gonzalez [11] uses smart-phone data in order to model the behavior of an urban population in Boston, in an attempt to assess the impact of efficient ridesharing services on the urban traffic, and specifically on the expected levels of congestion. This data-centric approach leads to a highly accurate modeling of the mobility patterns in the city. However, much like most of the recent work on this subject, the researchers have followed an aggregative modeling, that tries to find the static long-term definitive mobility patterns, purposely omitting any dynamic fluctuations.

In another study, researchers from the Microsoft Research Center [76] analyzed the ride data of 12,000 taxis during 110 days in order to model the mobility patterns of potential passengers. Using this probabilistic model the researchers were able to build a recommendation system for taxi drivers that would maximize their profits (yielding an overall 10% improvement in the overall profits) and a second recommendation system for passengers, advising them where to turn in order to maximize their chances of finding a vacant taxi (with 67% accuracy). A similar research can be found in [77].

A recent review of dynamic ridesharing systems [78] focused on the optimization problem of finding efficient matches between passengers and drivers. This ride-matching optimization problem determines vehicle routes and the assignment of passengers to vehicles considering the conflicting objectives of maximizing the number of serviced passengers, minimizing the operating cost, and minimizing passenger inconvenience. Another study [79] presented an algorithm that increases the potential destination choice for ride sharing schemes set by considering alternative destinations that are within given space-time budgets.

On a similar note, a recent study [80] analyzed the benefits of meeting points in ride-sharing systems, investigating the potential benefits of introducing meeting points in a ride-sharing system. With meeting points, riders can be picked up and dropped off either at their origin and destination or at a meeting point that is within a certain distance from their origin or destination. The increased flexibility results in additional feasible matches between drivers and riders, and allows a driver to be matched with multiple riders without increasing the number of stops the driver needs to make. A similar approach for the optimization of such meetings points was discussed in [81].

The challenge of rides-matching was also discussed in works such as [82], [83] or [84], which have demonstrated that 2,000 vehicles (15% of the taxi fleet in New York) with a capacity of 10 passengers (or 3,000 vehicles of capacity of 4 passengers) can serve 98% of the New York taxi demand within a mean waiting time of 2.8 minutes and mean trip delay of 3.5 minutes.

A *path merging* approach, which instead of merging rides to and from the same locations calculates new paths which go through the same locations of the original trips, in the same order, and thus improves the ability to merge rides, was discussed in [85].

In a recent theoretical study [86] where the combinatorial optimization of ridesharing matching problem was tackled using the proof of equivalence between classical centroid clustering problems and a special case of set partitioning called metric k-set partitioning, in which an efficient expectation maximization algorithm was used to achieve a 69% reduction in total vehicle distance, as compared with no ridesharing.

A fully decentralized reputation-based approach is discussed in [87], using a peer-to-peer architecture to provide self-assembling ride sharing infrastructure capable of functioning with no central authority or regulator.

2.3 Dataset and Methodology

Our analysis was performed using a dataset of 14,776,615 taxi rides collected in New York City over a period of one month (January 2013) (publicly accessible at the NYC Taxi Repository [19]). Each ride record consists of the following fields: pick-up time, pick-up longitude, pick-up latitude, drop-off longitude, drop-off latitude, number of passengers per ride, average velocity and overall trip duration. Times granularity is second-based and positional information has been collected via GPS technology by the data provider. From this raw data sample we omit records containing missing or erroneous GPS coordinates, as well as records that represent rides that started or ended outside Manhattan, yielded a cleaned dataset containing 12,784,243 rides.

As a first step in modeling the feasibility and efficiency of ride-sharing schemes using taxi rides in New York City, a comprehensive understanding of the data itself is required. How do the rides distribute over the various geographic locations? Are there patterns that emerge when observing the O-D matrix of the various rides? Can we use those in order to predict the destinations of passengers when they board a taxi at a certain location? The figures below attempt to answer some of the Power Low distributions) strongly implying on the potential of a network-centric approach as the method of choice with respect to the modeling of the dynamics of the data.

Some of the following illustrations analyzing the dataset's statistical properties were first presented in [88, 89]. These illustrations appear here to contribute to the reader's understanding of the nature of the data and the behavior dynamics it encapsulates.

Figure 2.1 reports the distribution of rides per day of the week and per hour of the day. As can be seen in the figure, the number of rides has a far-from-uniform time distribution. More specifically, the number of rides is higher in the middle of the week and is lower during the weekend. In addition, the daily rides distribution peaks, as expected, in the morning hours and around 6–7pm.

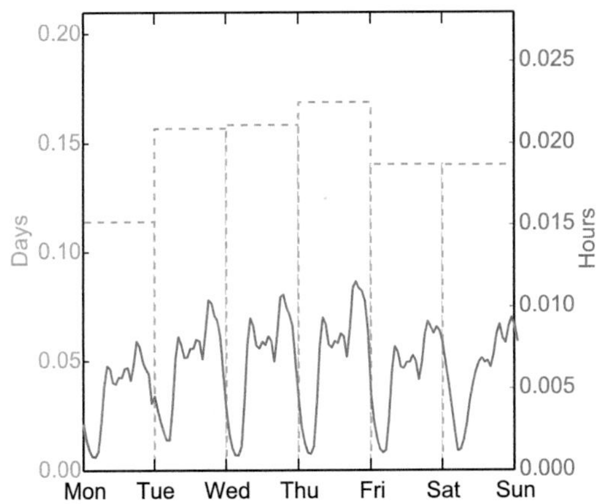

Figure 2.1: Probability Density Function (PDF) of the number of rides per day of week/hour of day. Afternoon peaks are centered on average around 7pm. This chart was taken from [89].

We use the set of taxi ride records to construct a "rides network" G_{T_1,T_2}, comprising of $|V|$ nodes representing equally sized squared regions of New York City, and a set of $|E|$ edges, such that each edge $(u,v) \in E$ corresponds to a connection between two regions $u,v \in V$ if and only if there exists at least one ride from region u to region v in the time-frame referred to by the network. Such a connection exists if and only if a ride started at some time t departing at u and reaching v, or vice versa, such that $T_1 \leq t \leq T_2$ is contained in the time period defined for the network G_{T_1,T_2}.

As we create edges only based on rides that were created during a certain period of time the network may change (and quite significantly so) for various values selected for T_1 and T_2. As the time period defined by these values increases the network is expected to contain more edges, with the densest network received for $G = G_{-\infty,\infty}$ being the network that is based on the complete aggregation of all the rides. In order to encapsulate the traffic properties of a certain point in time T we would observe the time period circumventing T. Similarly, in order to analyze the network dynamics, that is – the way it changes over time, we would analyze the evolution of the network properties for networks created in non-identical yet partially-overlapping time periods. This methodology is extensively used in Section 2.4.

For different granularity of city partitioning (reflecting through the use of different sizes of the square regions) different ride networks would be produced. However, Network Theory implies that changing this parameter would not affect the existence of various mathematic invariants such as the network's "Scale Free-ness" or its expected small diameter [90], but rather – mainly change the sparsity of the network and its number of nodes. During this work we have examined several sizes of squared-regions, ranging from rectangular regions of 0.0156 square miles in size, to 1 square mile, obtaining similar results. The analysis below is based on square tiles of 0.39 square mile (i.e. 1 square kilometer). In such a case, when taking $G = G_{-\infty,\infty}$, the network that aggregates all the rides, it comprises of 813 nodes and 58,014 edges. Figure 2.2 illustrates the geographical distribution of the nodes V_G on the map of New York.

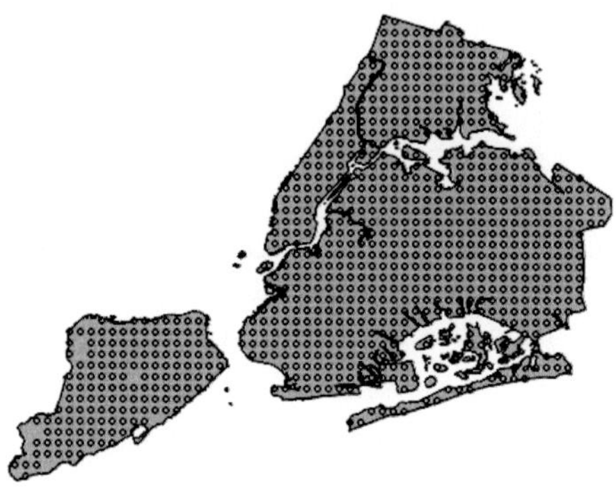

Figure 2.2: Illustration of the rides network G, portrayed on the map of NYC. It can be seen that the network has high density through the city, with a few empty spots in Staten Island. The figure is based on the result presented in [89].

Figure 2.3 illustrates the distribution of the number of trips on the various O-D routes in the taxi network. By *weight* we refer to the number of trips that took place through this edge and by *Frequency* we refer to the number of edges that have a specific weight. Note the small number of edges which have more than 500 rides (approximately 5,000 edges out of 58,000 edges). Similarly, over 47,000 edges has less than 50 rides passing through them. This observation coincides well with the fact that human mobility is known to follow a power low distribution [3].

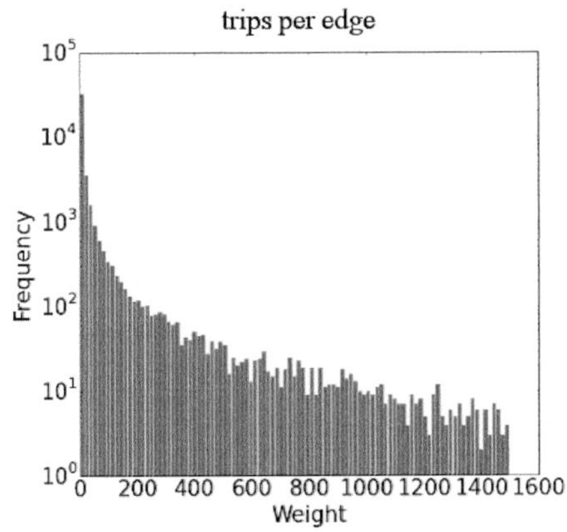

Figure 2.3: Edge weights of the taxi rides network G, denoting the number of trips per edge (namely, between every two nodes in the city). The X-axis denotes the number of trips per edge (representing a pair of origin-destination nodes), and the Y-axis (shown in a log-scale) represents the number of edges who have such number of trips.

As we analyze the network properties of the graph implied by the taxi rides, it is interesting to observe the characteristics of the degrees of the nodes of the network G. A 'degree' of a node $v \in V$ is the number of nodes v connected through edges in E, where such nodes represent the actual destinations passengers who boarded a taxi at location v chose to go to. Namely, a degree of a node v represents therefore the number of possible destinations a passenger boarding a taxi on location v may chose to go to. An important observation is that the popularity of a node v as reflected both by its in-degree (i.e. the number of origins passengers depart from in order to get to v) as well as by its out-degree (i.e. the number of destinations passengers leaving v may go to) is independent of the geographic size or shape of node v – as all nodes refer to equally-sized square regions.

Interestingly, analyzing the distribution of this property reveals that whereas there are some nodes with a high degree (probably corresponding to main train stations or large administration facilities) the vast majority of the nodes have a very low degree. In other words—for the vast majority of locations in New York, it is extremely easy to predict the destinations of a passenger starting his ride there (as a low degree implies a low number of possible destinations, and a high chance of guessing the correct one). This observation is quite remarkable, as it implies that taxi users are *much more predictable* than may seem. Indeed, it seems that when one boards a taxi one's destination can quite accurately be predicted.

Specifically, in 24% of the possible origins of a taxi ride in New York City, the number of possible destinations of a passenger leaving these origins is on average 5, and in 43% of the origins it is 10. A quick arithmetics yields that if at some point in time we would pick a random person just boarding a taxi anywhere in New York, we would have more than 7.5% of guessing precisely his or her destination. This probability is about three times *higher* than rolling a "Snake Eyes" (two 1's in a 6-sided dice). See Figure 2.4 for more details.

In this context it is also important to note that in this work we are less interested in the specific characterization of nodes having high (or low) degrees, but rather – in the dynamics those values represent over time, as discussed in detail in the following sections.

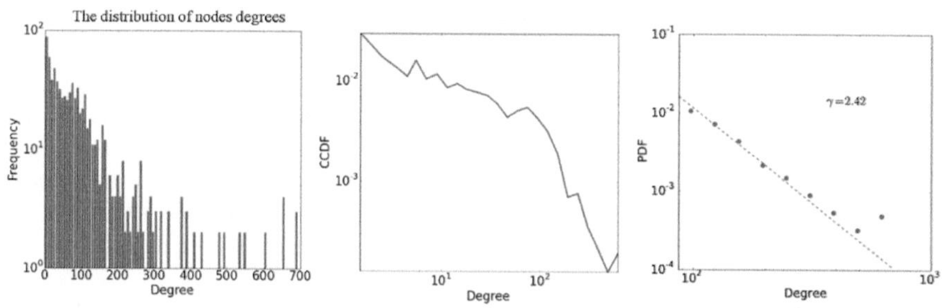

Figure 2.4: The distribution of nodes degrees in the taxi rides network, representing the number of possible destinations a passenger boarding a taxi at some location in the city may chose to go to. Note the surprisingly high number of origins with very low degrees – number of possible destinations. The nodes' degrees are unaffected by the size or shape of the actual geographic region they refer to as all nodes refer to equally-sized square patches of the city.

In order to analyze the 'sharability', or the ability to merge rides using the same vehicle at overlapping times, we applied a simplified version of the methodology used by Santi et al. [68] to calculate the potential benefits of ride sharing: Let $T_i = (o_i, d_i, t_i^o, t_i^d), i = 1...k$ be k trips where o_i denotes the origin of the trip, d_i the destination, and t_i^o, t_i^d the starting and ending times, respectively. We say that multiple trips T_i are shareable if there exists a route connecting all of their origins and destinations in any order where each o_i precedes the corresponding d_i.

Sharability, or 'ridesharing utilization' is expressed in terms of the number of rides that can be 'merged', as a function of the guaranteed quality of service, expressed through the number of latency minutes agreeable by the passengers — the maximum time delay in catching a ride and arriving at a destination, representing the maximum discomfort that a passenger can experience using the service. In other words, given a pre-defined level of discomfort passengers are willing to undertake (expressed in a prolonged wait-time), the ridesharing utilization depicts the portion of rides that are redundant and can be saved by merging with other rides to and from the same locations.

Our analysis aims at finding pairs of rides, which are represented in the network by the same edge (i.e., have the same origin and destination), that can be shared. For each edge, we examine its corresponding set of originating rides, and count the number of ride pairs that can be merged, taking into consideration the maximum time delay parameter.

The main difference between our approach and the one discussed in [68] is that we only merge rides that leave the same origin 'tile' and go to the same destination 'tile'. There are several advantages for this approach:

1. The routing-agnostic scheme is significantly less sensitive than the temporary changes in the infrastructure, such as detours, traffic jams, accidents, and so on.
2. Merging rides based only on their origin and destination makes our ridesharing policy entirely agnostic to the routing decision of the driver. Alternatively, the approach that is based on allowing rides to be merged even if they do not leave from the same origin, but are rather partially overlapping, depends on the assumption that the route of the "containing ride" indeed passes through the origin of the second ride. This assumption in turn depends on either perfectly guessing the routing decisions of the driver, or – dictating those decisions to the driver by the ridesharing service.
3. As a result, our routing-agnostic approach is also expected to be easier to implement in a real-life scenario, as it requires less cooperation from the drivers.
4. In addition, the increased simplicity of the routing-agnostic approach makes it easier to optimize from a computational point of view. The routing-aware approach discussed in [68] has a time complexity of $O(n^2 \log(n))$ when merging pairs of rides [91], becomes much harder when triple rides merging is allowed [92], and eventually becomes computationally unfeasible for larger numbers of rides-to-be-merged [68].
5. When comparing the merging efficiencies of our proposed routing-agnostic approach with the routing-aware one it is shown that whereas the latter is slightly more efficient when long wait-times are allowed (increasing our proposed 73% sharability to 93% for 5 minutes maximal delay), the improvement for shorter wait times becomes significantly smaller (this is illustrated by comparing Figure 2.6 below to Figure 2.3 in [68]).

Figure 2.5 shows the probability density function (pdf) of the number of rides per edge. As can be seen from the figure, the distribution is heavy tailed and seems to follow a power-law. In other words, most of the edges (i.e., pairs of origin-destination) induce a small number of rides while a small number of edges induce an extremely high number of rides.

Figure 2.6 presents the percentage of shareable rides as a function of the maximum time delay parameter. The results are encouraging: more than 70% of the rides can be shared when passengers can accept a delay of up to 5 minutes. As expected, the benefit of ride sharing increases when the passengers are willing to take a higher discomfort, and the percentage of shareable rides is more than 90% when passengers can wait 30 minutes or more.

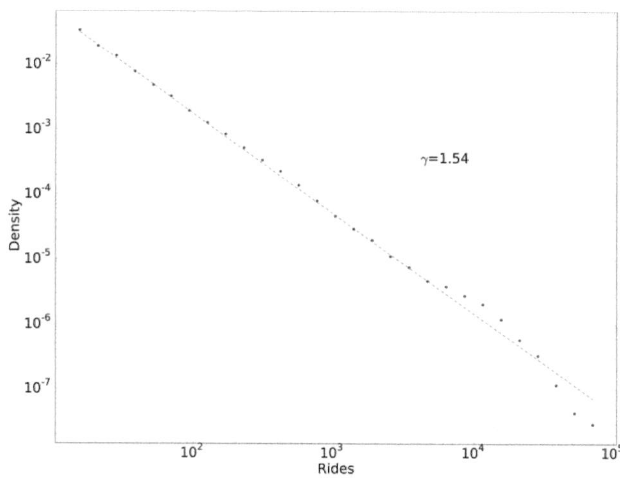

Figure 2.5: Probability Density Function of the number of rides per edge. The figure is based on the result presented in [89].

It should be noted that the simplified analysis illustrated in Figure 2.6 assumes that two rides that took place at the same time can always be merged, regardless of the number of passengers in each ride. Since the average number of passengers per ride is 1.7 and most of the rides involve a single passenger, the number of saved rides could have been even higher by merging more than 2 rides at a time. On the other hand, in some cases, even merging two rides at a time might have resulted in overcrowding of the vehicle.

In order to assess the effect of these two potential phenomena over our analysis we can observe the distribution of the number of passengers per trip in the data. While doing so, we artificially segregate trips made using private taxi caps (that can board up to 4 passengers) and trips made with larger vehicles (capable of boarding from 5 to 48 passengers):

- 49.22 percent of the trips have 1 passenger.
- 24.22 percent of the trips have 2 passengers.
- 15.72 percent of the trips have 3 passengers.
- 10.84 percent of the trips have 4 passengers.

We examine two approaches for the assessment of the actual theoretical ride-sharing utilization:

Greedy merging, assuming an even distribution of the number of passengers: In this approach analyze the merging process in a two phases greedy approach. In the first phase, we assume that all the original trips that can be merged are indeed merged, and are done so under the assumption that the number of passengers is distributed approximately uniformly, with respect to the various geographic locations. Then, the resulting merged trips are merged again, if possible. This analysis approach should result in a lower bound for the actual ride-sharing utilization, as in real life our ride-matching algorithm would aspire for maximizing the number of merged rides, where possible.

Optimal merging: In this approach we assume that whenever two rides are merged, the number of passengers they have receives the value that would result in the most efficient merging scheme possible (confined to the overall distribution of the numbers of passengers for rides). This analysis approach should result in an upper bound for the actual ride-sharing utilization, as in real life there will be times where the only way to merge rides would be in a sub-optimal way.
The following is a detailed analysis of both approaches:

Greedy merging: The expected distribution of the merged trips for the first phase would be :

- In 24.23 percent of the pairs, we would merge a trip that has 1 passenger with a trip that has 1 passenger. This results in a merged trip of 2 passengers.
- In 23.84 percent of the pairs, we would merge a trip that has 1 passenger with a trip that has 2 passengers. This results in a merged trip of 3 passengers. These trips cannot be merged, assuming the greedy 2-steps approach.
- In 15.48 percent of the pairs, we would merge a trip that has 1 passenger with a trip that has 3 passengers. This results in a merged trip of 4 passengers, that cannot be further merged.
- In 5.87 percent of the pairs, we would merge a trip that has 2 passengers with a trip that has 2 passengers. This results in a merged trip of 4 passengers, that cannot be further merged.
- In 30.58 percent of the pairs, we would not be able to merge the trips, as these would be pairs that either (a) have one of the trips with 4 passengers,

or (b) have a trip with 2 passengers and a trip with 3 passengers, or (c) have two trips having 3 passengers each.

The second phase will therefore be able to merge another $0.2423 \cdot 0.2423 \cdot 100 = 5.87$ percent of the original pairs, which reflects a $5.87 \cdot 2 = 11.74$ percent increase. Overall, this would sum up to $100 - 30.58 + 11.74 = 81.16$ percent of the naive potential utilization (namely, the utilization that is calculated under the assumption that all rides are merge-able, and that we do not merge more than two rides.

Optimal merging: Assuming an optimal merging scheme we can calculate the merging of the relevant New York City data as follows:

- 10.84 percent of the rides that have 4 passengers cannot be merged at all.
- 24.22 percent of the rides that have 2 passengers would be merged among themselves.
- 15.72 percent of the rides that have 3 passengers would be merged with a matching 15.72 percent of the rides that have 1 passenger.
- This would leave another (49.22 – 15.72 =) 33.5 percent of the rides, that have 1 passenger. These rides would be merged in a 4-to-1 ratio, virtually implying a $33.5 \cdot 1.5 = 50.25$ percent save.

Altogether, the actual optimal theoretical utilization would sum up to $24.22 + 15.72 + 15.72 + 50.25 = 105.91$ percent (namely, under the assumption of optimal merging. The benefit from merging 4 rides of a single passenger more than compensates the loss due to rides with 4 passengers.

Therefore, the actual theoretical utilization for the New York City taxi dataset, denoted as U, would be bounded by:

$$0.8116 \cdot \alpha \leq U \leq 1.0591 \cdot \alpha$$

such that α is the potential utilization that is calculated throughout this work, using the method that was described above, ignoring the effect of multiple merges, as well as the effect of over-population of rides.

2.4 Analyzing the Dynamic Ride-Sharing Network

In the previous Section we have described the taxi data that was used for this study, illustrated its various mathematical properties and discussed the way it can be analyzed for the purpose of assessing the potential ability of ride-sharing schemes to merge rides between similar locations (denoted as the *ride-sharing potential utilization*). In this Section we demonstrate the inability of static analytic approaches to efficiently model this utilization and suggest an alternative approach, that is based on the construction of multiple

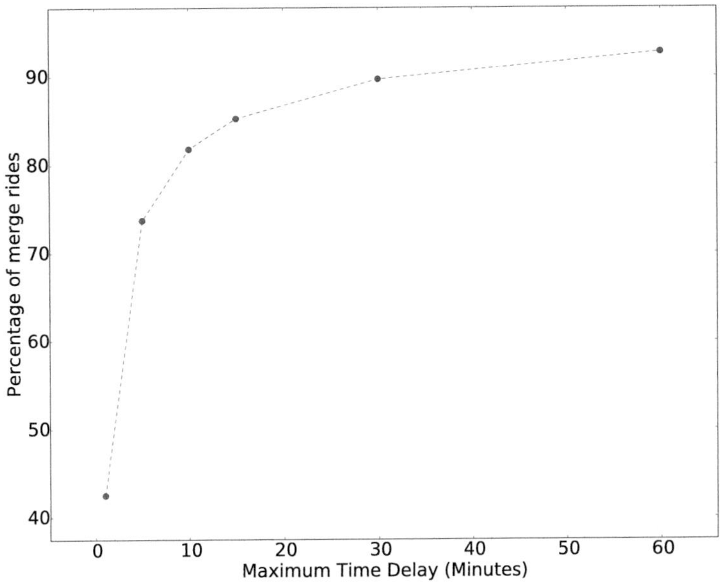

Figure 2.6: Percentage of merged rides (for the entire network). The figure is based on the result presented in [89].

network-snapshots, derived using a sliding-window based aggregation of the taxi rides. We show that this technique can serve as a valuable methodology for both (a) assessing the potential ride-sharing utilization of the current supply and demand scheme (as appears in Section 2.4.2), as well as (b) serve as a prediction method for estimating *changes* in this utilization, in the near future, up to a few hours (as shown in Section 2.4.3).

2.4.1 The Need for Dynamic Ridesharing Optimization and Prediction

Mainstream transportation analysis models (such as [93–97] and many more) approach the problem of transportation forecasting and analysis through the use of long-term data aggregation. Simply put, the dominating approach today sees the accurate approximation of the "steady state", or "average state", of the transportation system as the most efficient way to understand the behavior of the system, and to use this understanding in order to arrive at better decisions [98]. Such decisions are often concerned with the locations, type or size of new infrastructures that should be built, large-scale budgets' investment alternatives or long-term policy revisions [99].

When examining the rapidly expanding field of ridesharing this approach suffers an inherent limitation, as it is not adequate for the nature of decisions ridesharing operators and regulators are required to make. As ridesharing uses existing roads and metropolitan infrastructure, does not require setting fixed-place stations of fixed-paths, and often uses existing vehicles, it is mostly located "outside" the realm of these analysis methodologies. Furthermore, ridesharing introduces a new set of factors that traditional methods usually cannot easily cope with, such as dynamic changes in fares, which may significantly influence network properties such as global congestion [100].

Analyzing ridesharing using the existing models would be inefficient at best. Taking the static approach using a long-term aggregation of the supply and demand would inevitably result in a model that would be optimized for the *average states* of the rides network, ignoring its inherent volatility (that is caused due to daily and weekly patterns as well as irregular spikes created by events such as street-parties, sports events, among others).

Interestingly, as shown in Section 2.4.2, the dynamic rides network spends only an extremely *small portion* of the time in those average network states. Furthermore, our analysis demonstrates that overlooking the dynamic nature of the traffic scheme disregards the vast majority of the network states, as manifested in the O-D matrix, as well as its possible ridesharing utilization. Specifically, this phenomenon is demonstrated in Figure 2.12 that reveals that the system spends approximately 33% of the time in states that have a potential utilization of either 50% above the monthly average, or 50% below it.

Ignoring this dynamic nature of the urban rides system through the use of a static analysis model (which is today's mainstream approach) will be inherently limited in its efficiency. The key to unlocking the development of effective next generation ridesharing systems therefore lays in an analysis that is rooted in the understanding of its dynamic nature, and the way to use it in order to develop pro-active strategies that dynamically adapt their forecast using an ad-hoc analysis of the network's state.

A potential example for this approach can be found in [101], containing a computational study aimed for identifying environments in which the use of "dedicated drivers" are most useful. As urban supply and demand environments are constantly (and significantly) changing (as demonstrated in our analysis of the New York taxi data), it is therefore likely that a strategy that detects the *times* where the use of such drivers is most efficient and upon such detection – launches these drivers to supply the demand[1] – would achieve a superior performance compared to a static strategy that does not react to such changes.

Another example can be the work of [102] in which the size of a carsharing fleet is optimized in order to maximize the monetary operational savings. Again, such an approach reaches the global optimization assuming a static approach whereas the incorporation of the dynamic nature of the system could yield a

[1]This can be done using a dynamic change in the commission drivers are required to pay, giving such drivers a temporary priority in certain roads, or forbidding them from granting service on a regular basis expect from when their service is required.

significant. This could be done for example by allowing the fleet operators to dynamically use the services of a public service (such as Uber or Lyft), rented cars, or private drivers. Using such a service when needed will allow to reduce the ongoing basic cost.

2.4.2 Dynamic Network Analysis

As discussed in previous sections one of the main hurdles that prevents the wide adoption of ride-sharing might be the high volatility of its potential utilization, and its extreme unpredictability. In this section we propose to mitigate this problem by using a dynamic network that represents the evolving travel patterns in the city. That is, a multitude of rides-networks, representing data of fixed-length time periods, each of which starts at different points in times of equal distances. Such a "sliding window" approach is useful for tracking changes in various properties of this dynamic network, which we show are not only highly correlated with the potential ride-sharing utilization at the corresponding points in time, but can also *predict* the utilization few hours ahead of time.

We divide the rides dataset into hourly aggregated snapshots, creating $31 \times 24 = 744$ sub-networks, each denoted by $G_{T_n,T_{n+1}}$, such that T_n represents the n-th hour in the month. An illustration of one such sub-network is shown in Figure 2.7. Intuitively we see that most of the nodes are highly connected, but a considerable number of nodes are connected to only one other node in the network.

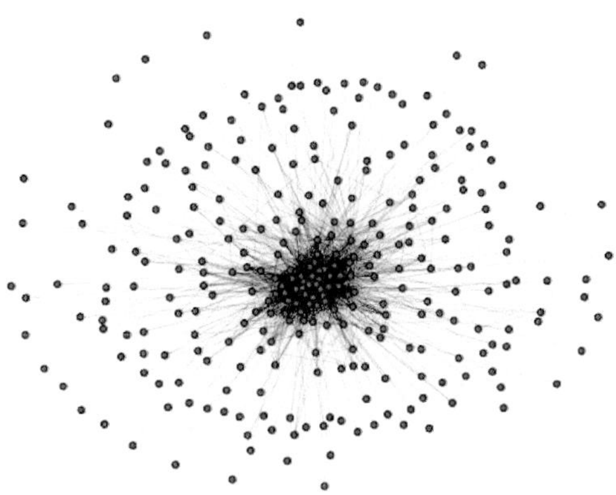

Figure 2.7: An illustration of the rides sub-network $G_{T_{144},T_{145}}$, denoting the structure that is implied by the aggregation of the rides between the 144-th and the 145-th hour of the month.

Similarly to Figure 2.6 in which the potential benefit of ride-sharing over the entire data was shown, we have performed the same calculation for every hourly network separately. Figure 2.8 presents the average potential ride-sharing utilization taken on all hourly networks, as a function of the maximal delay allowed (notice that this is in fact a lower bound, since we artificially prevent passengers from being merged with rides "outside" their hourly network). It can be seen that this produces a lower utilization than the previous calculation using the overall aggregation (approximately 10% decrease), caused by the fact that each pair of nodes has a lower probability of being connected.

We now extract a set of six common network properties for each traffic-network G_{T_i,T_j}, to be used as the features values representing each network. These features encapsulate various topological aspects of the network and enable us to project each hourly-collection of traffic data (containing a large and aprioricaly unknown number of rides) into a single coordinate in a 6-dimensional feature-space.

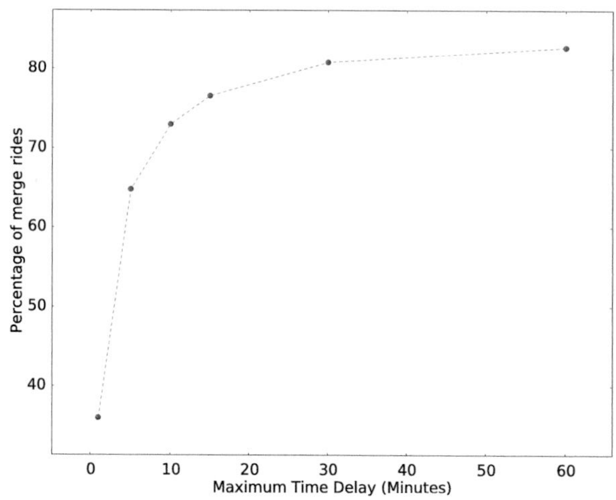

Figure 2.8: Potential of ride-sharing utilization, measured as the percentage of potentially merged rides (averaged over all sub-networks), as a function of maximal delay agreeable by the passengers. The figure is based on the result presented in [88].

2.4.2.1 *Number of Nodes*

The number of nodes in the network G, denoted as $|V|$, represent the number of unique pick-up and drop-off locations of rides made during this time window. Note that although all the networks refer to the same dataset, and the same geographic environment, different networks may have different values of $|V|$, since at different time-segments different locations may be "active".

2.4.2.2 Number of Edges

The number of edges in the network G, denoted as $|E|$, represent the number of unique pick-up to drop-off pairs of rides made during this time window. This is also the number of non-zero elements of the temporal O-D matrix that is derived from this network.

2.4.2.3 Network Density

The average degree of the network's nodes is defined as $\frac{|E|}{|V|}$. This property represents the average number of unique drop-off locations per pick-up location (and vice versa). Additionaly it is associated with the predictability of rides made during this time window, and is also related to the system's entropy.

2.4.2.4 Average Betweenness Centrality

Each node v in the network G has a calculate-able betweenness centrality score [103], representing the portion of "shortest paths" between all the node-pairs in the network, that pass through v. Formally, for a network node $v \in V$ this is defined as:

$$\sum_{s \neq v \neq t} \frac{\sigma_{s,t}(v)}{\sigma_{s,t}}$$

where $\sigma_{s,t}$ is the total number of shortest paths from node s to node t and $\sigma_{s,t}(v)$ is the number of those paths that pass through v.

Averaging these values yields an estimation of the network's efficiency, with respect to the number of nodes whose adequate availability is required in order to preserve the network's ability to maintain an efficient flow without increasing the length or durations of trips between arbitrary points [37, 104].

2.4.2.5 Average Closeness Centrality

The closeness centrality of a node [105] is a measure of centrality in a network, calculated as the sum of the length of the shortest paths between the node and all other nodes in the graph. Thus the more central a node is, the closer it is to all other nodes. For a node $v \in V$, the measure is defined as:

$$\frac{1}{\sum_x d(v,x)}$$

Averaging the closeness centrality over all the network nodes yields an estimation of its compactness, that is – how short it is to travel between an arbitrary pair of network nodes.

2.4.2.6 Average Eigenvalue Centrality

Eigenvalue centrality [106] (also called eigencentrality or eigenvector centrality) is a measure of the influence of a node in a network. It assigns relative scores to all nodes in the network based on the concept that connections to high-scoring nodes contribute more to the score of the node in question than equal connections to low-scoring nodes.

For a given graph G with an adjacency matrix A the centrality score of a node $v \in V$, denoted as $x(v)$, is defined as

$$\frac{1}{\lambda} \sum_{u \in M(v)} x_u$$

where $M(v)$ is a set of the neighbors of v and λ being the graph's largest positive real eigenvalue. This can be accurately estimated by taking the vth component in the eigenvector that corresponds to the largest positive real eigenvalue.

The use of eigenvalues to analyze propagation phenomena over networks can be see for example in [107], where its usability for predicting the epidemic potential of viruses is demonstrated.

We use a linear regression to fit these features for the calculated potential utilization, as well as a multiple linear regression to fit the potential utilization for the entire set of network properties. As can be seen in Figure 2.9 these features show a high correlation with the potential utilization for this hourly network (the figure reports the adjusted R squared to account for the different number of predictors).

2.4.3 Ride-Sharing Potential Prediction

In the previous section we have shown that the monthly rides can be partitioned into hourly aggregative snapshots, each with different characteristics (and specifically, network oriented ones), and different ride-sharing potentials. In addition, we have demonstrated the correlation between these network properties and the ride-sharing potentials of the rides the corresponding networks are implied from (as appears in Figure 2.9). In this section, we discuss whether this correlation can also be used for predictive purposes. Specifically, can we deduce from the current values of various network properties how the *change* in the ride-sharing potential compared to its current value.

In order to do so, we first analyze the evolution of various network properties of the hourly aggregative rides network $G_{T_n,T_{n+1}}$ over time. Figure 2.10 illustrates the evolution of the mean nodes degree of the rides network as a function of time (that is, the average over all of the network's nodes' degrees, for all the dynamic hourly networks). For the sake of clarity, we have increased the time granularity used in the analysis, so that the hourly networks are now generated in 5 minute intervals, thus significantly overlapping, and subsequently generating a

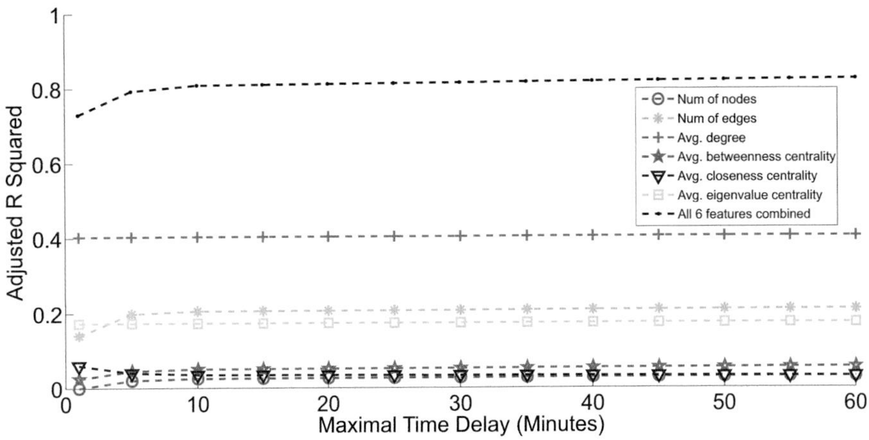

Figure 2.9: Adjusted R^2 of the correlation between seven features of the hourly rides network and the potential ride-sharing utilization for this network. Most features have low quality fit, but the combined mixture of all seven results has a remarkably high correlation ($R^2 = 0.82$). Features are (1) the number of nodes in the network, (2) the number of edges, (3) the averaged degree, (4) the averaged betweenness centrality, (5) the averaged closeness centrality, and (6) the averaged eigenvector centrality.

smoother and easier to read graph. The change from the monthly average of the mean degree as a function of time is portrayed, clearly showing a dominant daily pattern. However, on top of this pattern we can see significant hourly fluctuations, tens of percent in magnitude. This reveals the existence of strong volatility in the ride dynamics alongside the predicted daily and weekly dynamics.

A similar dynamics is observed when analyzing the evolution of the largest eigenvalue of the rides-network' adjacency matrix over time. The use of eigenvalues to analyze propagation phenomena over networks can be see for example in [107, 108], where its usability for predicting the epidemic potential of viruses (both human and computer-based) is demonstrated. Additional mathematical analysis on the role of eigenvalues in the analysis of network structures can be found in [109]. This property, known to encapsulate various behavioral characteristics of the people whose mobility patterns the network is depicting, displays a clear (and easy to predict and understand) daily pattern, on top of which significant and erratic spikes are added, as can be seen in Figure 2.11. These spikes seem to appear sporadically, lacking any clear patterns or internal regularity, implying again the need for understanding the dynamic aspects of the network.

Now, let us perform a similar analysis over the potential ride-sharing utilization, looking at its evolution over time. The results of this analysis, presented in Figure 2.12, clearly demonstrate a similar dynamics to the couple of network properties mentioned earlier. Specifically, it can be seen that

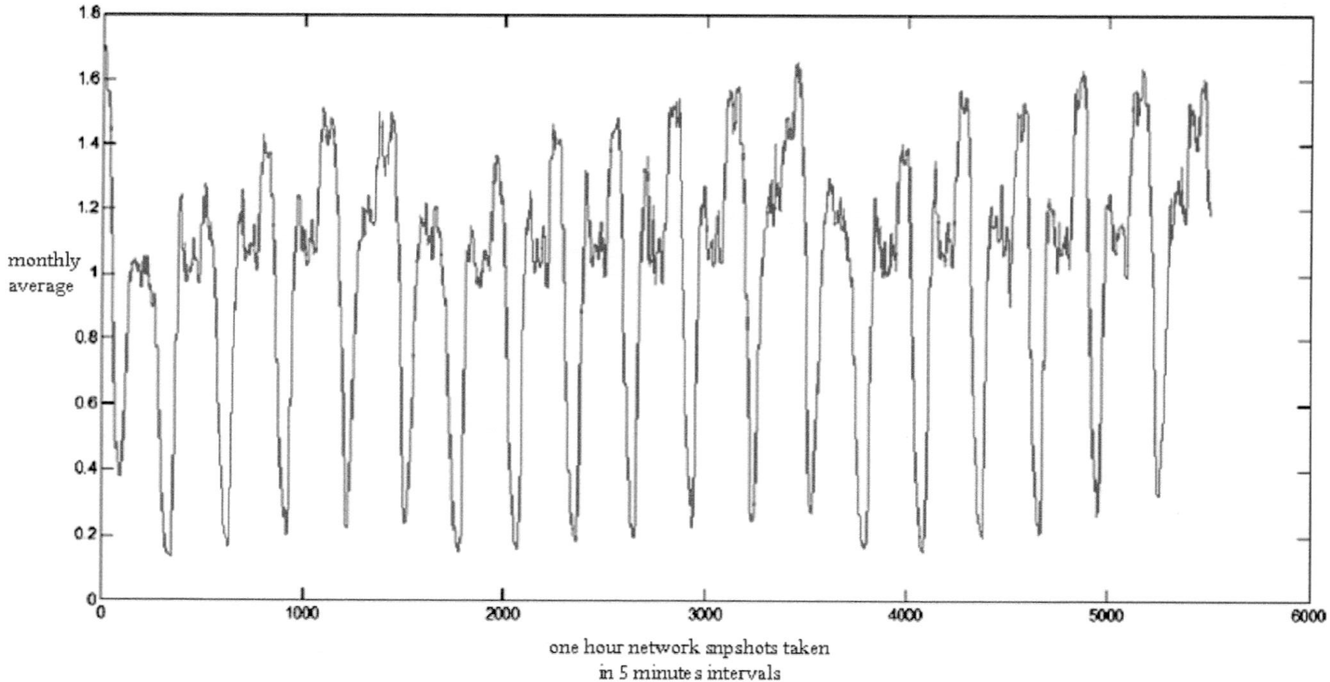

Figure 2.10: Dynamics of the mean degree of the rides network nodes. X-axis denotes the time, given in 5 minutes granularity. Y-axis denotes the change of the mean degree of the network compared to its monthly average.

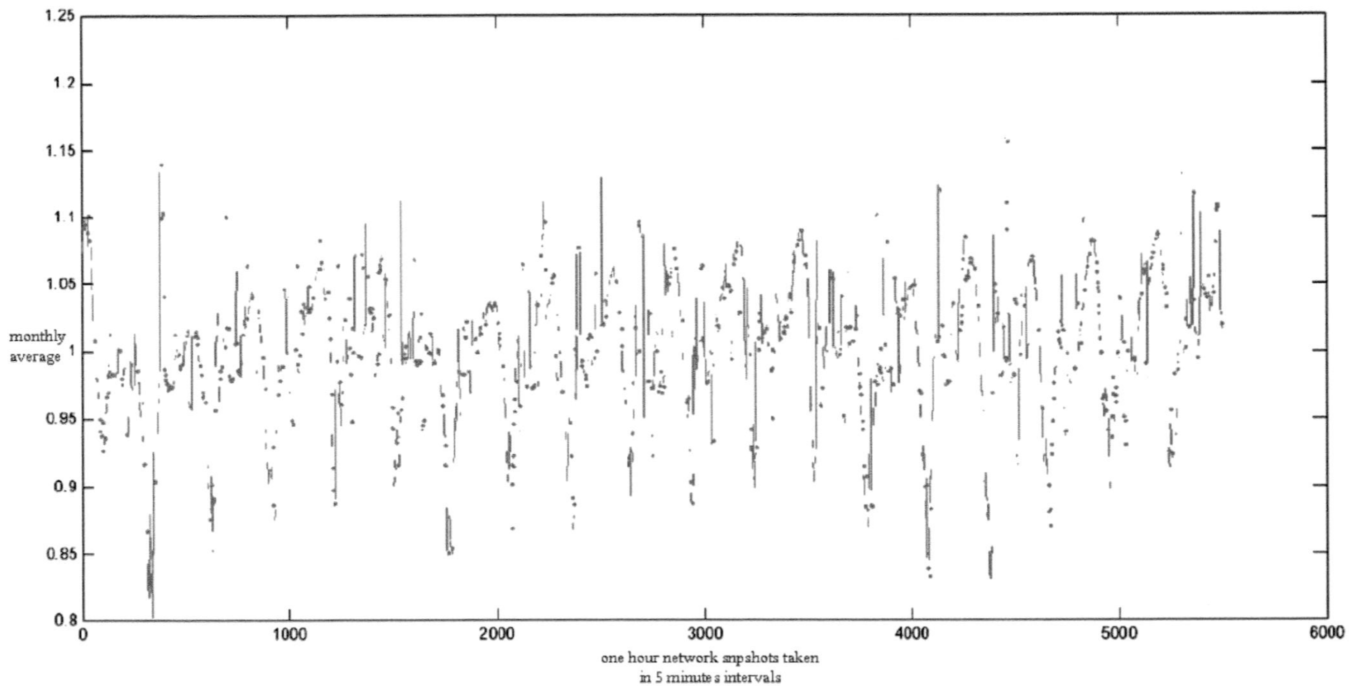

Figure 2.11: Dynamics of the largest eigenvalue of the rides-dynamic network over time. X-axis denotes the time, given in 5 minutes granularity. Y-axis denotes the change of the largest eigenvalue of the network compared to its monthly average.

alongside the dominating daily pattern (and weaker, but still easy to see, weekly one), there are clear changes in the potential utilization. These changes take various shapes and forms, from a sudden decrease in the daily peak (as can be seen around $X = 1400$), to changes in the intra-weekly peaks (the first week analyzed showing a 'U-shaped' form among its days, the second week showing an equal-peaks dynamics, and the third week showing an extremely high Monday and Tuesday, and weaker Wednesday, Thursday and Friday), and others. Surprisingly, the magnitude of these changes may even exceed the dominating daily pattern. For example, the change between the first Tuesday (around $X = 200$) and the third Tuesday ($X = 4100$) is 90% compared to the monthly average, whereas the average change in potential utilization between workdays and weekends is only 70%.

At this point we ask the following question: "can we find a statistical correlation between *current* values of the rides network properties and *future* values of the potential ride-sharing utilization?". This question is of interest, as such a correlation would allow us to predict future changes in the potential utilization, providing valuable tools for ride-sharing users, operators, and regulators.

We first address this question by comparing network properties values at time T with potential utilization at time $T+1$ (1 hour prediction). Figure 2.13 presents an example of such a comparison, in the form of a scatter plot showing for each point in time T a dot whose X-axis is the mean nodes degree of the network $G_{T,T+1}$ and whose Y-axis is the change in the potential utilization of the rides between $T+1$ and $T+2$ compared to the rides between T and $T+1$. That is, the change in the momentary ride-sharing utilization between "now" (time T) and "in an hour" (time $T+1$). It is easy to see that this representation reveals a clear and strong negative correlation between the two.

Trying to increase our lookahead and predict the change in the dynamic ride-sharing utilization from a 2 hours time-frame, Figure 2.14 illustrates the correlation between the value of the largest eigenvalue of the rides network at time T and the change in the potential utilization between time T (aggregated to $T+1$) and $T+2$ (aggregated to $T+3$). Again, a clear strong negative correlation is easily visible. For example, in times where the value of the largest eigenvalue of the rides network is smaller than 0.012 the potential ride-sharing utilization was statistically guaranteed (during the month of the observation) to significantly increase in the coming 2 hours. Similarly, the largest eigenvalue of 0.014 would indicate a significant decrease in the ride-sharing potential within the next 2 hours.

Figures 2.13 and 2.14 are based on the analysis of the first 3 weeks of January 2013. These observations were then validated using the last week of January, as can be seen in Figures 2.15 and 2.16.

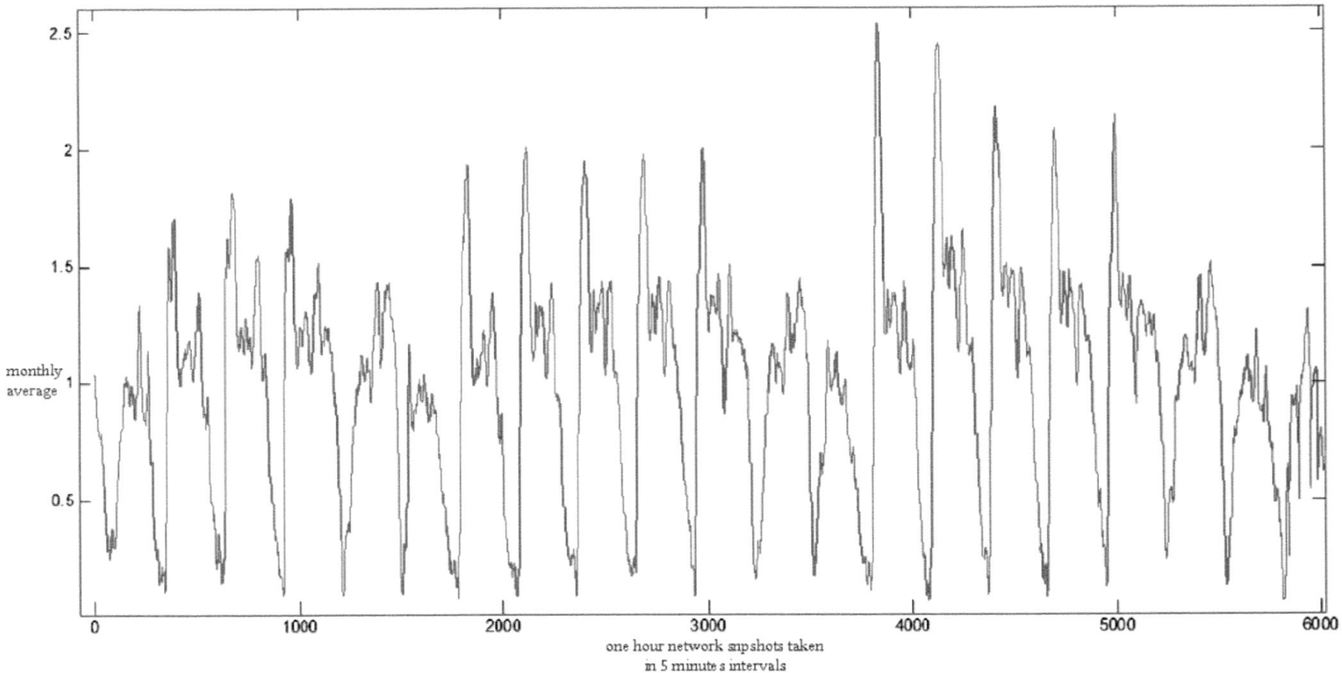

Figure 2.12: Dynamics of the potential ride-sharing utilization over time. X-axis denotes the time, given in 5 minutes granularity. Y-axis denotes the change of the potential utilization compared to its monthly average.

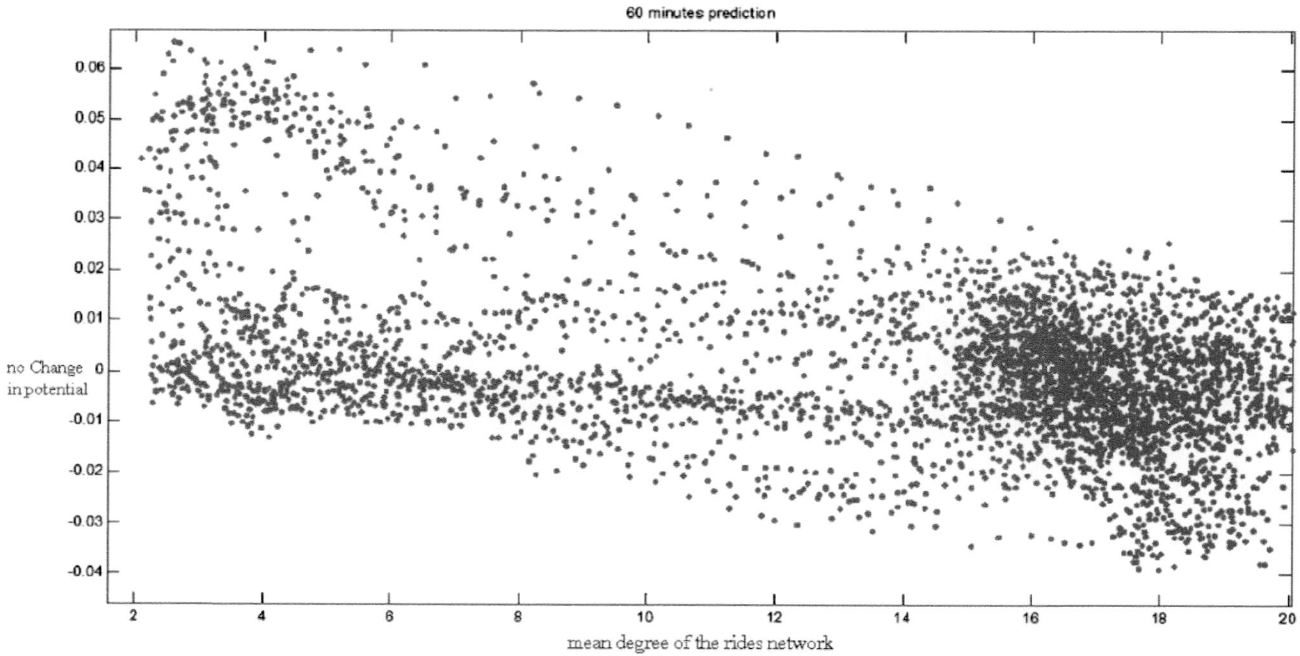

Figure 2.13: Change in potential ride-sharing utilization (Y-axis), 1 hour prediction, as a function of the mean degree of the rides network (X-axis).

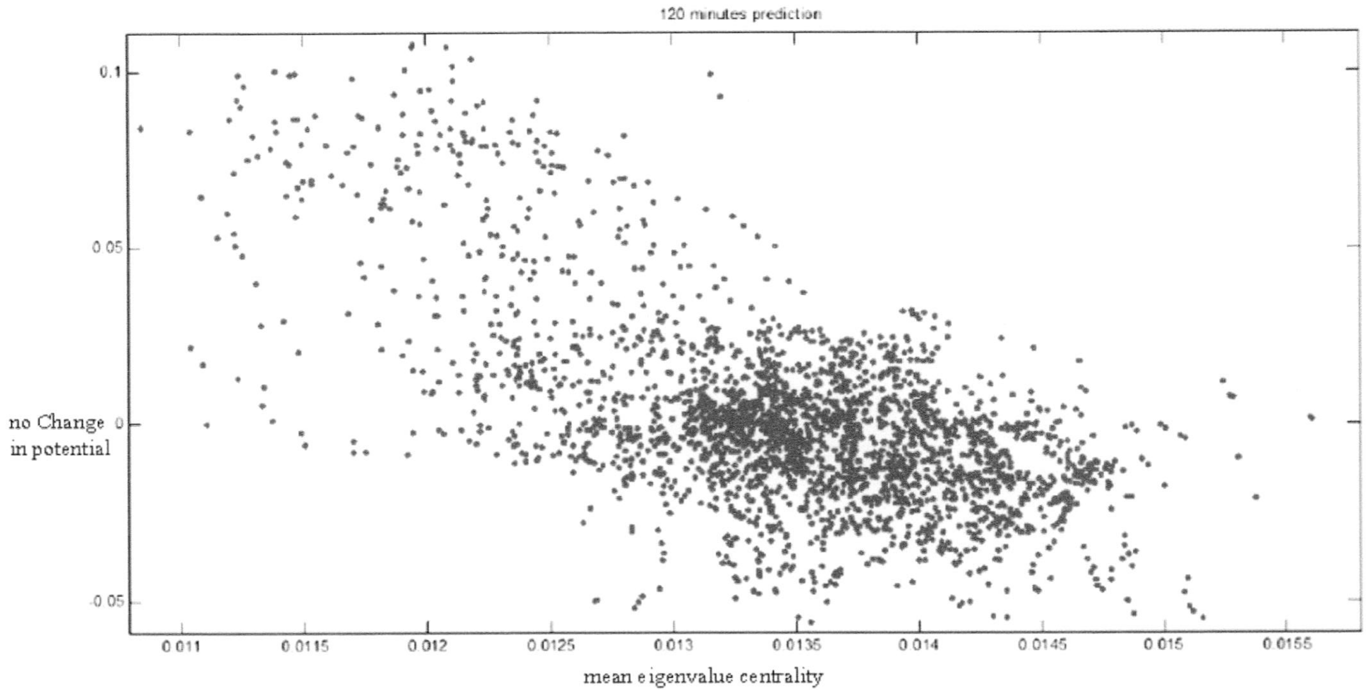

Figure 2.14: Change in potential ride-sharing utilization (Y-axis), 2 hours prediction, as a function of the largest eigenvalue of the rides network (X-axis).

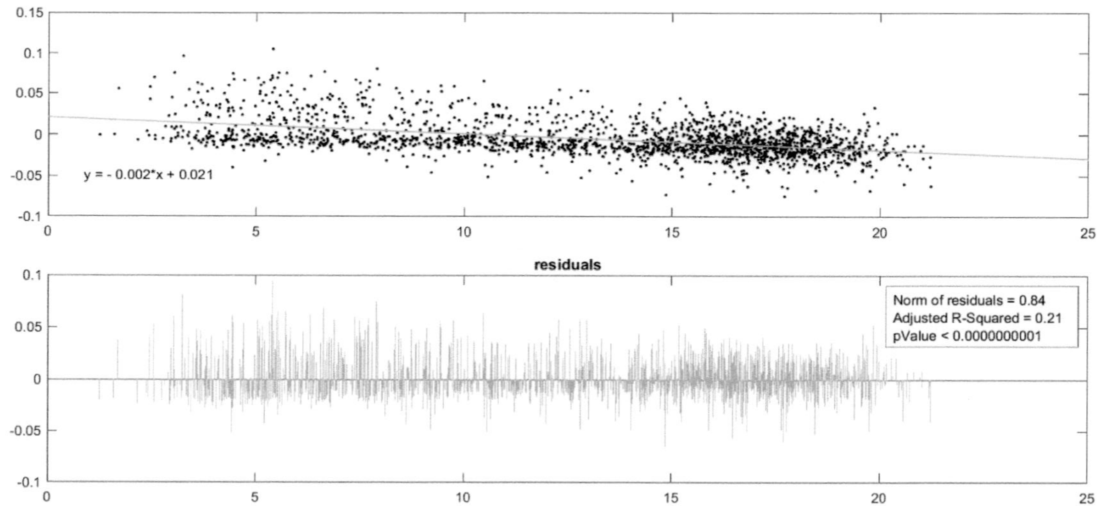

Figure 2.15: Change in potential ride-sharing utilization (Y-axis), 1 hour prediction, as a function of the mean degree of the rides network (X-axis), created for the last week of the data.

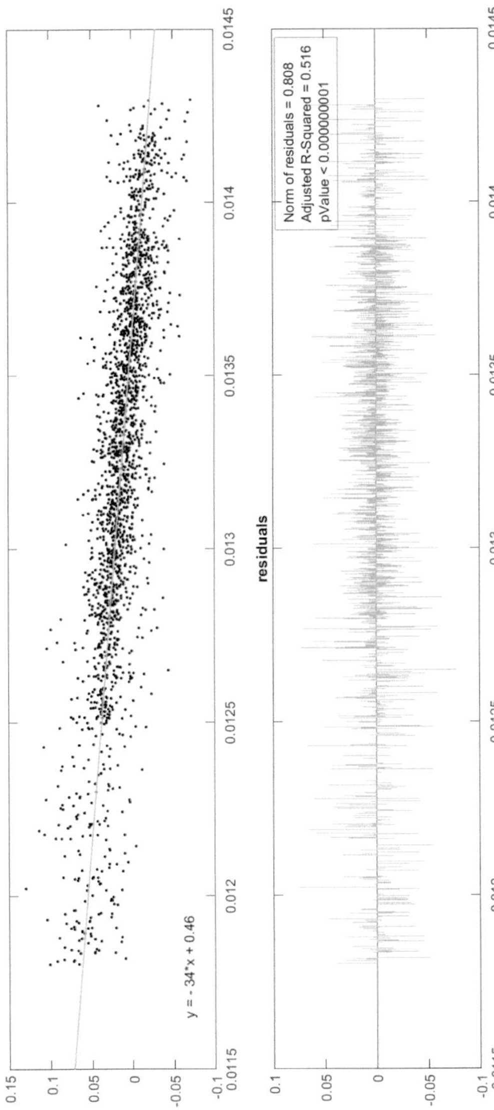

Figure 2.16: Change in potential ride-sharing utilization (Y-axis), 2 hours prediction, as a function of the largest eigenvalue of the rides network (X-axis), created for the last week of the data.

Once demonstrating the predictive power of the dynamic network's properties with respect to the network's future ride-sharing potential we can now construct a multiple linear regression model that would fit all of these 6 properties. We have created 18 models, for 2 values of distance tolerance (400 meters and 800 meters, denoting the pick-up and drop-off distances that still allow rides to be merged), 3 values of time tolerance (30 seconds, 2 minutes and 5 minutes, denoting the duration the passengers would be willing to wait for in order to merge their rides) and 3 values of prediction horizon (no prediction, 1 hour prediction and 2 hours prediction). The results include a scatter plot of the data, effects of the various properties, Anova analysis and other statistical analysis appear in Figures 2.17 through 2.34.

The effectiveness of the prediction as a function of the prediction horizon (i.e. the distance between the point in time where the prediction is calculated and the point in time this prediction refers to) is illustrated in Figures 2.35 through 2.40, showing the R^2 of the model (both ordinal and adjusted) as a function of the time horizon (between 0 and 12 hours), for several values of distance tolerance and time tolerance. It can clearly be seen that in general (and as expected) the accuracy of the model decreases with the increase in the prediction horizon used (that is, when the model tries to predict the behavior of the system further into the future).

The effect of each feature, depicted by the adjusted response plot for its various values, is presented in Figures 2.41 through 2.46, created for a scenario with distance tolerance of 800 meters, time tolerance of 5 minutes, and prediction horizon of 2 hours.

2.4.4 Usability and Applications

The main contribution of the analysis presented throughout this work is threefold:

The need for dynamic ride-sharing models: Traditionally, the main approach for analyzing the feasibility of ride-sharing schemes was based on long-term data aggregation, for calculating the average portion of the trip that can be merged. Our work, and specifically the analysis that is illustrated in Figure 2.12, clearly demonstrate that any model that would rely on such a static approach would be extremely inefficient, as there would be significant periods of time where the system would either have starvation (namely, rides that would not be able to be merged, and therefore, insufficient resources), or become saturated (namely, would have too many vehicles, with not enough rides to merge). This insight is expected to have immediate practical implications on the design and analysis of any ride-sharing system.

The usefulness of a network based approach ride-sharing utilization modeling: In Section 2.4.2 we offer an alternative approach, using a network-oriented analysis. The main contribution of this discussion is the clear demonstration that network properties can be used in order to efficiently explain the dynamics of ride-sharing utilization (see for example Figure 2.9). Future works will therefore either be able to use the network properties that have been defined, or employ alternative ones. As this is the first work to take a network-oriented approach for ride-sharing utilization modeling, the Adjusted R^2 that is presented in Figure 2.9 can also serve as a point of reference for future works.

The ability to forecast the future potential ride-sharing utilization: After demonstrating the ability of projecting the data to a network-based representation to explain the dynamics of the ride-sharing utilization, we have further shown the ability of this approach to also *predict* changes in the utilization a few hours before they take place. Two examples are shown in Figures 2.13 and 2.14, and then validated in Figures 2.15 and 2.16 respectively. A comprehensive analysis of the prediction capabilities of the model can be found in Appendix A.

The ability to predict changes in ride-sharing utilization under future traffic supply and demand conditions would enable a wide variety of new applications, including among others:

- Municipal authorities could dynamically change the values of incentives for subsidizing ride-sharing services.
- Municipal authorities could dynamically change the tolls collected for using roads in various parts of the city.
- Ride-sharing service operators could more efficiently control the size of their operational fleet, by dynamically optimizing the amount of available drivers at every point in time.
- The ability to dynamically optimize the resources used by ride-sharing service operators is expected to both trigger a reduction of service fees for end-users, as well as improve the service level in terms of wait times.
- End-users would be able to get a better prediction regarding availability and cost of ride-sharing services in future times.

2.5 Summary and Future Work

As the popularity of ride-sharing systems grow, its user-base gradually transforms from early adopters to mainstream consumers. While the former are characterized by a keen affection for innovative solutions powered by cutting edge technologies, aiming to disrupt the governing paradigm in the field, the latter are often interested mainly in the advantages these services can offer them with the smallest change in their habits as possible. With respect to ride sharing these new users are willing to sustain far less wait-times and are extremely susceptible to inconvenience than their preceding tech-savvy innovation-hungry early users. The key to a scalable mature ride sharing infrastructure is found in the level of service such systems provide, mainly measured by the availability of vehicles when they are needed. Alas, availability maximization is immediately linked to a reduction in the financial savings that the service can offer. In other words, a further expansion of ride-sharing is being constrained among others by the ability to offer high utilization, defined as the ability to "merge" similar rides in a way that would not require the passengers to sustain more than a minimal delay in their trips.

This optimization problem was extensively discussed in previous literature (comprehensive literature review can be found in Section 2.2). However, the conventional approach to this problem assumed a static environment which needs to be optimized. By finding the optimal number of cars, or optimal pricing policy, the efficiency (or potential) of the system was assumed to be calculable in a robust way — a key component in the decision of operators where to deploy new systems, in the design of relevant urban legislations by municipal policy makers, and of course in the likelihood of passengers to use these services.

In this work we discussed the dynamic nature of ride sharing systems. Specifically, we were interested in whether ride-sharing utilization is stable over time (which coincides with the implicit assumption of most previous works in this field) or if it undergo significant and often rapid changes (which would imply the inherent inefficiency of schemes assuming a static nature). We modeled the ride sharing utilization using the known New York Taxi dataset and clearly showed that it is highly dynamic, and that any system that would be designed for the "average" utilization would be highly inefficient.

We then showed assuming a dynamic approach the taxi data can be modeled as a sequence of data-snapshots, resulting in a dynamic traffic-network model. Several recent works have shown that network features can effectively be used to predict a variety of events and properties, e.g., emergency situations, individuals personalities and spending behaviors [110, 111]. We used a similar technique in order to project the taxi data as into a feature space comprised of topological features of the dynamic network implied by this traffic. This (dynamic) feature space is then used to model the dynamics of ride-sharing utilization over time.

Using this approach we were able to demonstrate a clear correlation between the utilization of the ride sharing system over time and several topological features of the network it creates. In addition, we demonstrated that the potential benefit of ride sharing expressed as the percentage of rides that can be shared with limited discomfort for riders can also be predicted a few hours in advance. Such prediction can be used as a tool for an accurate short-term forecasting of the ride-sharing potential in cities and metropolitan areas.

Researchers in [8, 68, 112] and others have focused on addressing the computational challenges of trip-matching (an NP-hard optimization problem) in real-time and developed heuristics to quantify potential ride-sharing demand. These algorithms re-route trips in order to match them with similar, overlapping trips, explicitly capturing demand for ridesharing relative to passenger willingness to experience prolonged travel time. However, finding an optimal solution to this problem is not computationally plausible (even under extreme limitations of the problem's space [113]), and even the calculation of approximation heuristics would be computationally intense when done ad-hoc. Therefore, the ability to use current traffic dynamics in order to predict properties of an efficient near-future ride-sharing scheme — such as the method we propose in this work — can be used to make this process significantly more efficient [114, 115].

Future work should focus on the analysis of the correlation we find in this paper, trying to detect traces of possible causalities. Are network properties merely correlated with ride-sharing utilization, or do they possess an active influence over it? Evidences of the latter would enable us to offer urban designers and policy makers an innovative tool for encouraging and facilitating the adoption of ride-sharing systems. Alternatively, incentives and fees could be better moderated, used as "remedies" in the case of a change in the travel patterns, in order to balance it and maintain a sustainable ride-sharing paradigm. Another approach could be the pipelining of the dynamic ride-sharing utilization forecast as the input of models intended to predict the benefits of ride-sharing on the overall traffic [116].

Recent works have demonstrated the benefit of tracking the network's dynamics in order to improve collaborative decision making [117, 118]. A possible continuation of the current work can analyze ride-sharing optimization as a case of decentralized decision making process, using the technique that is presented here.

As the prediction of future ride-sharing potential is ultimately needed for optimization purposes (of the overall travel time, congestion or any other utilization metric) of a dynamic coverage problem, comparing the performance of any proposed method to the theoretical results that are available for various types of such decentralized collaborative coverage challenges can also be of value (see [119–126] and specifically [127] and [128], and for a detailed discussion regarding the importance of various geometric properties of the geography see [129, 130]).

Finally, as our suggested approach is agnostic to the actual route taken by the drivers it would be interesting to see whether the introduction of ride-sharing affects additional factors such as detours (that for a merged ride may become cost-effective) and usage of toll-routes.

References

[1] A. Caragliu, C. Del Bo and P. Nijkamp, Smart cities in Europe, *Journal of Urban Technology*, 18(2): 65–82, 2011.

[2] H. Chourabi, T. Nam, S. Walker, J.R. Gil-Garcia, S. Mellouli, et al. Understanding smart cities: An integrative framework. In: *Proceedings of the 45th Hawaii International Conference on System Science (HICSS), IEEE*, pp. 2289–2297, 2012.

[3] M.C. Gonzalez, C.A. Hidalgo and A.-L. Barabási. Understanding individual human mobility patterns, *Nature*, 453(7196): 779–782, 2008.

[4] F. Calabrese, M. Diao, G. Di Lorenzo, J. Ferreira Jr, C. Ratti, et al. Understanding individual mobility patterns from urban sensing data: A mobile phone trace example. *Transportation Research part C: emerging technologies*, 26: 301–313, 2013.

[5] A. Noulas, S. Scellato, R. Lambiotte, M. Pontil, C. Mascolo, et al. A tale of many cities: universal patterns in human urban mobility, *PloS one*, 7(5): e37027: 2012.

[6] waze.com, `http://www.waze.com`, [Online; accessed 15-October-2014].

[7] R. Chin, Solving transport headaches in the cities of 2050, BBC Future [Online; accessed 15-October-2014] (June 2013).

[8] B. Cici, A. Markopoulou, E. Frias-Martinez and N. Laoutaris, Assessing the potential of ride-sharing using mobile and social data: a tale of four cities. In: *Proceedings of the 2014 ACM International Joint Conference on Pervasive and Ubiquitous Computing*. ACM, 2014, pp. 201–211.

[9] T. Altshuler, Y. Shiftan, R. Katoshevski, N. Oliver, A.S. Pentland, et al. Mobile phones for on-demand public transportation. In: *NetSci*, 2014.

[10] D. Graziotin, An analysis of issues against the adoption of dynamic carpooling. *arXiv preprint arXiv*:1306.0361, 2013.

[11] L. Alexander and M. Gonzalez, Assessing the impact of real-time ridesharing on urban traffic using mobile phone data, 2015.

[12] S. Hobrink, Explaining regional adoption differentials in dutch car sharing markets. 2014.

[13] L. Coenen, P. Benneworth and B. Truffer, Toward a spatial perspective on sustainability transitions. *Research Policy*, 41(6): 968–979, 2012.

[14] S. Shaheen, A. Cohen and M. Chung, North american carsharing: 10-year retrospective. *Transportation Research Record: Journal of the Transportation Research Board*, 2110, 35–44, 2009.

[15] lyft.com, `http://www.lyft.com`.

[16] uber.com, `http://www.uber.com`.

[17] side.cr (currently closed, and acquired by General Motors), `http://www.side.cr`.

[18] D. Efthymiou, C. Antoniou and P. Waddell, Factors affecting the adoption of vehicle sharing systems by young drivers. *Transport Policy*, 29: 64–73, 2013.

[19] NYC Open Data Website, http://data.ny.gov/. Accessed, September, 2014.

[20] T. Altshuler, Y. Altshuler, R. Katoshevski and Y. Shiftan, Modeling and prediction of ride-sharing utilization dynamics. *Journal of Advanced Transportation*, 2019.

[21] R. Albert and A.-L. Barabási, Statistical mechanics of complex networks. *Reviews of Modern Physics*, 74(1): 47, 2002.

[22] A.-L. Barabási, The elegant law that governs us all, 2017.

[23] M.E. Newman, The structure and function of complex networks. *SIAM Review*, 45(2): 167–256, 2003.

[24] Y. Altshuler, A.S. Pentland and G. Gordon, Social behavior bias and knowledge management optimization. In: *Social Computing, Behavioral-Cultural Modeling, and Prediction*, Springer, pp. 258–263, 2015.

[25] P. M. Krafft, J. Zheng, W. Pan, N. Della Penna, Y. Altshuler, et al. Human collective intelligence as distributed bayesian inference. *arXiv preprint arXiv*:1608.01987, 2016.

[26] D. Lazer, A.S. Pentland, L. Adamic, S. Aral, A.-L. Barabási, et al. Computational social science. *Science*, 323(5915): 721–723, 2009. arXiv:http://www.sciencemag.org/content/323/5915/721.full.pdf, doi:10.1126/science.1167742.
URL `http://www.sciencemag.org/content/323/5915/721.short`

[27] Y. Altshuler, W. Pan and A. Pentland, Trends prediction using social diffusion models. *arXiv.org*, 2011.

[28] Y. Altshuler, W. Pan and A. Pentland, Trends prediction using social diffusion models. In: *International Conference on Social Computing, Behavioral-Cultural Modeling and Prediction*, Springer, pp. 97–104, 2012.

[29] S. Somin, G. Gordon and Y. Altshuler, Network analysis of erc20 tokens trading on ethereum blockchain. In: *Unifying Themes in Complex Systems IX: Proceedings of the Ninth International Conference on Complex Systems*, Springer, p. 439, 2018.

[30] Y. Altshuler, M. Fire, E. Shmueli, Y. Elovici, A. Bruckstein, et al. The social amplifier—reaction of human communities to emergencies. *Journal of Statistical Physics*, 152(3): (2013) 399–418.

[31] E. Shmueli, V.K. Singh, B. Lepri and A.S. Pentland, Sensing, understanding, and shaping social behavior. *IEEE Transactions on Computational Social Systems*, 1(1): 22–34, 2014. doi:10.1109/TCSS.2014.2307438.

[32] A. Almaatouq, A. Alabdulkareem, M. Nouh, E. Shmueli, M. Alsaleh, et al. Twitter: Who gets caught? observed trends in social micro-blogging spam. In: *Proceedings of the 2014 ACM Conference on Web Science*, WebSci '14, ACM, pp. 33–41, 2014. doi:10.1145/2615569.2615688. URL `http://doi.acm.org/10.1145/2615569.2615688`

[33] M. Zwang, S. Somin, A.S. Pentland and Y. Altshuler, Detecting bot activity in the ethereum blockchain network, 2018. arXiv:1810.01591.

[34] M.C. Gonzalez, C.A. Hidalgo and A.-L. Barabasi, Understanding individual human mobility patterns, *Nature*, 453(7196): 779–782, 2008. URL `http://dx.doi.org/10.1038/nature06958`

[35] Y. Altshuler, R. Puzis, Y. Elovici, S. Bekhor, A.S. Pentland, et al. On the rationality and optimality of transportation networks defense: a network centrality approach. *Securing Transportation Systems*, 35–63.

[36] J. Armstrong and A. Khan, Modelling urban transportation emissions: role of gis. *Computers, Environment and Urban Systems*, 28(4): 421–433, 2004.

[37] R. Puzis, Y. Altshuler, Y. Elovici, S. Bekhor, Y. Shiftan, et al. Augmented betweenness centrality for environmentally-aware traffic monitoring in transportation networks. *Journal of Intelligent Transportation Systems*, 17(0): 91-–105, 2013.

[38] B. Waclaw, Statistical mechanics of complex networks, Arxiv preprint arXiv: 0704.3702 (2007).

[39] S. Wassermann and K. Faust, Social network analysis: Methods and applications, New York, 1994.

[40] Y. Altshuler, M. Fire, E. Shmueli, Y. Elovici, A. Bruckstein, et al. Detecting anomalous behaviors using structural properties of social

networks. In: *Social Computing, Behavioral-Cultural Modeling and Prediction*, Springer, pp. 433–440, 2013.

[41] M. Granovetter, The strength of weak ties. *American journal of Sociology*, 1360–1380, 1973.

[42] R. Albert, H. Jeong and A. Barabási, Error and attack tolerance of complex networks. *Nature*, 406(6794): 378–382, 2000.

[43] J. Onnela, J. Saramäki, J. Hyvönen, G. Szabó, D. Lazer, et al. Structure and tie strengths in mobile communication networks. *Proceedings of the National Academy of Sciences*, 104(18): 7332, 2007.

[44] R. Pastor-Satorras and A. Vespignani, Epidemic spreading in scale-free networks. *Physical Review Letters*, 86(14): 3200–3203, 2001.

[45] R. Pastor-Satorras and A. Vespignani, Evolution and structure of the internet: A statistical physics approach, Cambridge Univ Pr, 2007.

[46] Y. Altshuler, M. Fire, N. Aharony, Y. Elovici, A. Pentland, et al. How many makes a crowd? On the correlation between groups' size and the accuracy of modeling, In: *International Conference on Social Computing, Behavioral-Cultural Modeling and Prediction*, Springer, pp. 43–52, 2012.

[47] Y. Altshuler, N. Aharony, M. Fire, Y. Elovici, A. Pentland, et al. Incremental learning with accuracy prediction of social and individual properties from mobile-phone data, CoRR, 2011.

[48] Y. Altshuler, E. Shmueli, G. Zyskind, O. Lederman, N. Oliver, et al. Campaign optimization through behavioral modeling and mobile network analysis. *Computational Social Systems, IEEE Transactions on*, 1(2): 121–134, 2014.

[49] Y. Altshuler, E. Shmueli, G. Zyskind, O. Lederman, N. Oliver, et al. Campaign optimization through mobility network analysis. *Geo-intelligence and Visualization Through Big Data Trends*, 33–74, 2015.

[50] Y. Altshuler, M. Fire, N. Aharony, Z. Volkovich, Y. Elovici, et al. Trade-offs in social and behavioral modeling in mobile networks. In: *Social Computing, Behavioral-Cultural Modeling and Prediction*, Springer, pp. 412–423, 2013.

[51] Y. Altshuler, N. Aharony, A. Pentland, Y. Elovici, M. Cebrian, et al. Stealing reality: When criminals become data scientists (or vice versa). *Intelligent Systems, IEEE*, 26(6): 22–30, 2011. doi:10.1109/MIS.2011.78.

[52] C. Jernigan and B.F. Mistree, Gaydar: Facebook friendships expose sexual orientation, 14(10), 2009.

[53] Y.-Y. Liu, J.C. Nacher, T. Ochiai, M. Martino and Y. Altshuler, Prospect theory for online financial trading. *PloS one*, 9 (10): e109458, 2014.

[54] W. Pan, Y. Altshuler and A. Pentland, Decoding social influence and the wisdom of the crowd in financial trading network. In: Privacy, *Security, Risk and Trust (PASSAT), 2012 International Conference on and 2012*

International Confernece on Social Computing (SocialCom), IEEE, pp. 203–209, 2012.

[55] A. Barabasi, The origin of bursts and heavy tails in human dynamics. *Nature*, 435(7039): 207–211, 2005.

[56] J. Candia, M. González, P. Wang, T. Schoenharl, G. Madey, et al. Uncovering individual and collective human dynamics from mobile phone records. *Journal of Physics A: Mathematical and Theoretical*, 41: 224015, 2008.

[57] S. Somin, G. Gordon and Y. Altshuler, Social signals in the ethereum trading network. *arXiv preprint arXiv*: 1805.12097, 2018.

[58] E. Shmueli, Y. Altshuler and A.S. Pentland, Temporal dynamics of scale-free networks. In: *Social Computing, Behavioral-Cultural Modeling and Prediction*, Springer, pp. 359–366, 2014.

[59] L. Alexander, S. Jiang, M. Murga and M.C. González, Origin–destination trips by purpose and time of day inferred from mobile phone data. *Transportation Research Part C: Emerging Technologies*, pp. 240–250, 2015.

[60] G. Palla, A. Barabasi, T. Vicsek, Quantifying social group evolution. *Nature*, 446(7136) 664–667, 2007.

[61] D. Brockmann, L. Hufnagel and T. Geisel, The scaling laws of human travel. *Nature*, 439(7075): 462–465, 2006.

[62] Y.-A. de Montjoye, C.A. Hidalgo, M. Verleysen and V.D. Blondel, Unique in the crowd: The privacy bounds of human mobility, *Scientific Reports*, 3, 2013.

[63] C. Song, Z. Qu, N. Blumm and A. Barabási, Limits of predictability in human mobility. *Science*, 327(5968): 1018–1021, 2010.

[64] Y. Elovici and Y. Altshuler, Introduction to security and privacy in social networks. *Security and Privacy in Social Networks*, 1–6, 2012.

[65] M. Furuhata, M. Dessouky, F. Ordóñez, M.-E. Brunet, X. Wang, et al. Ridesharing: The state-of-the-art and future directions. *Transportation Research Part B: Methodological*, 57, 28–46, 2013.

[66] K. Ghoseiri, Dynamic rideshare optimized matching problem, 2012.

[67] L. Rayle, S. Shaheen, N. Chan, D. Dai and R. Cervero, App-based, on-demand ride services: Comparing taxi and ridesourcing trips and user characteristics in san francisco university of california transportation center (uctc). *Tech. rep.*, UCTC-FR-2014-08, 2014.

[68] P. Santi, G. Resta, M. Szell, S. Sobolevsky, S.H. Strogatz, et al. Quantifying the benefits of vehicle pooling with shareability networks. *Proceedings of the National Academy of Sciences*, 111(37): 13290–13294, 2014.

[69] J. Miller and J.P. How, Predictive positioning and quality of service ridesharing for campus mobility on demand systems. *arXiv preprint arXiv*: 1609.08116 (2016).

[70] M. Ota, H. Vo, C. Silva and J. Freire, Stars: Simulating taxi ride sharing at scale. *IEEE Transactions on Big Data*, 2016.

[71] R. Tachet, O. Sagarra, P. Santi, G. Resta, M. Szell, et al. Scaling law of urban ride sharing, arXiv preprint arXiv:1610.09921 (2016).

[72] M. Stiglic, N. Agatz, M. Savelsbergh amd M. Gradišar, Enhancing urban mobility: Integrating ride-sharing and public transit. *Available at SSRN*, 2805342, 2016.

[73] S. Banerjee, R. Johari and C. Riquelme, Dynamic pricing in ridesharing platforms. *ACM SIGecom Exchanges*, 15(1): 65–70, 2016.

[74] K. Bimpikis, O. Candogan and S. Daniela, Spatial pricing in ride-sharing networks, 744-769, 2016.

[75] C. Dai, Ridesharing recommendation: Whether and where should i wait? In: *International Conference on Web-Age Information Management*, Springer, pp. 151–163, 2016.

[76] L.Z. X. X. G.S. Nicholas Jing Yuan, Yu Zheng, Where to find my next passenger? In: *Ubicomp*, 2011, 2011.

[77] X. Li, G. Pan, Z. Wu, G. Qi, S. Li, et al. Prediction of urban human mobility using large-scale taxi traces and its applications. *Frontiers of Computer Science*, 6(1): 111–121, 2012.

[78] N. Agatz, A. Erera, M. Savelsbergh and X. Wang, Optimization for dynamic ride-sharing: A review. *European Journal of Operational Research*, 223(2): 295–303, 2012.

[79] Y. Wang, R. Kutadinata and S. Winter, Activity-based ridesharing: increasing flexibility by time geography. In: *Proceedings of the 24th ACM SIGSPATIAL International Conference on Advances in Geographic Information Systems, ACM*, p. 1, 2016.

[80] M. Stiglic, N. Agatz, M. Savelsbergh and M. Gradisar, The benefits of meeting points in ride-sharing systems. *Transportation Research Part B: Methodological*, 82, 36–53, 2015.

[81] P. Goel, L. Kulik, K. Ramamohanarao, Optimal pick up point selection for effective ride sharing. *IEEE Transactions on Big Data*, 2016.

[82] L. Alarabi, B. Cao, L. Zhao, M.F. Mokbel and A. Basalamah, A demonstration of sharek: an efficient matching framework for ride sharing systems. In: *Proceedings of the 24th ACM SIGSPATIAL International Conference on Advances in Geographic Information Systems, ACM*, p. 95, 2016.

[83] M. Stiglic, N. Agatz, M. Savelsbergh and M. Gradisar, Making dynamic ride-sharing work: The impact of driver and rider flexibility.

Transportation Research Part E: Logistics and Transportation Review, 91: 190–207, 2016.

[84] J. Alonso-Mora, S. Samaranayake, A. Wallar, E. Frazzoli, D. Rus, et al. On-demand high-capacity ride-sharing via dynamic trip-vehicle assignment. *Proceedings of the National Academy of Sciences*, 201611675, 2017.

[85] E. D'Andrea, D. Di Lorenzo, B. Lazzerini, F. Marcelloni, F. Schoen, et al. Path clustering based on a novel dissimilarity function for ride-sharing recommenders. In: *Smart Computing (SMARTCOMP)*, 2016 *IEEE International Conference on, IEEE*, pp. 1–8, 2016.

[86] C. Wu, E. Kamar and E. Horvitz, Clustering for set partitioning with a case study in ridesharing. In: *Intelligent Transportation Systems (ITSC)*, 2016 *IEEE 19th International Conference on, IEEE*, pp. 1384–1388, 2016.

[87] D. Sánchez, S. Martínez and J. Domingo-Ferrer, Co-utile p2p ridesharing via decentralization and reputation management. *Transportation Research Part C: Emerging Technologies*, 73 (2016) 147–166.

[88] T. Altshuler, Y. Shiftan, R. Katoshevski, N. Oliver, A. Pentland, et al. The network dimension of ride sharing. In: *International Conference on Computational Social Science (ICCSS)*, 22–26, 2015.

[89] E. Shmueli, I. Mazeh, L. Radaelli, A. S. Pentland, Y. Altshuler er al. Ride sharing: A network perspective. In: *Social Computing, Behavioral-Cultural Modeling*, and Prediction, Springer, pp. 434–439, 2015.

[90] A.-L. Barabasi and R. Albert, Emergence of scaling in random networks, Science 286 (5439): 509–512, 1999.

[91] Z. Galil, Efficient algorithms for finding maximum matching in graphs, *ACM Computing Surveys (CSUR)*, 18(1): 23–38, 1986.

[92] B. Chandra and M. M. Halldorsson, Greedy local improvement and weighted set packing approximation. *Journal of Algorithms* 39(2): 223–240, 2001.

[93] M.E. Ben-Akiva and S.R. Lerman, Discrete choice analysis: theory and application to travel demand, Vol. 9, MIT press, 1985.

[94] S. Erlander and N.F. Stewart, The gravity model in transportation analysis: theory and extensions, Vol. 3, Vsp, 1990.

[95] C.-H. Wen and F.S. Koppelman, The generalized nested logit model. *Transportation Research Part B: Methodological*, 35(7): 627–641, 2001.

[96] W. J. G., Some theoretical aspects of road traffic research. *Proceedings of the Institution of Civil Engineers*, 1(1952): 325–378, 1952.

[97] M.G. McNally, The four step model, Handbook of transport modelling, 1: 35–41, 2007.

[98] T.L. Friesz, Transportation network equilibrium, design and aggregation: key developments and research opportunities. *Transportation Research Part A: General*, 19(5): 413–427, 1985.

[99] M.D. Meyer and E.J. Miller, Urban transportation planning: a decision-oriented approach, 1984.

[100] H. Xu, F. Ordóñez and M. Dessouky, A traffic assignment model for a ridesharing transportation market. *Journal of Advanced Transportation*, 49(7): 793–816, 2015.

[101] A. Lee and M. Savelsbergh, Dynamic ridesharing: Is there a role for dedicated drivers? *Transportation Research Part B: Methodological*, 81: 483–497, 2015.

[102] D.J. Fagnant and K.M. Kockelman, Dynamic ride-sharing and optimal fleet sizing for a system of shared autonomous vehicles. In: *Transportation Research Board 94th Annual Meeting,* No. 15–1962, 2015.

[103] L.C. Freeman, A set of measures of centrality based on betweenness. *Sociometry*, 40(1): 35–41, 1977.

[104] Y. Altshuler, R. Puzis, Y. Elovici, S. Bekhor and A. Pentland, Augmented betweenness centrality for mobility prediction in transportation networks. In: *International Workshop on Finding Patterns of Human Behaviors in NEtworks and MObility Data, NEMO11*, 2011.

[105] K. Stephenson and M. Zelen, Rethinking centrality: Methods and examples. *Social Networks*, 11(1): 1–37, 1989.

[106] P. Bonacich, Factoring and weighting approaches to status scores and clique identification. *Journal of Mathematical Sociology*, 2(1): 113–120, 1972.

[107] B.A. Prakash, D. Chakrabarti, M. Faloutsos, N. Valler, C. Faloutsos, et al. Got the flu (or mumps)? check the eigenvalue!. *arXiv preprint arXiv*: 1004.0060 (2010).

[108] D. Chakrabarti, Y. Wang, C. Wang, J. Leskovec, C. Faloutsos, et al. Epidemic thresholds in real networks. *ACM Transactions on Information and System Security*, 10(4): 1–26, 2008. doi:http://doi.acm.org/10.1145/1284680.1284681.

[109] S. Boccaletti, V. Latora, Y. Moreno, M. Chavez, D.-U. Hwang, et al. Complex networks: Structure and dynamics. *Physics Reports*, 424(4): 175–308, 2006.

[110] Y.-A. de Montjoye, J. Quoidbach, F. Robic and A.S. Pentland, Predicting personality using novel mobile phone-based metrics. In: *Social Computing, Behavioral-Cultural Modeling and Prediction, Springer*, pp. 48–55, 2013.

[111] V.K. Singh, L. Freeman, B. Lepri and A.S. Pentland, Predicting spending behavior using socio-mobile features. In: *Proceedings of the International Conference on Social Computing (SocialCom), IEEE*, pp. 174–179, 2013.

[112] S. Ma, Y. Zheng, O. Wolfson and T-share: A large-scale dynamic taxi ridesharing service. In: *Proceedings of the 29th International Conference on Data Engineering (ICDE), IEEE*, pp. 410–421, 2013.

[113] M.R. Garey, D.S. Johnson and L. Stockmeyer, Some simplified np-complete graph problems. *Theoretical Computer Science*, 1(3): 237–267, 1976.

[114] F.A. Kuipers and P.F. Van Mieghem, Conditions that impact the complexity of qos routing. *IEEE/ACM Transactions on Networking (TON)*, 13(4): 717–730, 2005.

[115] Y. Naveh, Y. Richter, Y. Altshuler, D.L. Gresh, D.P. Connors, et al. Workforce optimization: Identification and assignment of professional workers using constraint programming. *IBM Journal of Research and Development*, 51(3.4) 263–279, 2007.

[116] O. Bahat and S. Bekhor, Incorporating ridesharing in the static traffic assignment model. *Networks and Spatial Economics*, 1–25, 2015.

[117] Y.-A. de Montjoye, A. Stopczynski, E. Shmueli, A. Pentland and S. Lehmann, The strength of the strongest ties in collaborative problem solving. *Scientific Reports*, 4, 2014.

[118] D. Lazer and A. Friedman, The network structure of exploration and exploitation. *Administrative Science Quarterly*, 52(4): 667–694, 2007.

[119] Y. Altshuler, A. Pentland and A.M. Bruckstein, Introduction to swarm search. In: *Swarms and Network Intelligence in Search*, Springer, pp. 1–14, 2018.

[120] Y. Altshuler, A. Pentland and A.M. Bruckstein. *Swarms and Network Intelligence in Search*, Springer, 2018.

[121] Y. Altshuler, I. Wagner and A. Bruckstein, Collaborative exploration in grid domains. In: *Sixth International Conference on Informatics in Control, Automation and Robotics (ICINCO)*, 13–21, 2009.

[122] Y. Altshuler, V. Yanovsky, A. Bruckstein and I. Wagner, Efficient cooperative search of smart targets using uav swarms. ROBOTICA, 26: 551–557, 2008.

[123] J. Svennebring and S. Koenig, Building terrain-covering ant robots: A feasibility study. *Autonomous Robots* 16(3): 313–332: 2004.

[124] S. Koenig, B. Szymanski and Y. Liu, Efficient and inefficient ant coverage methods. *Annals of Mathematics and Artificial Intelligence*, 31: 41–76: 2001.

[125] Y. Altshuler, I. Wagner, V. Yanovski and A. Bruckstein, Multi-agent cooperative cleaning of expanding domains. *International Journal of Robotics Research*, 30: 1037–1071, 2010.

[126] Y. Altshuler, A. Pentland and A.M. Bruckstein, Collaborative patrolling swarms in stochastically expanding environments. In: *Swarms and Network Intelligence in Search, Springer*, pp. 155–185, 2018.

[127] Y. Altshuler, A. Pentland and A.M. Bruckstein, Optimal dynamic coverage infrastructure for large-scale fleets of reconnaissance uavs. In: *Swarms and Network Intelligence in Search*, Springer, pp. 207–238, 2018.

[128] Y. Altshuler, A.M. Bruckstein, Static and expanding grid coverage with ant robots: Complexity results. *Theoretical Computer Science*, 412(35): 4661–4674, 2011.

[129] Y. Altshuler, I. Wagner and A. Bruckstein, On swarm optimality in dynamic and symmetric environments, 7: 11, 2008.

[130] Y. Altshuler, I. Wagner and A. Bruckstein, Shape factor's effect on a dynamic cleaners swarm. In: *Third International Conference on Informatics in Control, Automation and Robotics (ICINCO), the Second International Workshop on Multi-Agent Robotic Systems (MARS)*, pp. 13–21, 2006.

Appendix A: Prediction Results

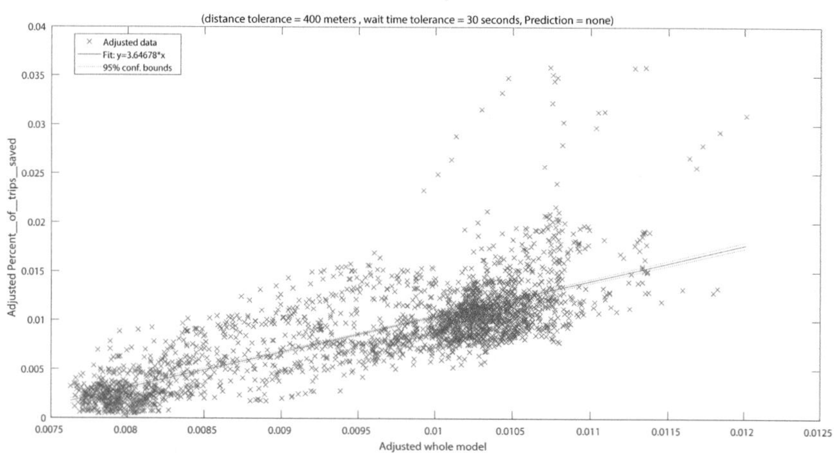

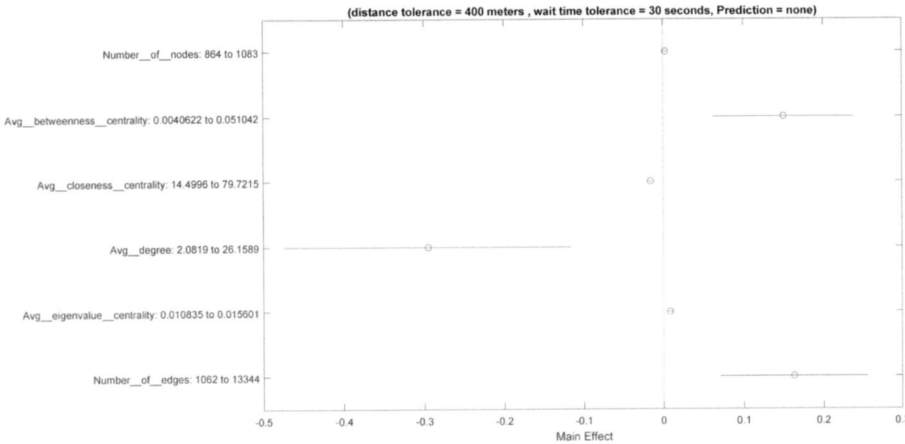


```
Estimated Coefficients:
```

	Estimate	SE	tStat	pValue
(Intercept)	-0.026063	0.010258	-2.5407	0.011126
Number__of__nodes	1.0874e-05	7.6323e-06	1.4247	0.15436
Avg__betweenness__centrality	3.196	0.94927	3.3667	0.00077247
Avg__closeness__centrality	-0.00025104	1.3216e-05	-18.994	3.9737e-75
Avg__degree	-0.012254	0.0037849	-3.2377	0.0012212
Avg__eigenvalue__centrality	1.7564	0.39193	4.4813	7.7627e-06
Number__of__edges	1.3273e-05	3.8399e-06	3.4566	0.00055645

```
Number of observations: 2440, Error degrees of freedom: 2433
Root Mean Squared Error: 0.00307
R-squared: 0.589,  Adjusted R-Squared 0.588
F-statistic vs. constant model: 581, p-value = 0

ANOVA_ANALYSIS =
```

	SumSq	DF	MeanSq	F	pValue
Total	0.055862	2439	2.2904e-05		
Model	0.032903	6	0.0054838	581.12	0
Residual	0.022959	2433	9.4367e-06		

Figure 2.17: Multilinear regression model predicting the ride-sharing utilization using the dynamic network's properties.

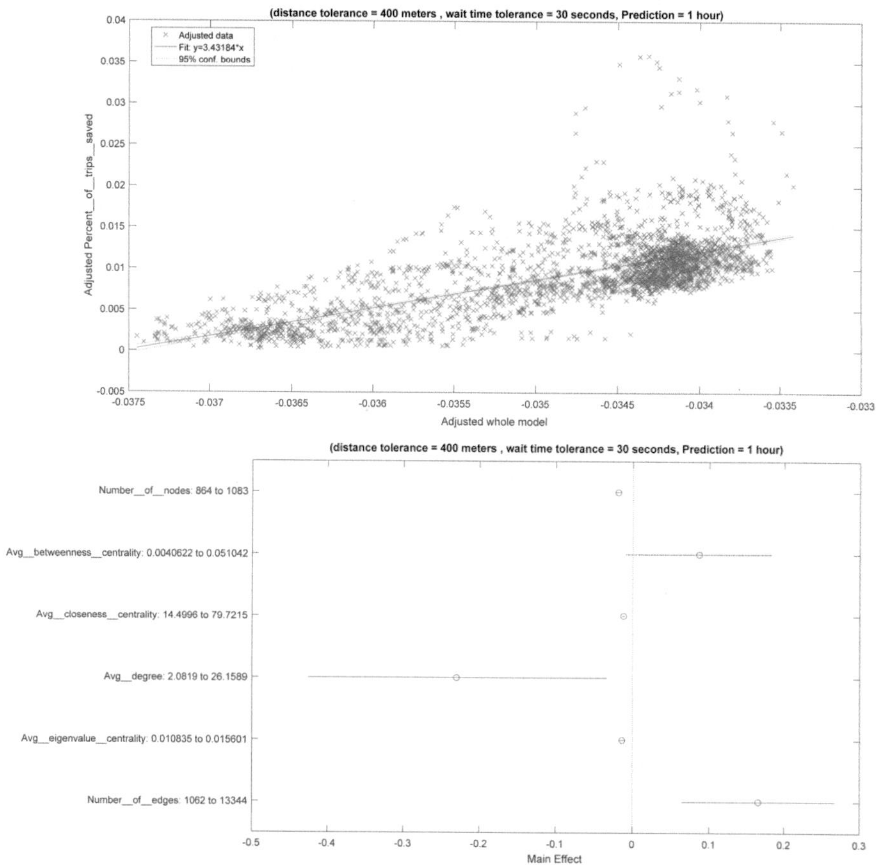


```
Estimated Coefficients:
```

	Estimate	SE	tStat	pValue
(Intercept)	0.12878	0.011271	11.426	1.7369e-29
Number__of__nodes	-8.5893e-05	8.3765e-06	-10.254	3.4997e-24
Avg__betweenness__centrality	1.8516	1.0415	1.7777	0.075571
Avg__closeness__centrality	-0.00018101	1.4562e-05	-12.431	1.9343e-34
Avg__degree	-0.0095375	0.0041534	-2.2963	0.021741
Avg__eigenvalue__centrality	-2.8895	0.43166	-6.6939	2.6888e-11
Number__of__edges	1.3508e-05	4.2145e-06	3.2052	0.0013673

```
Number of observations: 2437, Error degrees of freedom: 2430
Root Mean Squared Error: 0.00337
R-squared: 0.493,  Adjusted R-Squared 0.492
F-statistic vs. constant model: 394, p-value = 0

ANOVA_ANALYSIS =
```

	SumSq	DF	MeanSq	F	pValue
Total	0.054512	2436	2.2378e-05		
Model	0.026891	6	0.0044818	394.29	0
Residual	0.027621	2430	1.1367e-05		

Figure 2.18: Multilinear regression model predicting the ride-sharing utilization using the dynamic network's properties.

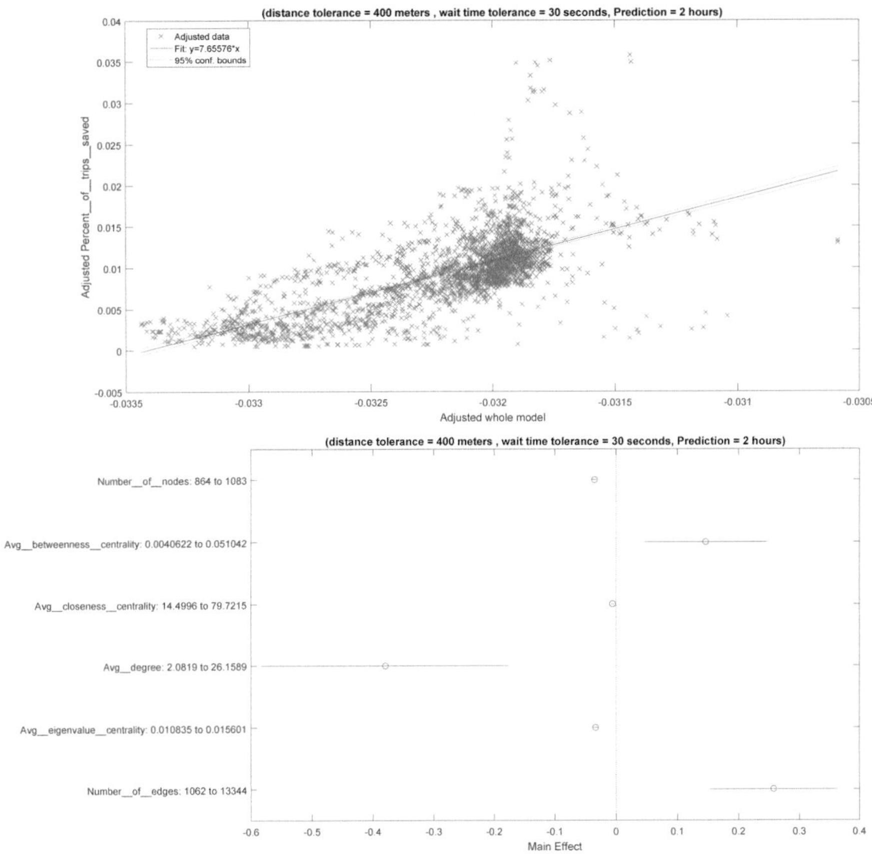

Estimated Coefficients:

	Estimate	SE	tStat	pValue
(Intercept)	0.25583	0.011669	21.924	2.3641e-97
Number__of__nodes	-0.00016059	8.6769e-06	-18.507	1.2197e-71
Avg__betweenness__centrality	3.1095	1.0789	2.8821	0.0039851
Avg__closeness__centrality	-8.4426e-05	1.5012e-05	-5.6241	2.0791e-08
Avg__degree	-0.015776	0.0043026	-3.6666	0.00025105
Avg__eigenvalue__centrality	-6.9958	0.44458	-15.736	3.3485e-53
Number__of__edges	2.0941e-05	4.3658e-06	4.7966	1.7113e-06

Number of observations: 2434, Error degrees of freedom: 2427
Root Mean Squared Error: 0.00349
R-squared: 0.46, Adjusted R-Squared 0.459
F-statistic vs. constant model: 345, p-value = 2.42e-320

ANOVA_ANALYSIS =

	SumSq	DF	MeanSq	F	pValue
Total	0.054816	2433	2.253e-05		
Model	0.025219	6	0.0042032	344.68	2.4165e-320
Residual	0.029596	2427	1.2195e-05		

Figure 2.19: Multilinear regression model predicting the ride-sharing utilization using the dynamic network's properties.

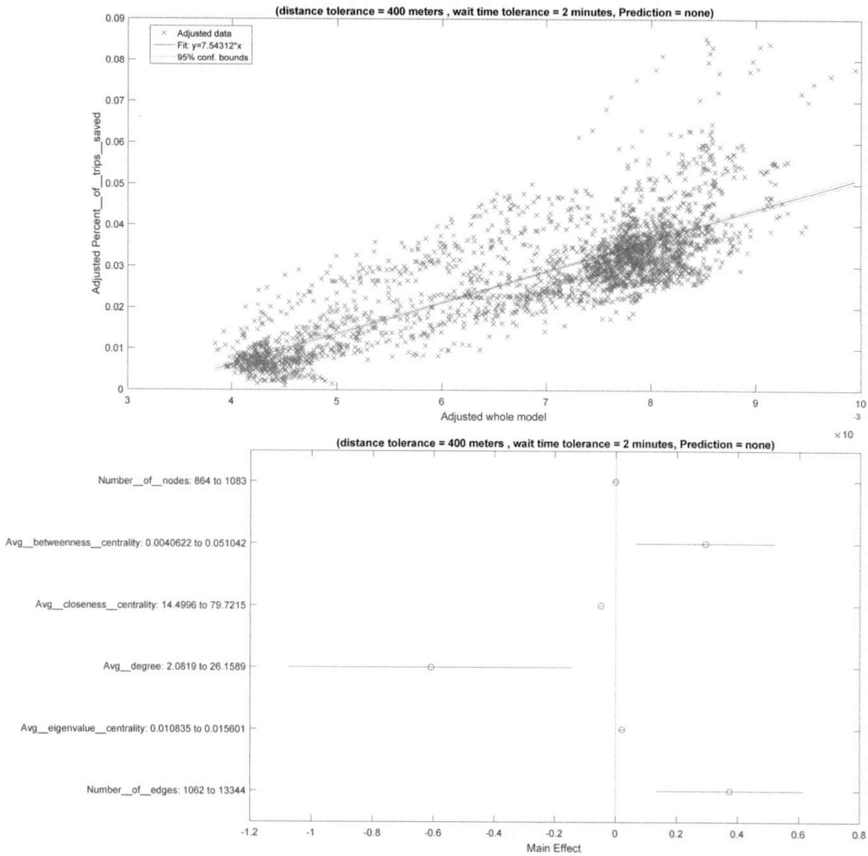

Estimated Coefficients:

	Estimate	SE	tStat	pValue
(Intercept)	-0.024022	0.026666	-0.90085	0.36776
Number__of__nodes	-8.6047e-06	1.9866e-05	-0.43314	0.66495
Avg__betweenness__centrality	6.227	2.4724	2.5187	0.011844
Avg__closeness__centrality	-0.00075145	3.4186e-05	-21.981	6.9988e-98
Avg__degree	-0.025278	0.0098607	-2.5635	0.010422
Avg__eigenvalue__centrality	4.257	1.0088	4.22	2.5325e-05
Number__of__edges	3.0418e-05	1.0005e-05	3.0401	0.0023896

Number of observations: 2455, Error degrees of freedom: 2448
Root Mean Squared Error: 0.008
R-squared: 0.663, Adjusted R-Squared 0.662
F-statistic vs. constant model: 803, p-value = 0

ANOVA_ANALYSIS =

	SumSq	DF	MeanSq	F	pValue
Total	0.46544	2454	0.00018966		
Model	0.30858	6	0.05143	802.64	0
Residual	0.15686	2448	6.4075e-05		

Figure 2.20: Multilinear regression model predicting the ride-sharing utilization using the dynamic network's properties.

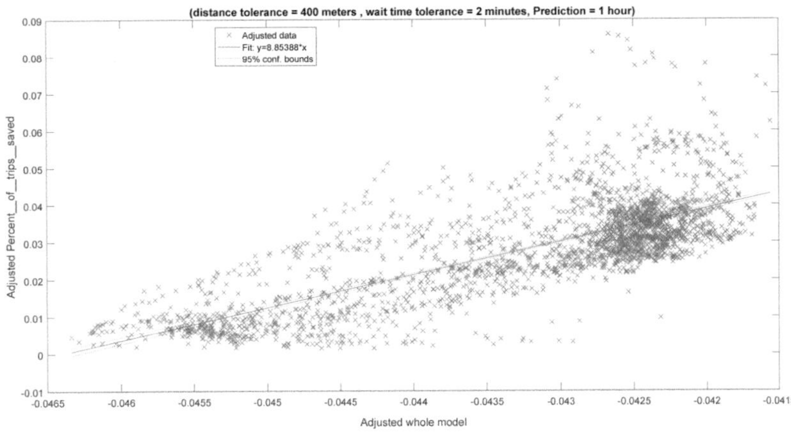

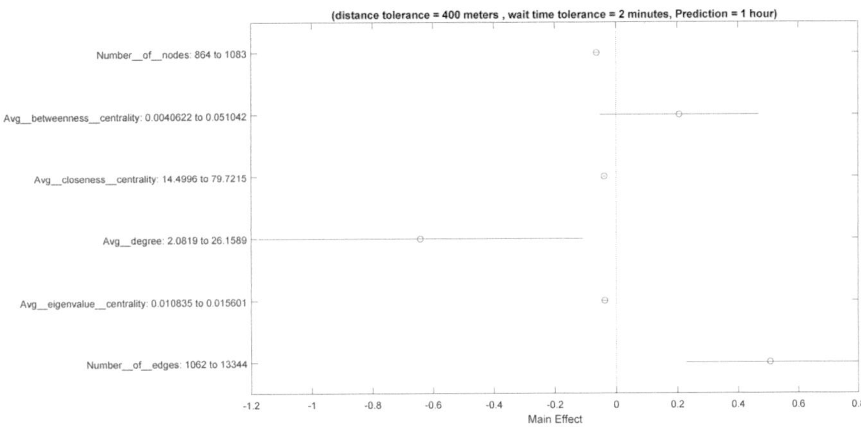

Estimated Coefficients:

	Estimate	SE	tStat	pValue
(Intercept)	0.41087	0.030574	13.439	8.9037e-40
Number__of__nodes	-0.00029226	2.2777e-05	-12.831	1.599e-36
Avg__betweenness__centrality	4.4153	2.8347	1.5576	0.11946
Avg__closeness__centrality	-0.00059769	3.9196e-05	-15.249	3.1646e-50
Avg__degree	-0.026662	0.011306	-2.3583	0.018437
Avg__eigenvalue__centrality	-7.6743	1.1566	-6.6352	3.9763e-11
Number__of__edges	4.1122e-05	1.1472e-05	3.5847	0.00034407

Number of observations: 2455, Error degrees of freedom: 2448
Root Mean Squared Error: 0.00918
R-squared: 0.557, Adjusted R-Squared 0.556
F-statistic vs. constant model: 513, p-value = 0

ANOVA_ANALYSIS =

	SumSq	DF	MeanSq	F	pValue
Total	0.46549	2454	0.00018969		
Model	0.25929	6	0.043215	513.05	0
Residual	0.2062	2448	8.4232e-05		

Figure 2.21: Multilinear regression model predicting the ride-sharing utilization using the dynamic network's properties.

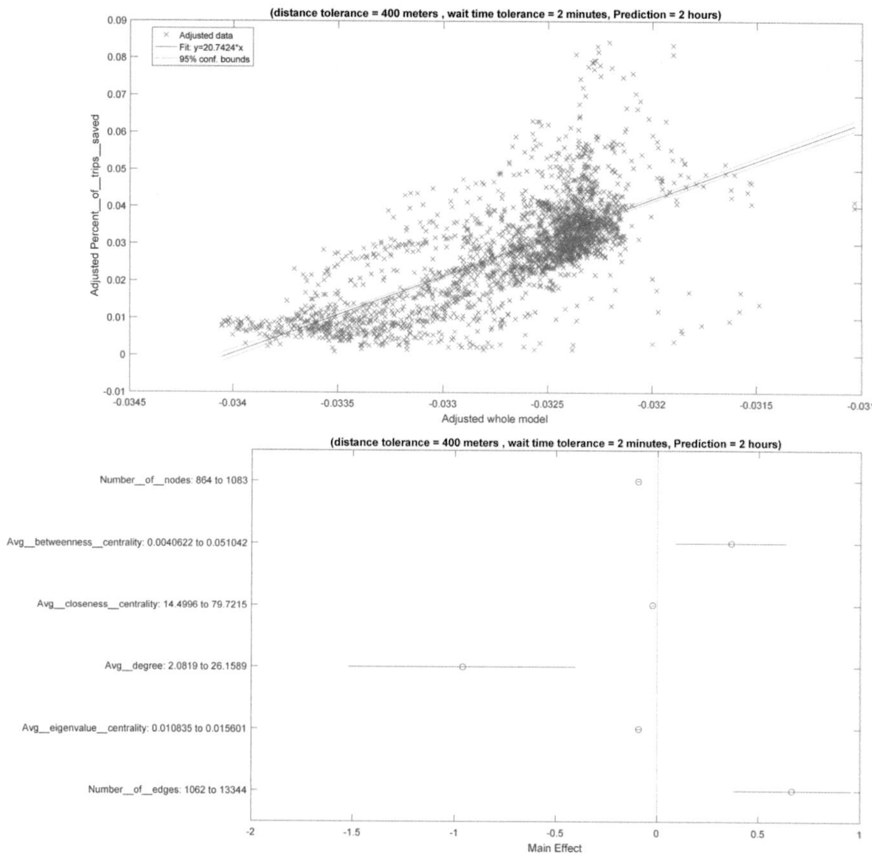


```
Estimated Coefficients:
                                    Estimate          SE         tStat        pValue
                                   __________    __________    _______    __________

    (Intercept)                       0.70576      0.031989     22.063    1.5528e-98
    Number__of__nodes             -0.00043985    2.3831e-05    -18.457    2.5497e-71
    Avg__betweenness__centrality       7.7561        2.9659     2.6151     0.0089749
    Avg__closeness__centrality    -0.00034689     4.101e-05    -8.4587    4.5808e-17
    Avg__degree                     -0.039914      0.011829    -3.3743    0.00075165
    Avg__eigenvalue__centrality       -19.238        1.2101    -15.897    3.1308e-54
    Number__of__edges              5.4288e-05    1.2002e-05      4.523    6.3871e-06


Number of observations: 2455, Error degrees of freedom: 2448
Root Mean Squared Error: 0.0096
R-squared: 0.516,  Adjusted R-Squared 0.515
F-statistic vs. constant model: 435, p-value = 0

ANOVA_ANALYSIS =

                 SumSq      DF       MeanSq        F        pValue
                _______    ____    _________    ______    ______

    Total       0.46627    2454      0.00019
    Model       0.24054       6      0.04009    434.77       0
    Residual    0.22573    2448    9.221e-05
```

Figure 2.22: Multilinear regression model predicting the ride-sharing utilization using the dynamic network's properties.

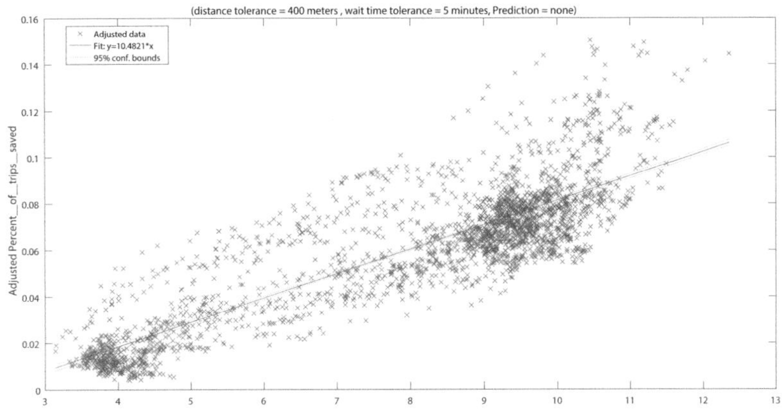

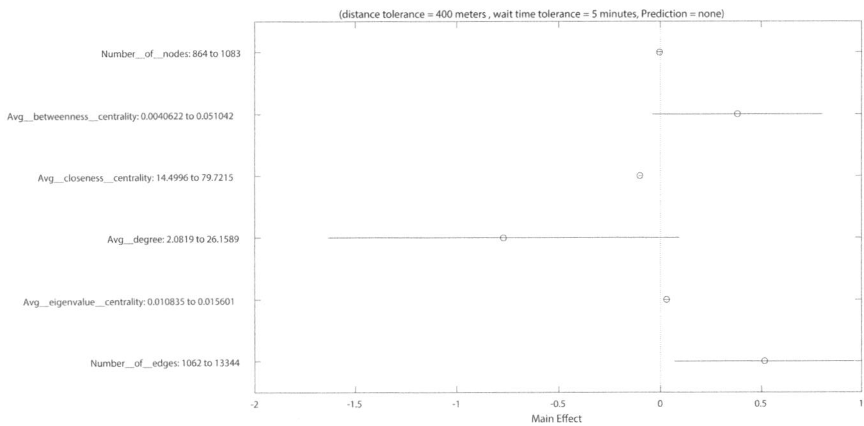

Estimated Coefficients:

	Estimate	SE	tStat	pValue
(Intercept)	-0.023402	0.049455	-0.47319	0.63612
Number__of__nodes	-1.933e-05	3.6844e-05	-0.52466	0.59987
Avg__betweenness__centrality	8.1552	4.5853	1.7786	0.075438
Avg__closeness__centrality	-0.0015276	6.3402e-05	-24.094	2.802e-115
Avg__degree	-0.031999	0.018288	-1.7498	0.080285
Avg__eigenvalue__centrality	6.5853	1.8709	3.5199	0.00043966
Number__of__edges	4.1957e-05	1.8556e-05	2.2611	0.023841

```
Number of observations: 2455, Error degrees of freedom: 2448
Root Mean Squared Error: 0.0148
R-squared: 0.73,  Adjusted R-Squared 0.729
F-statistic vs. constant model: 1.1e+03, p-value = 0

ANOVA_ANALYSIS =
```

	SumSq	DF	MeanSq	F	pValue
Total	1.9966	2454	0.00081362		
Model	1.4571	6	0.24285	1101.9	0
Residual	0.53953	2448	0.0002204		

Figure 2.23: Multilinear regression model predicting the ride-sharing utilization using the dynamic network's properties.

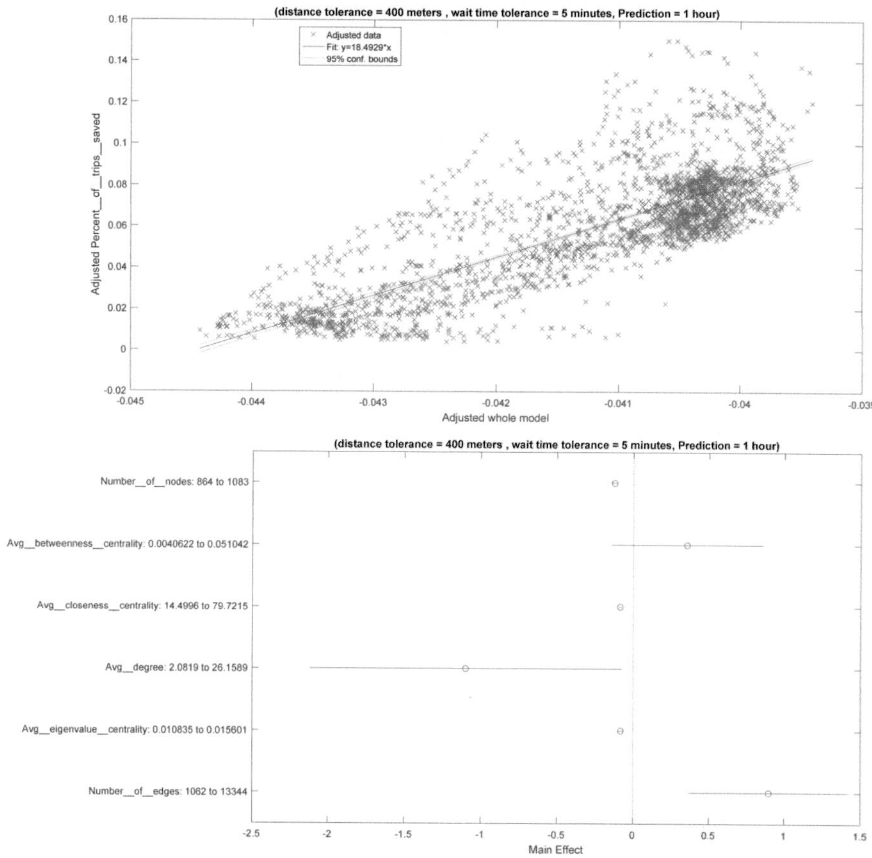

Estimated Coefficients:

	Estimate	SE	tStat	pValue
(Intercept)	0.82177	0.058462	14.056	3.2169e-43
Number_of_nodes	-0.00056557	4.3554e-05	-12.985	2.4598e-37
Avg_betweenness_centrality	7.5943	5.4204	1.401	0.16133
Avg_closeness_centrality	-0.0012879	7.495e-05	-17.184	1.4197e-62
Avg_degree	-0.045583	0.021619	-2.1085	0.035088
Avg_eigenvalue_centrality	-16.862	2.2117	-7.624	3.4958e-14
Number_of_edges	7.2904e-05	2.1936e-05	3.3236	0.00090187

Number of observations: 2455, Error degrees of freedom: 2448
Root Mean Squared Error: 0.0175
R-squared: 0.624, Adjusted R-Squared 0.624
F-statistic vs. constant model: 678, p-value = 0

ANOVA_ANALYSIS =

	SumSq	DF	MeanSq	F	pValue
Total	2.0075	2454	0.00081805		
Model	1.2535	6	0.20892	678.35	0
Residual	0.75396	2448	0.00030799		

Figure 2.24: Multilinear regression model predicting the ride-sharing utilization using the dynamic network's properties.

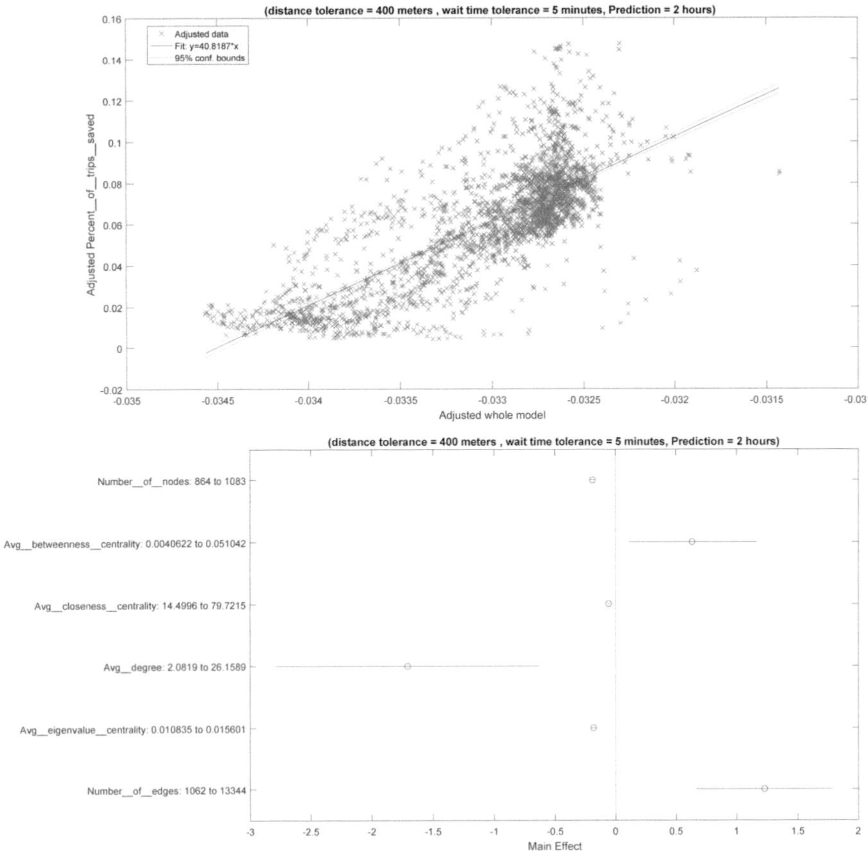

Estimated Coefficients:

	Estimate	SE	tStat	pValue
(Intercept)	1.4086	0.0618	22.793	1.9116e-104
Number__of__nodes	-0.0008726	4.6041e-05	-18.952	7.4474e-75
Avg__betweenness__centrality	13.531	5.73	2.3614	0.018284
Avg__closeness__centrality	-0.00088309	7.9229e-05	-11.146	3.5287e-28
Avg__degree	-0.070953	0.022853	-3.1047	0.0019262
Avg__eigenvalue__centrality	-38.511	2.3379	-16.472	6.7669e-58
Number__of__edges	9.9818e-05	2.3188e-05	4.3047	1.738e-05

Number of observations: 2455, Error degrees of freedom: 2448
Root Mean Squared Error: 0.0186
R-squared: 0.578, Adjusted R-Squared 0.577
F-statistic vs. constant model: 560, p-value = 0

ANOVA_ANALYSIS =

	SumSq	DF	MeanSq	F	pValue
Total	1.9988	2454	0.00081451		
Model	1.1563	6	0.19271	559.94	0
Residual	0.84252	2448	0.00034417		

Figure 2.25: Multilinear regression model predicting the ride-sharing utilization using the dynamic network's properties.

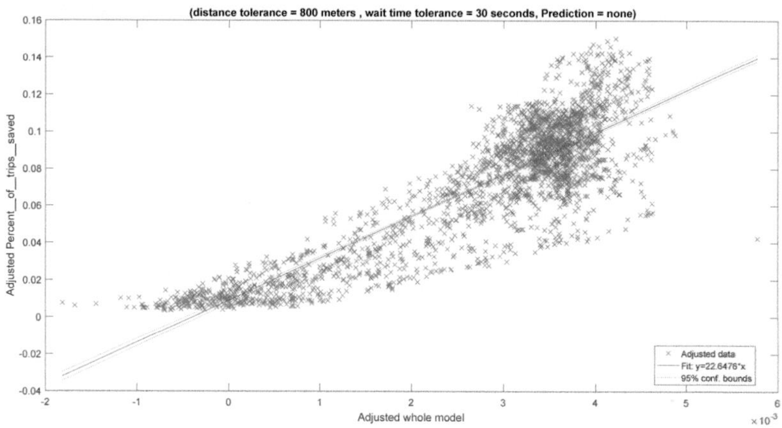

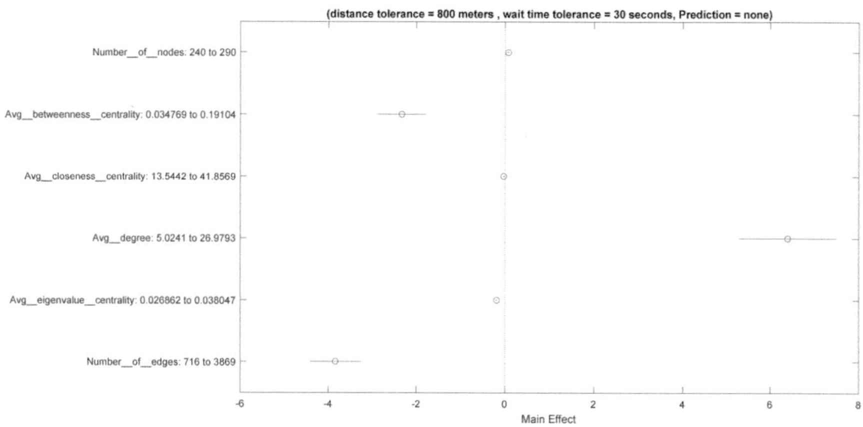

Estimated Coefficients:

	Estimate	SE	tStat	pValue
(Intercept)	0.0090474	0.071194	0.12708	0.89889
Number__of__nodes	0.0014356	0.00021643	6.6332	4.0185e-11
Avg__betweenness__centrality	-14.963	1.8334	-8.1612	5.2123e-16
Avg__closeness__centrality	-0.0009661	0.00032024	-3.0168	0.0025803
Avg__degree	0.29089	0.025602	11.362	3.3613e-29
Avg__eigenvalue__centrality	-16.998	0.98392	-17.276	3.0715e-63
Number__of__edges	-0.0012151	9.2937e-05	-13.075	7.9079e-38

Number of observations: 2490, Error degrees of freedom: 2483
Root Mean Squared Error: 0.0177
R-squared: 0.761, Adjusted R-Squared 0.761
F-statistic vs. constant model: 1.32e+03, p-value = 0

ANOVA_ANALYSIS =

	SumSq	DF	MeanSq	F	pValue
Total	3.2576	2489	0.0013088		
Model	2.4805	6	0.41342	1320.9	0
Residual	0.77712	2483	0.00031298		

Figure 2.26: Multilinear regression model predicting the ride-sharing utilization using the dynamic network's properties.

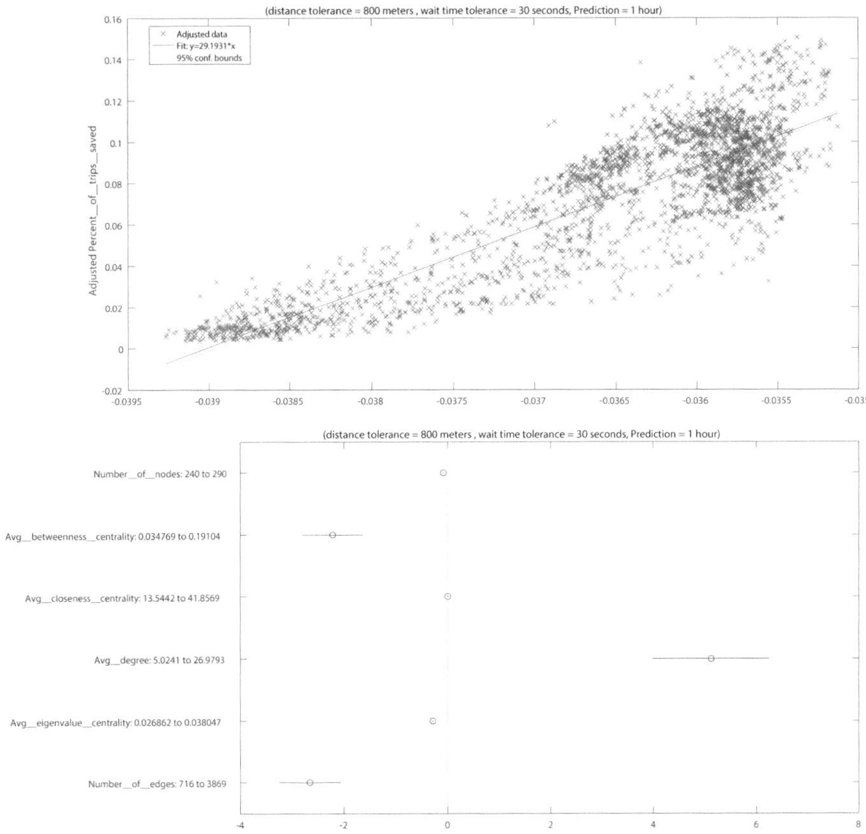

Estimated Coefficients:

	Estimate	SE	tStat	pValue
(Intercept)	1.1391	0.073557	15.485	1.0529e-51
Number__of__nodes	-0.0016839	0.00022362	-7.5302	7.0487e-14
Avg__betweenness__centrality	-14.176	1.8943	-7.4833	1.0001e-13
Avg__closeness__centrality	0.0001357	0.00033087	0.41015	0.68173
Avg__degree	0.23293	0.026452	8.8058	2.3867e-18
Avg__eigenvalue__centrality	-25.519	1.0166	-25.103	4.0157e-124
Number__of__edges	-0.00084118	9.6022e-05	-8.7603	3.5338e-18

Number of observations: 2490, Error degrees of freedom: 2483
Root Mean Squared Error: 0.0183
R-squared: 0.746, Adjusted R-Squared 0.745
F-statistic vs. constant model: 1.21e+03, p-value = 0

ANOVA_ANALYSIS =

	SumSq	DF	MeanSq	F	pValue
Total	3.2621	2489	0.0013106		
Model	2.4325	6	0.40541	1213.5	0
Residual	0.82957	2483	0.0003341		

Figure 2.27: Multilinear regression model predicting the ride-sharing utilization using the dynamic network's properties.

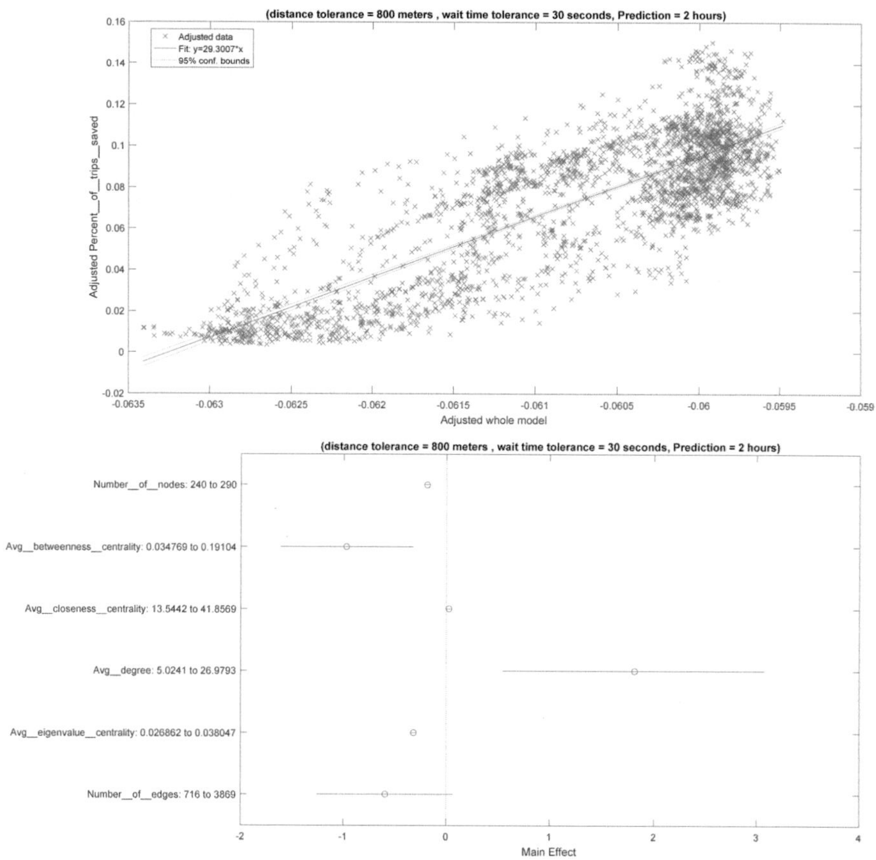

Estimated Coefficients:

	Estimate	SE	tStat	pValue
(Intercept)	1.8532	0.081866	22.637	2.6549e-103
Number__of__nodes	-0.0037658	0.00024888	-15.131	1.5281e-49
Avg__betweenness__centrality	-6.2244	2.1083	-2.9524	0.003183
Avg__closeness__centrality	0.00084982	0.00036824	2.3078	0.021093
Avg__degree	0.082561	0.02944	2.8044	0.0050799
Avg__eigenvalue__centrality	-28.632	1.1314	-25.306	6.7398e-126
Number__of__edges	-0.00018889	0.00010687	-1.7675	0.07726

Number of observations: 2490, Error degrees of freedom: 2483
Root Mean Squared Error: 0.0203
R-squared: 0.677, Adjusted R-Squared 0.676
F-statistic vs. constant model: 866, p-value = 0

ANOVA_ANALYSIS =

	SumSq	DF	MeanSq	F	pValue
Total	3.1783	2489	0.0012769		
Model	2.1507	6	0.35846	866.17	0
Residual	1.0276	2483	0.00041384		

Figure 2.28: Multilinear regression model predicting the ride-sharing utilization using the dynamic network's properties.

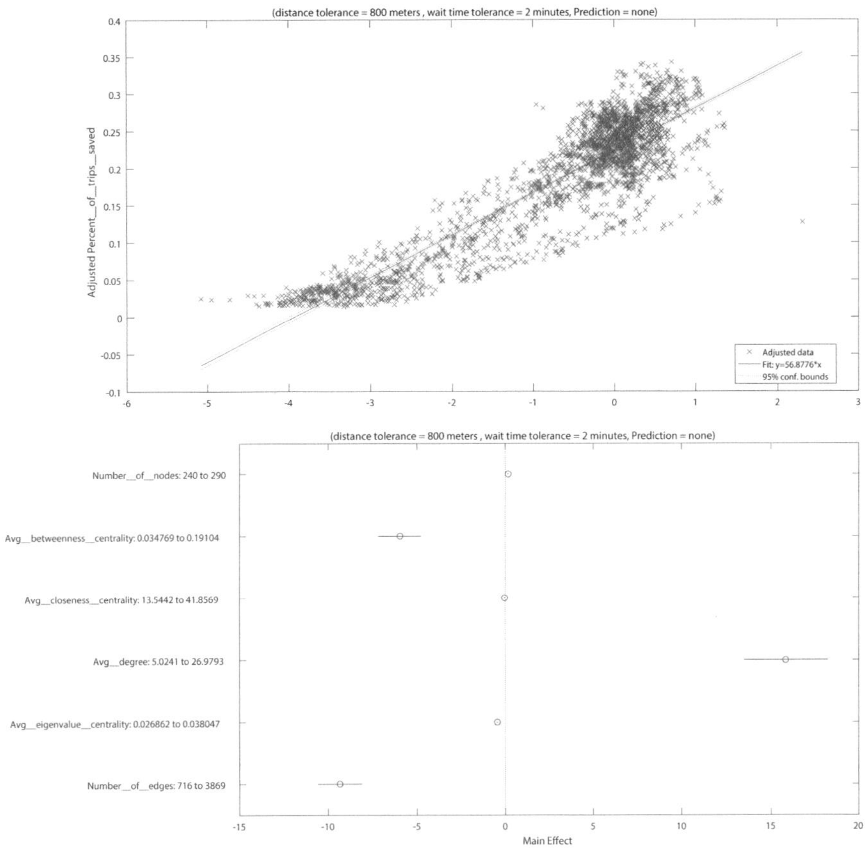

Estimated Coefficients:

	Estimate	SE	tStat	pValue
(Intercept)	0.22375	0.15331	1.4595	0.14455
Number__of__nodes	0.0028521	0.00046607	6.1195	1.0873e-09
Avg__betweenness__centrality	-38.255	3.9481	-9.6895	8.0908e-22
Avg__closeness__centrality	-0.001842	0.00068959	-2.6712	0.0076081
Avg__degree	0.72187	0.055131	13.094	6.2587e-38
Avg__eigenvalue__centrality	-42.084	2.1188	-19.863	1.3341e-81
Number__of__edges	-0.0029629	0.00020013	-14.805	1.3708e-47

Number of observations: 2490, Error degrees of freedom: 2483
Root Mean Squared Error: 0.0381
R-squared: 0.804, Adjusted R-Squared 0.804
F-statistic vs. constant model: 1.7e+03, p-value = 0

ANOVA_ANALYSIS =

	SumSq	DF	MeanSq	F	pValue
Total	18.405	2489	0.0073946		
Model	14.802	6	2.4669	1699.8	0
Residual	3.6036	2483	0.0014513		

Figure 2.29: Multilinear regression model predicting the ride-sharing utilization using the dynamic network's properties.

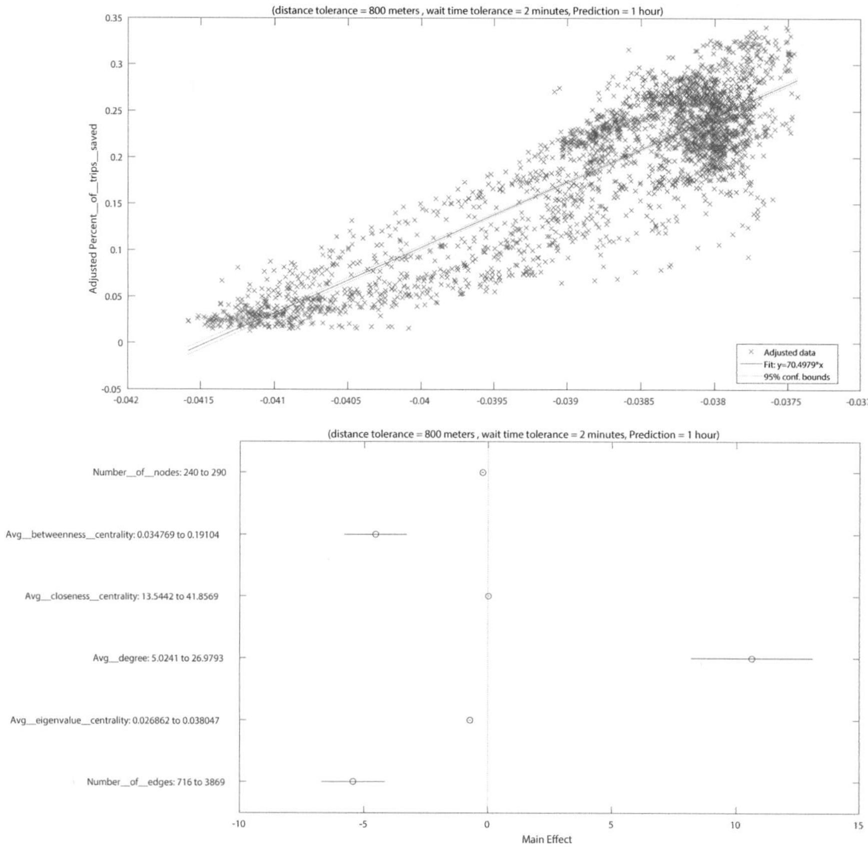


```
Estimated Coefficients:
                                    Estimate        SE         tStat       pValue
                                   __________   __________   ________   __________

    (Intercept)                        2.9227      0.15873     18.413    4.496e-71
    Number__of__nodes               -0.004423   0.00048255    -9.1658    1.002e-19
    Avg__betweenness__centrality      -29.097       4.0878    -7.1181   1.4262e-12
    Avg__closeness__centrality     0.00086988   0.00071399     1.2183      0.22321
    Avg__degree                       0.48447     0.057081     8.4874   3.5786e-17
    Avg__eigenvalue__centrality       -64.211       2.1937    -29.271  4.6691e-162
    Number__of__edges              -0.0017278   0.00020721    -8.3384   1.2315e-16


Number of observations: 2490, Error degrees of freedom: 2483
Root Mean Squared Error: 0.0394
R-squared: 0.788,  Adjusted R-Squared 0.787
F-statistic vs. constant model: 1.54e+03, p-value = 0

ANOVA_ANALYSIS =

                SumSq     DF       MeanSq        F       pValue
                ______    ____    _________    ______    ______

    Total       18.221    2489    0.0073206
    Model       14.358       6        2.393    1538.1      0
    Residual     3.863    2483    0.0015558
```

Figure 2.30: Multilinear regression model predicting the ride-sharing utilization using the dynamic network's properties.

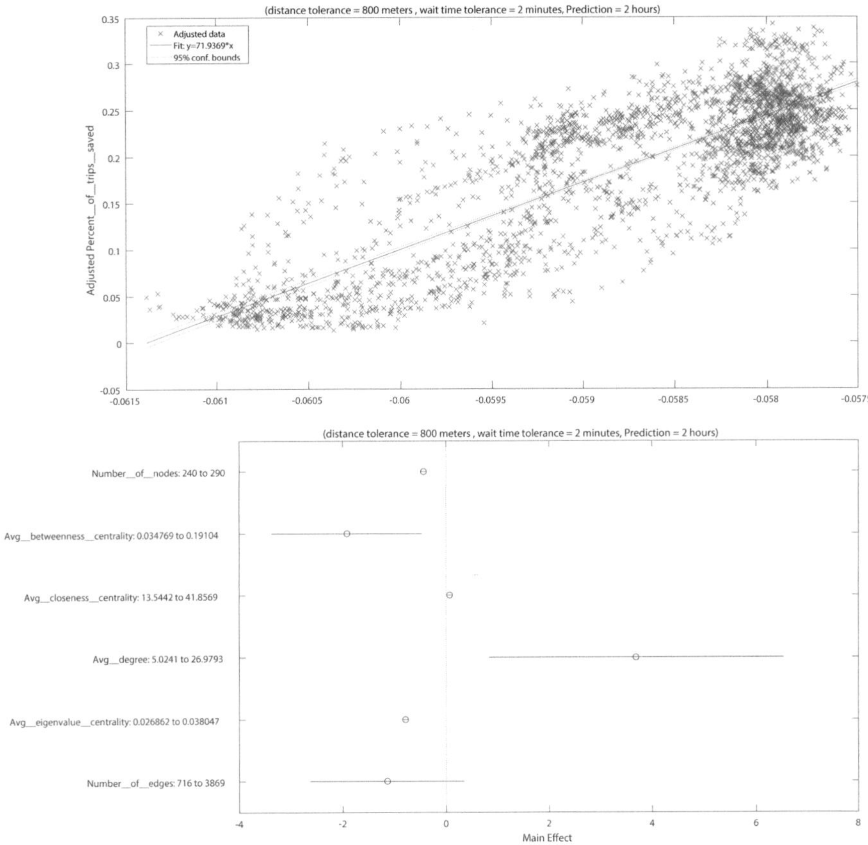

Estimated Coefficients:

	Estimate	SE	tStat	pValue
(Intercept)	4.4163	0.18429	23.964	2.461e-114
Number_of_nodes	-0.0087159	0.00056025	-15.557	3.8028e-52
Avg_betweenness_centrality	-12.259	4.746	-2.5831	0.0098488
Avg_closeness_centrality	0.0023073	0.00082895	2.7834	0.0054195
Avg_degree	0.16768	0.066272	2.5301	0.011463
Avg_eigenvalue_centrality	-70.884	2.5469	-27.831	1.32e-148
Number_of_edges	-0.00036267	0.00024057	-1.5075	0.1318

Number of observations: 2490, Error degrees of freedom: 2483
Root Mean Squared Error: 0.0458
R-squared: 0.708, Adjusted R-Squared 0.707
F-statistic vs. constant model: 1e+03, p-value = 0

ANOVA_ANALYSIS =

	SumSq	DF	MeanSq	F	pValue
Total	17.813	2489	0.0071568		
Model	12.606	6	2.101	1001.8	0
Residual	5.2072	2483	0.0020972		

Figure 2.31: Multilinear regression model predicting the ride-sharing utilization using the dynamic network's properties.

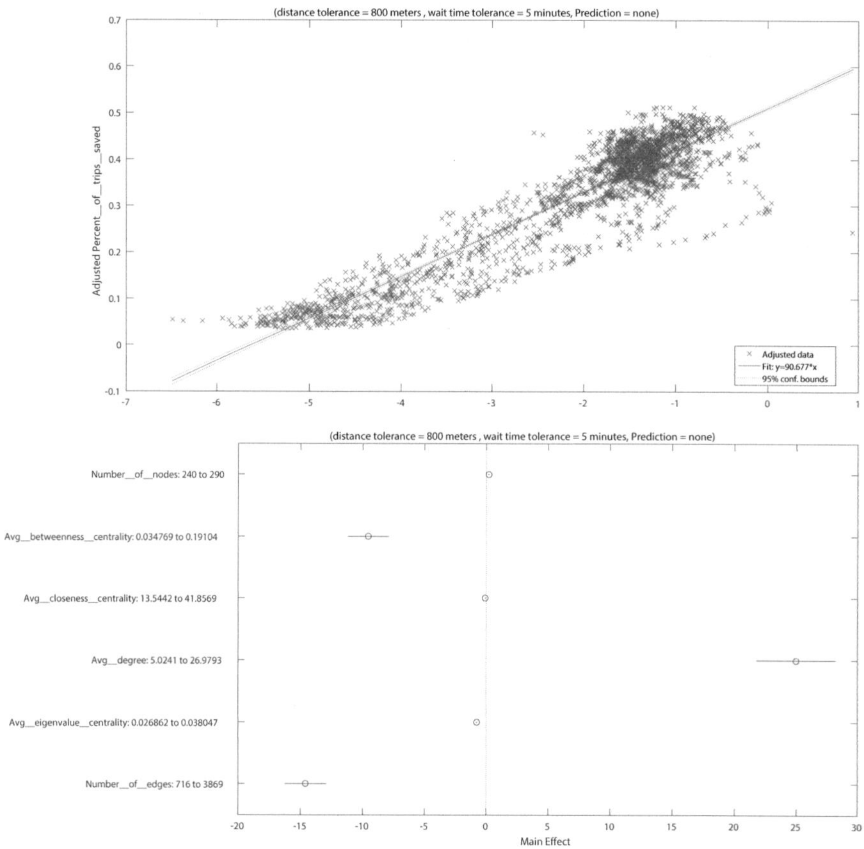

Estimated Coefficients:

	Estimate	SE	tStat	pValue
(Intercept)	0.5106	0.20836	2.4506	0.01433
Number__of__nodes	0.0040236	0.00063342	6.3521	2.5174e-10
Avg__betweenness__centrality	-61.095	5.3658	-11.386	2.5857e-29
Avg__closeness__centrality	-0.0020481	0.00093721	-2.1853	0.028962
Avg__degree	1.138	0.074928	15.188	6.8953e-50
Avg__eigenvalue__centrality	-66.996	2.8796	-23.266	1.7562e-108
Number__of__edges	-0.0046299	0.00027199	-17.022	1.5195e-61

Number of observations: 2490, Error degrees of freedom: 2483
Root Mean Squared Error: 0.0518
R-squared: 0.848, Adjusted R-Squared 0.848
F-statistic vs. constant model: 2.31e+03, p-value = 0

ANOVA_ANALYSIS =

	SumSq	DF	MeanSq	F	pValue
Total	43.806	2489	0.0176		
Model	37.15	6	6.1917	2309.7	0
Residual	6.6562	2483	0.0026807		

Figure 2.32: Multilinear regression model predicting the ride-sharing utilization using the dynamic network's properties.

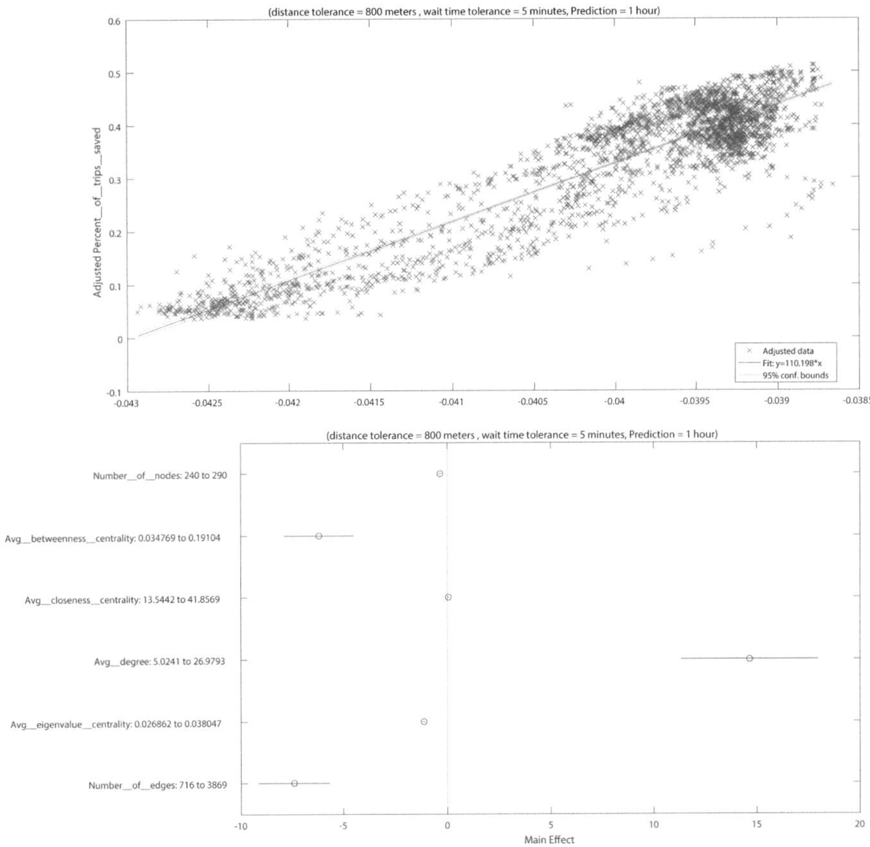

Estimated Coefficients:

	Estimate	SE	tStat	pValue
(Intercept)	4.7363	0.21495	22.035	1.9946e-98
Number__of__nodes	-0.0072379	0.00065345	-11.076	7.2563e-28
Avg__betweenness__centrality	-39.555	5.5355	-7.1457	1.1717e-12
Avg__closeness__centrality	0.0017503	0.00096685	1.8103	0.070375
Avg__degree	0.66746	0.077297	8.635	1.032e-17
Avg__eigenvalue__centrality	-102.85	2.9706	-34.623	1.1327e-214
Number__of__edges	-0.002347	0.00028059	-8.3642	9.9509e-17

Number of observations: 2490, Error degrees of freedom: 2483
Root Mean Squared Error: 0.0534
R-squared: 0.836, Adjusted R-Squared 0.835
F-statistic vs. constant model: 2.1e+03, p-value = 0

ANOVA_ANALYSIS =

	SumSq	DF	MeanSq	F	pValue
Total	43.093	2489	0.017313		
Model	36.009	6	6.0015	2103.6	0
Residual	7.0838	2483	0.0028529		

Figure 2.33: Multilinear regression model predicting the ride-sharing utilization using the dynamic network's properties.

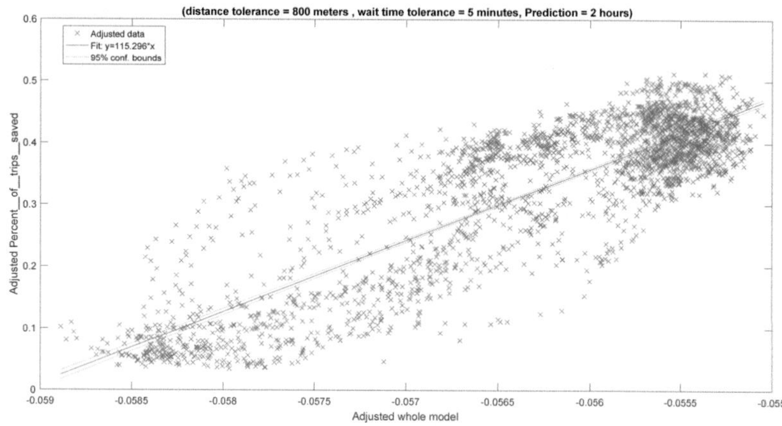

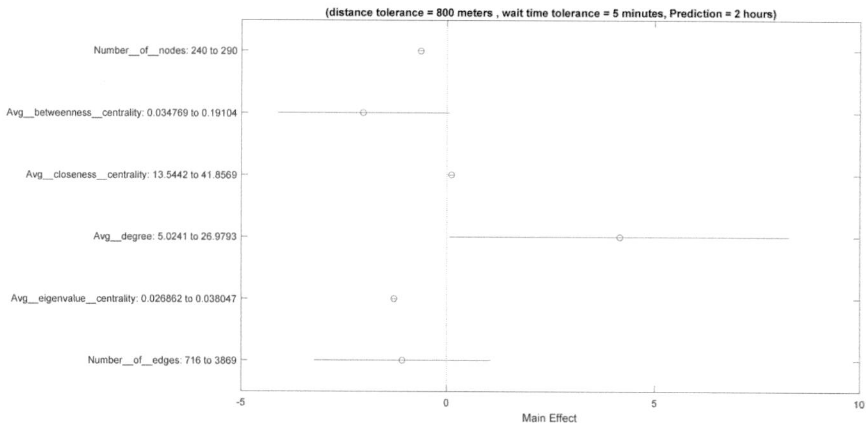

Estimated Coefficients:

	Estimate	SE	tStat	pValue
(Intercept)	6.814	0.26534	25.68	3.5345e-129
Number__of__nodes	-0.012945	0.00080667	-16.048	3.214e-55
Avg__betweenness__centrality	-13.006	6.8334	-1.9032	0.057124
Avg__closeness__centrality	0.0039826	0.0011935	3.3368	0.0008599
Avg__degree	0.18988	0.095421	1.99	0.046703
Avg__eigenvalue__centrality	-114.56	3.6671	-31.24	5.6128e-181
Number__of__edges	-0.00034396	0.00034638	-0.993	0.32081

Number of observations: 2490, Error degrees of freedom: 2483
Root Mean Squared Error: 0.0659
R-squared: 0.745, Adjusted R-Squared 0.744
F-statistic vs. constant model: 1.21e+03, p-value = 0

ANOVA_ANALYSIS =

	SumSq	DF	MeanSq	F	pValue
Total	42.302	2489	0.016995		
Model	31.507	6	5.2511	1207.8	0
Residual	10.795	2483	0.0043476		

Figure 2.34: Multilinear regression model predicting the ride-sharing utilization using the dynamic network's properties.

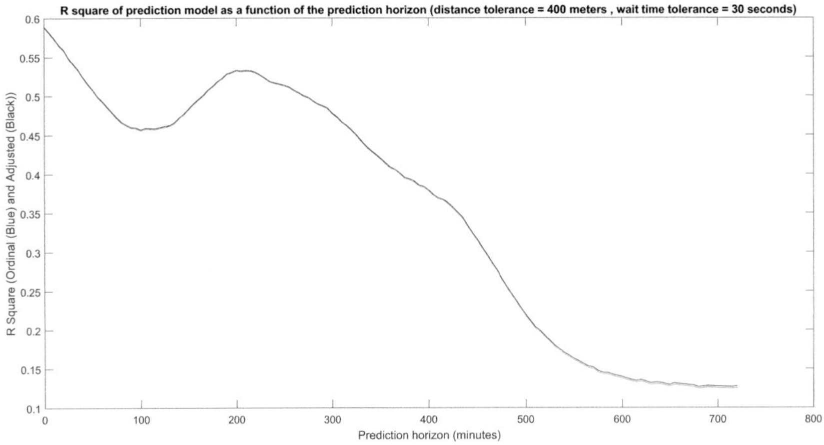

Figure 2.35: The accuracy of the multilinear regression model, as measured by its R^2, as a function of the prediction horizon 1.

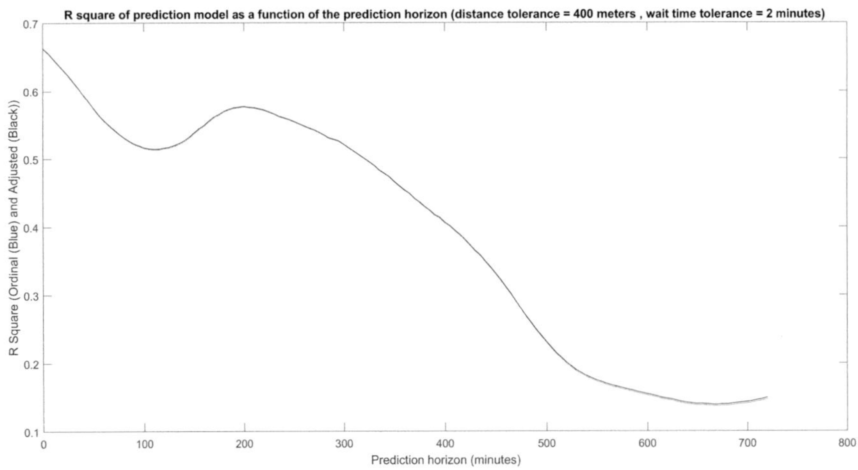

Figure 2.36: The accuracy of the multilinear regression model, as measured by its R^2, as a function of the prediction horizon 2.

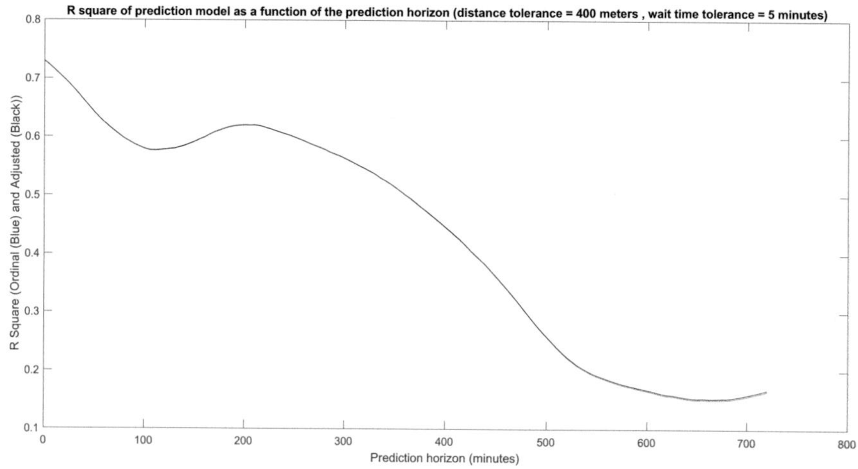

Figure 2.37: The accuracy of the multilinear regression model, as measured by its R^2, as a function of the prediction horizon 3.

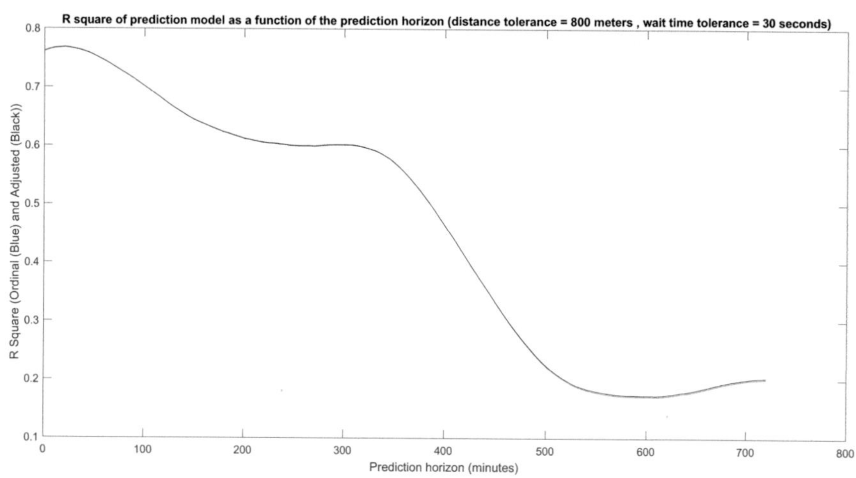

Figure 2.38: The accuracy of the multilinear regression model, as measured by its R^2, as a function of the prediction horizon 4.

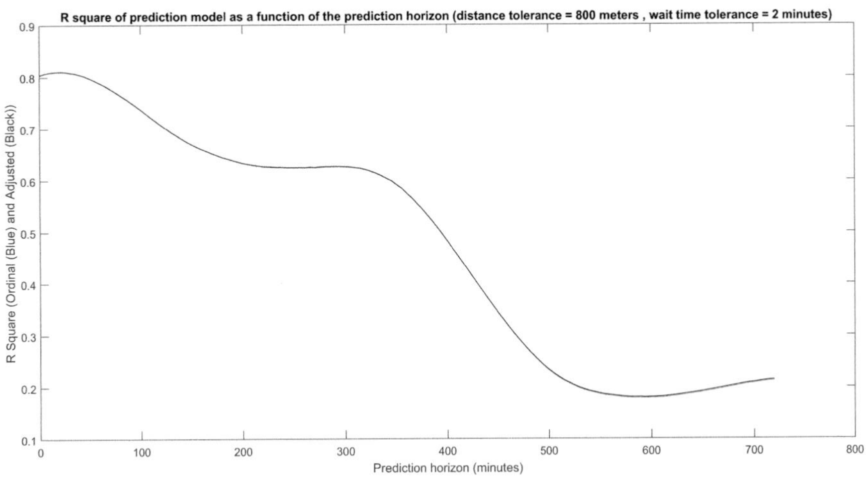

Figure 2.39: The accuracy of the multilinear regression model, as measured by its R^2, as a function of the prediction horizon 5.

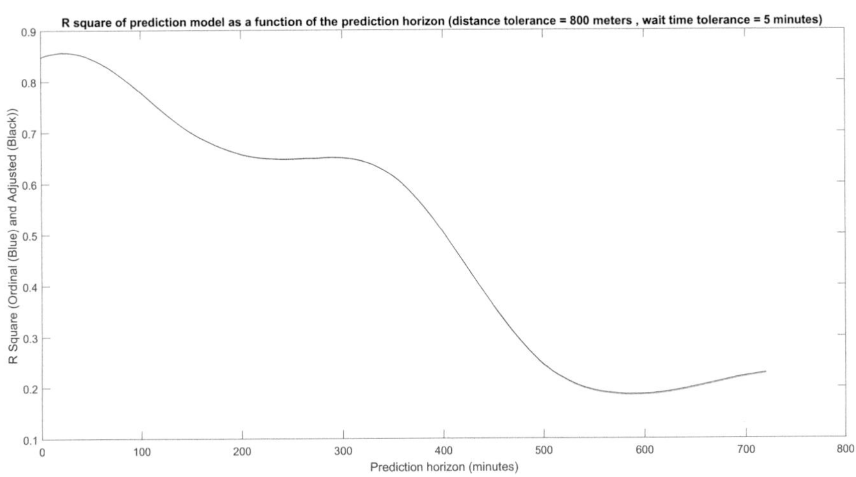

Figure 2.40: The accuracy of the multilinear regression model, as measured by its R^2, as a function of the prediction horizon 6.

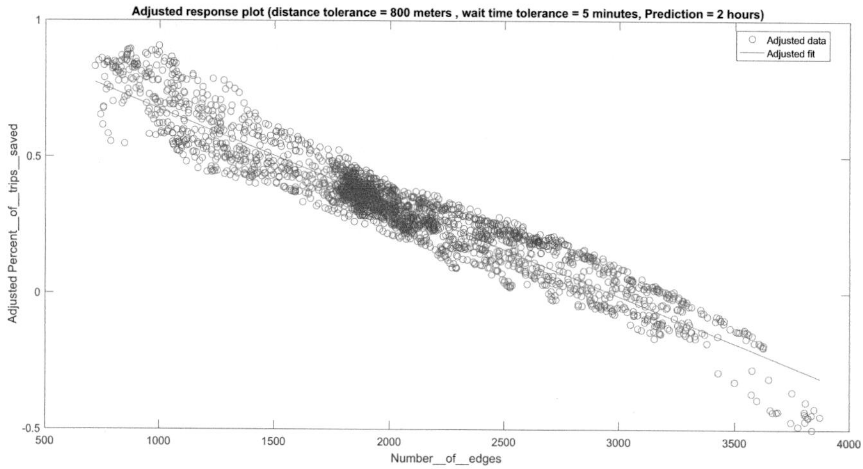

Figure 2.41: The adjusted response plot for the number of edges of the dynamic network, created for distance tolerance = 800 meters, wait time tolerance = 5 minutes, prediction horizon = 2 hours.

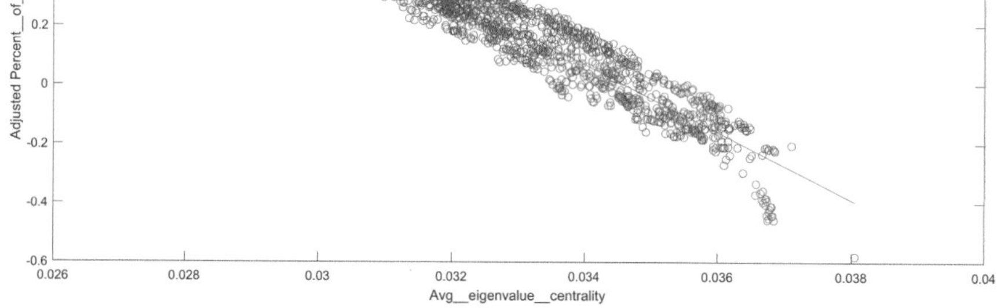

Figure 2.42: The adjusted response plot for the average eigenvalue centrality of the dynamic network, created for distance tolerance = 800 meters, wait time tolerance = 5 minutes, prediction horizon = 2 hours.

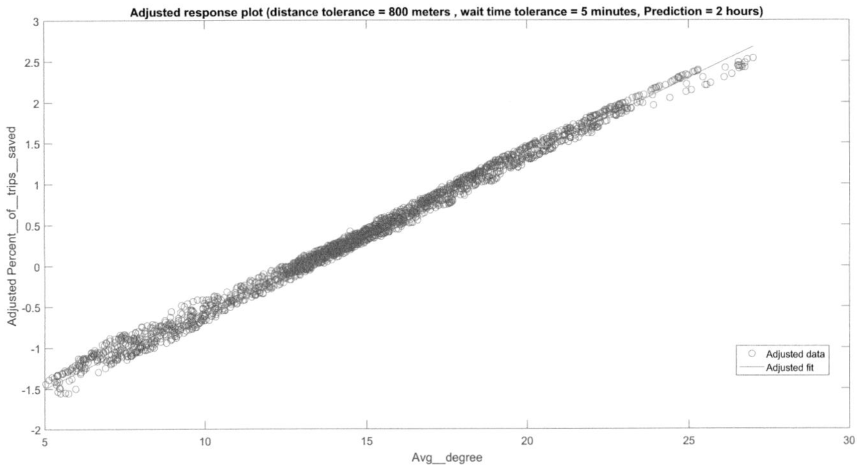

Figure 2.43: The adjusted response plot for the average degree of the dynamic network, created for distance tolerance = 800 meters, wait time tolerance = 5 minutes, prediction horizon = 2 hours.

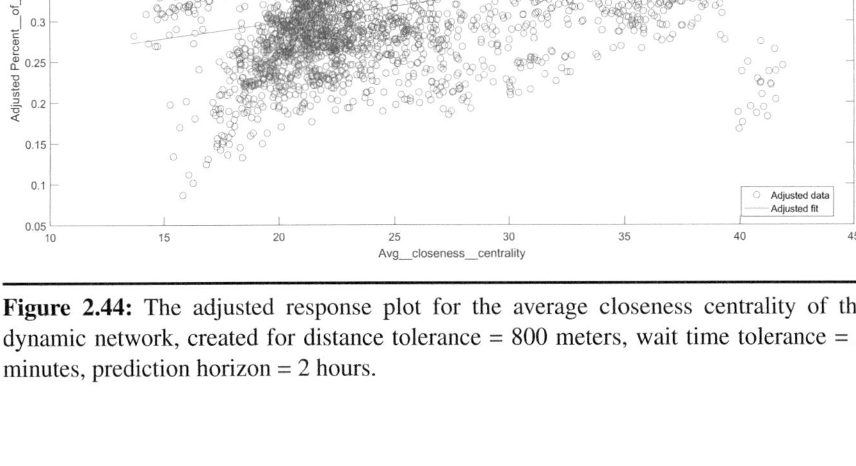

Figure 2.44: The adjusted response plot for the average closeness centrality of the dynamic network, created for distance tolerance = 800 meters, wait time tolerance = 5 minutes, prediction horizon = 2 hours.

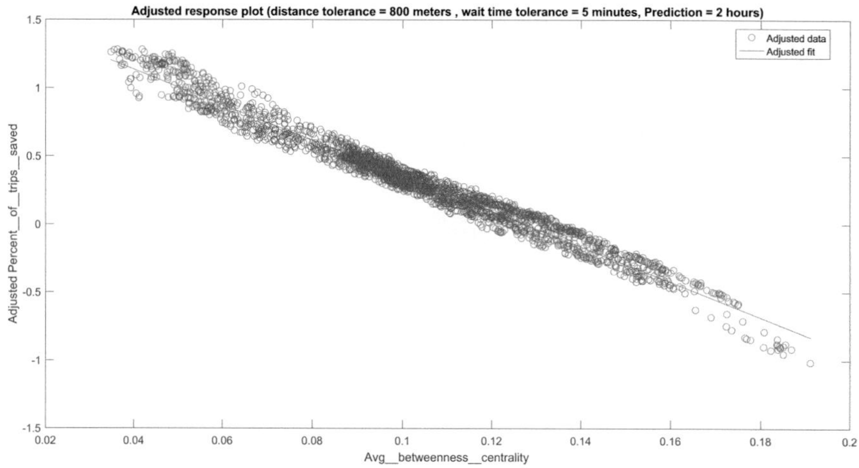

Figure 2.45: The adjusted response plot for the average betweenness centrality of the dynamic network, created for distance tolerance = 800 meters, wait time tolerance = 5 minutes, prediction horizon = 2 hours.

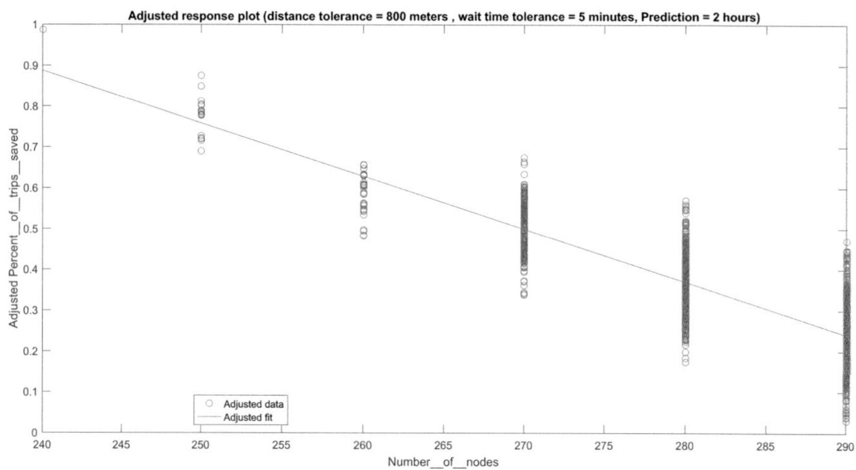

Figure 2.46: The adjusted response plot for the number of nodes of the dynamic network, created for distance tolerance = 800 meters, wait time tolerance = 5 minutes, prediction horizon = 2 hours.

Chapter 3

The Dynamics of Autonomous Drone Swarms in Surveillance: Exploring Efficiency, Feasibility, and Complexity

Yaniv Altshuler*[0]

Massachusetts Institute of Technology, Media Lab

3.1 Introduction

One of the most interesting challenges for a swarm of robotic drones is the design and analysis of a multi-robotics system for searching areas (whose dimensions and shape are either known or unknown) [22, 27, 32, 52, 57, 63, 66, 68] or see [4] for a survey of search and evasion strategies. Interesting works to mention in this scope are those of [53, 69], where

[0]xThis chapter is based on work previously discussed in parts in [10, 15, 20, 23–25] and published in [26].

a swarm of ant-like robots is used for repeatedly covering an unknown area, using a real time search method called *node counting*. By using this method, the robots are shown to be able to efficiently perform such a coverage mission, and analytic bounds for the coverage time are discussed.

While in most works the targets of the search mission were assumed to be idle, recent works considered dynamic targets, meaning — targets which after being detected by the searching robots, respond by performing various evasive maneuvers intended to prevent their interception. This problem, dating back to World War II operations research (see e.g. [62, 71]), requires the robotic agents to cope with a search area that expands while being scanned. The first documented example for search in dynamic domains discussed a planar search problem, considering the scanning of a corridor between parallel borders. This problem was solved in [56] in order to determine optimal strategies for aircraft searching for moving ships in a channel.

In this chapter we discuss the problem concerning a given (yet potentially unknown) region that needs to be searched (or *"cleaned"*) by a swarm of autonomous robotic agents, namely — to have all of its 'tiles', or 'pixels' visited at least once by at least a single member of the swarm, and that this "cleaning" would be guaranteed to be completed within a finite (and as short as possible) time. We further assume that the region that needs to be searched *"expands"* over time, namely—that parts of it that were already "cleaned" (or searched) can become "contaminated" (or "un-searched") again. Under this assumption, the completion of the cleaning mission is achieved only when all of the region's pixels are in "clean" status (e.g. were visited by the drones, and did not become contaminated again yet). This dynamic variant of the problem assumes a deterministic evolution of the environment, simulating a spreading *contamination*, or *fire* [6].

We assume that the swarm is collaborative, namely—that its various members are equipped with the same (and properly synchronized) software, and that each robot can acquire only the information which is available in its immediate vicinity, with the only way of inter-robot communication is leaving traces on the common ground and sensing the traces left by other robots. Similar works can be found in [3, 43, 59, 61, 64]. This problem is also strongly related to the problem of distributed search after mobile and evading target(s) [19, 34, 55] or the problems discussed under the names of "Cops and Robbers" or "Lions and Men" pursuits [42, 44, 47, 48].

Theoretical lower bounds for the problem are presented, as well as an impossibility result. A collaborative search protocol designed for a swarm of drones is presented, as well as a wealth of experimental results testing its performance and their comparison to the lower bounds.

In the spirit of [21, 35] we consider simple robots with only a bounded amount of memory (i.e. *finite-state-machines*). First, we discuss the collaborative coverage of static grids. We demonstrate that the best completion time known until recently ($O(n^2)$, achievable for example using the *LRTA** search algorithm [58]) can be improved to guarantee grid coverage in $O(\frac{1}{k}n^{1.5}+n)$.

We proceed to discuss the problem of covering an expanding domain, namely — a region in which "covered" tiles that are adjacent to "uncovered" tiles become "uncovered" every once in a while. Note that the grid is infinite, namely although the initial size of the region is n, it can become much greater over time. We show that using any conceivable algorithm, and using as sophisticated and potent robotic agents as possible, the minimal number of robots below which covering such a region is impossible equals $\Omega(\sqrt{n})$. We then show that when the region expands sufficiently slow, specifically — every $O(\frac{c_0}{\gamma_1})$ time steps (where c_0 is the circumference of the region and where γ_1 is a geometric property of the region, which ranges between $O(1)$ and $O(\ln n)$), a group of $\Theta(\sqrt{n})$ robots can successfully cover the region. Furthermore, we demonstrate that in this case a cover time of $O(n^2 \ln n)$ can be guaranteed.

These results are the first analytic results concerning the complexity of the number of robots required to cover an expanding grid, as well as for the time such a coverage requires.

3.2 Related Work

In general, most of the techniques used for the task of a distributed coverage use some sort of cellular decomposition. For example, in [66] the area to be covered is divided between the agents based on their relative locations. In [37] a different decomposition method is being used, which is analytically shown to guarantee a complete coverage of the area. Another interesting work is presented in [2], discussing two methods for cooperative coverage (one probabilistic and the other based on an exact cellular decomposition). All of the works mentioned above, however, rely on the assumption that the cellular decomposition of the area is possible. This in turn, requires the use of memory resources, used for storing the dynamic map generated, the boundaries of the cells, and more'. As the initial size and geometric features of the area are generally not assumed to be known in advance, agents equipped with merely a constant amount of memory will not be able to use such algorithms.

Surprisingly, while some existing works concerning distributed (and decentralized) coverage present analytic proofs for the ability of the system to guarantee the completion of the task (for example, in [2, 31, 37]), most of them lack analytic bounds for the coverage time (although in many cases an extensive amount of empirical results of this nature are made available by extensive simulations).

An interesting work discussing a decentralized coverage of terrains is presented in [75]. This work examines domains with non-uniform traversability.

Completion times are given for the proposed algorithm, which is a generalization of the forest search algorithm [76]. In this work, though, the region to be searched is assumed to be known in advance — a crucial assumption for the search algorithm, which relies on a cell-decomposition procedure.

A search for analytic results concerning the completion time of ant-robots covering an area in the grid revealed only a handful of works. The main result in this regard is that of [53, 69], where a swarm of ant-like robots is used for repeatedly covering an unknown area, using a real time search method called *node counting*. By using this method, the robots are shown to be able to efficiently perform such a coverage mission (using integer markers that are placed on the graph's nodes), and analytic bounds for the coverage time are discussed. Based on a more general result for strongly connected undirected graphs shown in [54, 70], the cover time of teams of ant robots (of a given size) that use node counting is shown to be $t_k(n) = O(n^{\sqrt{n}})$, when $t_k(n)$ is the cover time of a region of size n using k robots. It should be mentioned though, that in [69] the authors clearly state that it is their belief that the coverage time for robots using node counting in grids is much smaller. This evaluation is also demonstrated experimentally. However, no analytic evidence for this was available thus far.

Another algorithm to be mentioned in this scope is the *LRTA** search algorithm. This algorithm was first introduced in [58] and its multi-robotics variant is shown in [54] to guarantee cover time of undirected connected graphs in polynomial time. Specifically, on grids this algorithm is shown to guarantee coverage in $O(n^2)$ time (again, using integer markers).

Vertex-Ant-Walk, a variant of the node counting algorithm is presented in [74], is shown to achieve a coverage time of $O(n\delta_G)$, where δ_G is the graph's diameter. Specifically, the cover time of regions in the grid is expected to be $O(n^2)$ (however for various "round" regions, a cover time of approximately $O(n^{1.5})$ can be achieved). This work is based on a previous work in which a cover time of $O(n^2\delta_G)$ was demonstrated [72]. Additional analysis of the effects of various topological and geometric features of the search space on the ability of a collaborative swarm to efficiently guarantee successful completion can be found in [7–9, 29]. Taking a network perspective for the optimization of monitoring, patrolling or decentralized deployment is discussed in [5,28,41,65,67].

Another work called *Exploration as Graph Construction*, provides a coverage of degree bounded graphs in $O(n^2)$ time, can be found in [40]. Here a group of limited ant robots explore an unknown graph using special "markers".

Interestingly, a similar performance can be obtained by using the simplest algorithm for multi robots navigation, namely — random walk. Although in general undirected graphs a group of k random walking robots may require up to $O(n^3)$ time, in degree bounded undirected graphs such robots would achieve a much faster covering, and more precisely, $O\left(\frac{|E|^2 log^3 n}{k^2}\right)$ [36]. For regular graphs or degree bounded planar graphs a coverage time of $O(n^2)$ can be achieved [60], although in such cases there is also a lower bound for the coverage time, which equals $\Omega(n(\log n)^2)$ [51].

We next show that the problem of collaborative coverage in static grid domains can be completed in $O(\frac{1}{k}n^{1.5}+n)$ time and that collaborative coverage of dynamic grid domains can be achieved in $O(n^2 \ln n)$.

3.3 The Dynamic "Cooperative Cleaners" Problem

The definition of the dynamic cooperative cleaners problem is very similar to that of the static variant, albeit with several required extensions. For the sake of readability, following is the complete self-contained definition of the problem.

We shall assume that the time is discrete.

Definition 3.1 Let the undirected graph $G(V,E)$ denote a two dimensional integer grid $\mathbf{Z}^2$, whose vertices (or "*tiles*") have a binary property called '*contamination*'. Let $cont_t(v)$ denote the contamination state of the tile v at time t, taking either the value "*on*" (for "dirty" or "contaminated") or "*off*" (for "clean").

For two vertices $v,u \in V$, the edge (v,u) may belong to E at time t only if both of the following hold:

- v and u are $4Neighbors$ in G.
- $cont_t(v) = cont_t(u) = on$.

This however is a necessary but not a sufficient condition.

The edges of E represent the connectivity of the contaminated region. At $t=0$ all the contaminated tiles are connected, namely:

$$(v,u) \in E_0 \iff (v,u \text{ are } 4Neighbors \text{ in } G) \wedge (cont_0(v) = cont_0(u) = on)$$

Edges may be added to E only as a result of a contamination spread and can be removed only while contaminated tiles are cleaned by the agents (see below).

Definition 3.2 Let $F_t(V_{F_t}, E_t)$ be the contaminated sub-graph of G at time t, i.e.:

$$V_{F_t} = \{v \in G \mid cont_t(v) = on\}$$

We assume that F_0 is a single simply-connected component (the actions of the agents will be so designed that this property will be preserved).

Let a group of k agents that can move on the grid G (moving from a tile to its neighbor in one time step) be placed at time t_0 on F_0, at point $p_0 \in V_{F_t}$.

Each agent is equipped with a sensor capable of telling the *contamination* status of all tiles in the digital sphere of diameter 7, which surrounds the agent. An agent is also aware of other agents which are located in these tiles, and all the agents agree on a common direction. Each tile may contain any number of agents simultaneously. This information will later be required by the agents' cleaning protocol. Each agent is equipped with a memory of size $O(\log k)$ bits (used later as an agents counter).

When an agent moves to a tile v, it has the possibility of cleaning this tile (i.e. causing $cont(v)$ to become *off*. Once an agent cleaned a tile v, all the edges touching v are removed, namely:

$$E_{t+1} = E_t \setminus \{(v,u) \mid (v,u) \in E_t \ \wedge \ cont(v) = off\}$$

The agents do not have any prior knowledge of the shape or size of the sub-graph F_0 except that it is a single and simply connected component.

The contaminated region F_t is assumed to be surrounded at its boundary by a "rubber-like" elastic barrier, dynamically reshaping itself to fit the evolution of the contaminated region over time (the barrier is derived from the edges of E_t, as demonstrated in Figure 3.1). This process can be thought of as a water filled rubber balloon — while the shape of the water contained in it may change, as well as its volume, the balloon keeps bounding it tightly. This barrier is intended to guarantee the preservation of the simple connectivity of F_t, in order to allow the agents to perform their cleaning activities. The need for such a barrier is discussed in Section 3.12.

When an agent cleans a contaminated tile, the barrier is withdrawn, in order to free the void previously occupied by the cleaned tile. This is demonstrated in Figure 3.2.

Definition 3.3 Let d denote the number of time steps between contamination spreads. That is, if $t = nd$ for some positive integer n, then:

$$\forall v \in F_t \ \forall u \in 4Neighbors(v), \ cont_{t+1}(u) = on$$

While the contamination spreads, the elastic barrier stretches accordingly. First, all the clean tiles located to the right of a contaminated tile get contaminated. Then, the clean tiles located below the contaminated tiles, followed by the clean tiles located to the left of such tiles get contaminated. Finally, clean tiles located above a contaminated tile are affected. The barrier itself is implicitly derived from the edges connecting the tiles. Namely, it is the boundary which surrounds the connected contaminated tiles. This process is illustrated in Figures 3.3 and 3.4, and can be formalized as follows:

Definition 3.4 Elastic barrier's expansion as a result of a contamination spread:

1. Let $\widehat{V}_t = \emptyset$ be a set of vertices and let $\widehat{E}_t = \emptyset$ be a set of edges
2. For vertex u located at (x_u, y_u) let:

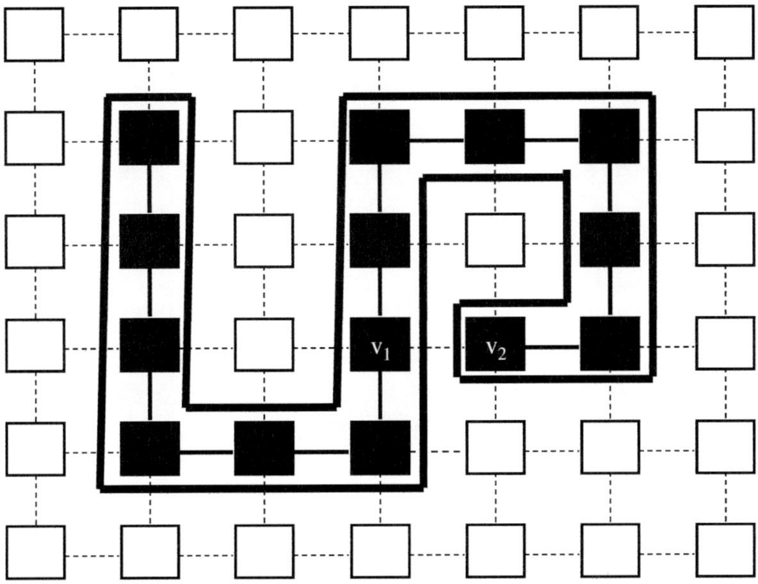

Figure 3.1: An illustration of the barrier surrounding the contaminated region, derived from the edges of E_t. Notice that when an edge connecting two contaminated tiles are not in E_t, those tiles are "on the other side" of the barrier. For example, observe the vertices v_1 and v_2, which are *4Neighours* in F_t (namely, the Manhattan distance between them in G equals one), however — as $\neg((v_1, v_2) \in E_t)$ the geodesic distance between them is significantly greater.

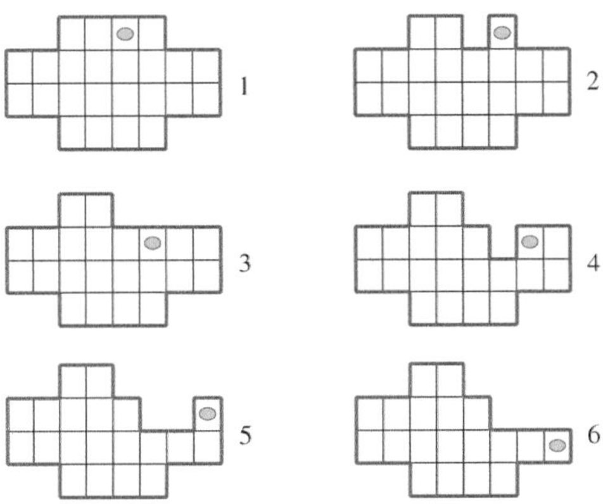

Figure 3.2: An illustration of the evolution of the elastic barrier as a result of the movement and cleaning of an agent, for example according to the **SWEEP** protocol (as described in Figures 3.18 and 3.19.

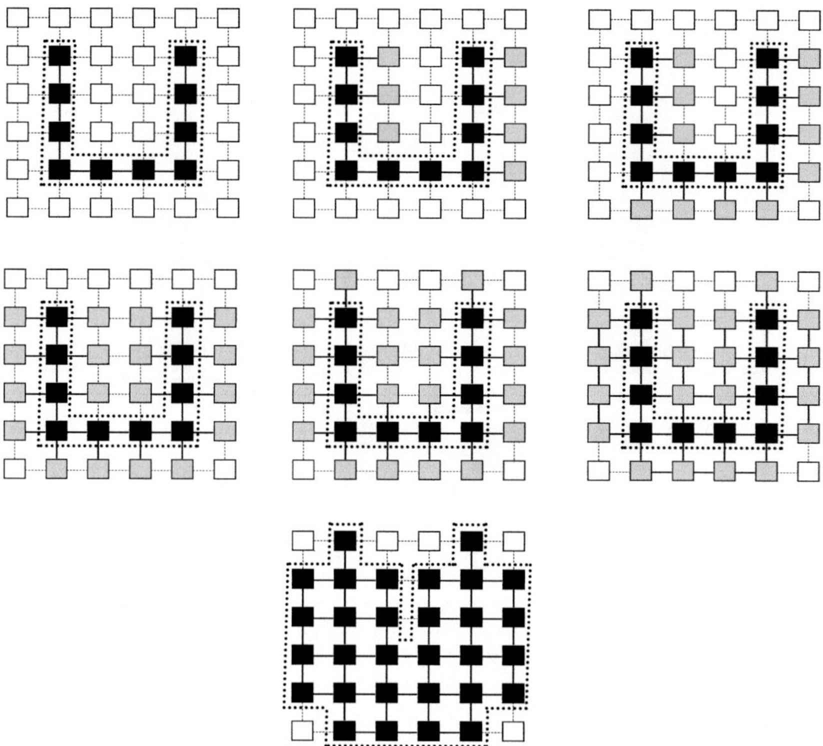

Figure 3.3: An illustration of the barrier expansion process as a result of a contamination spread. The example starts from the top left chart, going right.

- $Neighbor_{right}(u)$ denote the vertex located at (x_u+1, y_u)
- $Neighbor_{left}(u)$ denote the vertex located at (x_u-1, y_u)
- $Neighbor_{up}(u)$ denote the vertex located at (x_u, y_u+1)
- $Neighbor_{down}(u)$ denote the vertex located at (x_u+1, y_u-1)

3. Let $d_G(v,u)$ denote the number of edges between v and u in the graph G
4. Add new vertices to the right—
   (a) $V_{temp} = \left\{v \mid \neg(v \in V_{F_t} \cup \widehat{V}_t) \wedge (\exists u \in V_{F_t},\ v = Neighbor_{right}(u))\right\}$
   (b) $\widehat{E}_t \leftarrow \widehat{E}_t \cup \{(v,u) \mid (v \in V_{temp}) \wedge (v = Neighbor_{right}(u))\}$
   (c) $\widehat{E}_t \leftarrow \widehat{E}_t \cup \left\{(v,u) \;\middle|\; \begin{array}{c}(v \in \widehat{V}_t) \wedge (v = Neighbor_{right}(u)) \wedge \\ (u \in V_{F_t}) \wedge (d_{G(V_{F_t} \cup \widehat{V}_t, E_t \cup \widehat{E}t)}(v,u) \le 3)\end{array}\right\}$
   (d) $\widehat{V}_t \leftarrow \widehat{V}_t \cup V_{temp}$
5. Add new vertices to the bottom—
   (a) $V_{temp} = \left\{v \mid \neg(v \in V_{F_t} \cup \widehat{V}_t) \wedge (\exists u \in V_{F_t},\ v = Neighbor_{down}(u))\right\}$
   (b) $\widehat{E}_t \leftarrow \widehat{E}_t \cup \{(v,u) \mid (v \in V_{temp}) \wedge (v = Neighbor_{down}(u))\}$

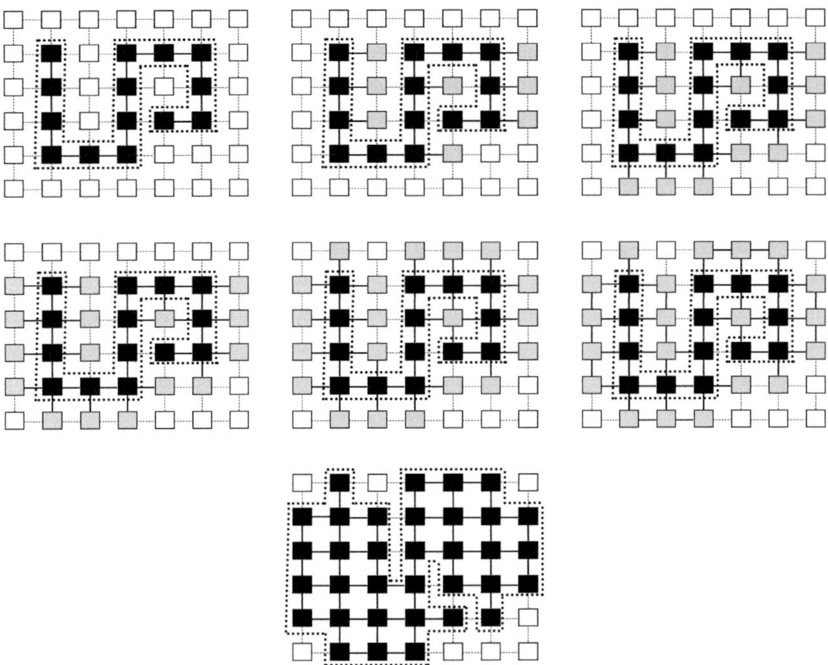

Figure 3.4: An illustration of the barrier expansion process as a result of a contamination spread. The example starts from the top left chart, going right.

(c) $\widehat{E_t} \leftarrow \widehat{E_t} \cup \left\{ (v,u) \;\middle|\; \begin{array}{c} (v \in \widehat{V_t}) \wedge (v = Neighbor_{down}(u)) \wedge \\ (u \in V_{F_t}) \wedge (d_{G(V_{F_t} \cup \widehat{V_t}, E_t \cup \widehat{Et})}(v,u) \leq 3) \end{array} \right\}$

(d) $\widehat{V_t} \leftarrow \widehat{V_t} \cup V_{temp}$

6. Add new vertices to the left—
   (a) $V_{temp} = \left\{ v \mid \neg(v \in V_{F_t} \cup \widehat{V_t}) \wedge (\exists u \in V_{F_t},\ v = Neighbor_{left}(u)) \right\}$
   (b) $\widehat{E_t} \leftarrow \widehat{E_t} \cup \{(v,u) \mid (v \in V_{temp}) \wedge (v = Neighbor_{leftt}(u))\}$
   (c) $\widehat{E_t} \leftarrow \widehat{E_t} \cup \left\{ (v,u) \;\middle|\; \begin{array}{c} (v \in \widehat{V_t}) \wedge (v = Neighbor_{left}(u)) \wedge \\ (u \in V_{F_t}) \wedge (d_{G(V_{F_t} \cup \widehat{V_t}, E_t \cup \widehat{Et})}(v,u) \leq 3) \end{array} \right\}$
   (d) $\widehat{V_t} \leftarrow \widehat{V_t} \cup V_{temp}$
7. Add new vertices to the top—
   (a) $V_{temp} = \left\{ v \mid \neg(v \in V_{F_t} \cup \widehat{V_t}) \wedge (\exists u \in V_{F_t},\ v = Neighbor_{up}(u)) \right\}$
   (b) $\widehat{E_t} \leftarrow \widehat{E_t} \cup \{(v,u) \mid (v \in V_{temp}) \wedge (v = Neighbor_{up}(u))\}$
   (c) $\widehat{E_t} \leftarrow \widehat{E_t} \cup \left\{ (v,u) \;\middle|\; \begin{array}{c} (v \in \widehat{V_t}) \wedge (v = Neighbor_{up}(u)) \wedge \\ (u \in V_{F_t}) \wedge (d_{G(V_{F_t} \cup \widehat{V_t}, E_t \cup \widehat{Et})}(v,u) \leq 3) \end{array} \right\}$
   (d) $\widehat{V_t} \leftarrow \widehat{V_t} \cup V_{temp}$

8. $V_{F_t} \leftarrow V_{F_t} \cup \widehat{V}_t$
9. $E_t \leftarrow E_t \cup \widehat{E}_t$
10. $E_t \leftarrow E_t \cup \{(v,u) \mid (v,u \in V_{F_t}) \wedge (v = 4Neighbors(u)) \wedge (d_{F_t}(v,u) \leq 3)\}$

For the agents who travel along the tiles of F, the barrier signals the boundary of the contaminated region. When an agent detects a contaminated tile which is "on the other side" of the barrier it treats this tile as though it was a clean tile (this aspect is important to remember when reviewing the agents' behavior while following the cleaning protocol). For example, examining the region which appears in Figure 3.5 and assuming an agent located in the "X" marked tile, then while looking "upwards", this agent acts as though the tile above it is clean (as the contaminated tiles are located behind the barrier and are thus masked as clean ones).

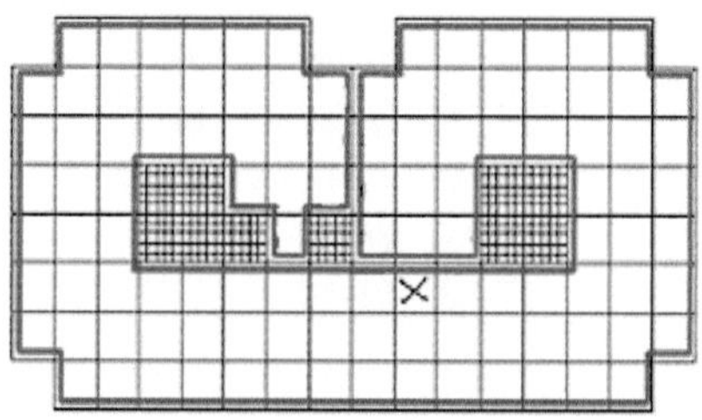

Figure 3.5: Observe how the stretching of the barrier is generating "double-fronts" after a contamination spread. Note that had the contaminated parts been allowed to merge, the simple connectivity of the region would have not been kept. In addition, observe how an agent located in the tile marked by an "X" is still kept on the boundary of the contaminated region, due to this mechanism.

As in the case of the static variant, it is the agents' goal to clean G by eliminating the contamination entirely, meaning that the agents must ensure that:

$$\exists t_{success} \; s.t \; F_{t_{success}} = \emptyset$$

In addition, it is desired that time $t_{success}$ will be minimal.

Note that as previously stated, we still impose the strict decentralization restriction of relying on no central control.

3.4 A Protocol-Agnostic Lower Bound for the Cleaning Time

Due to the dynamic nature of the problem, the shape of the contaminated region can change dramatically during the cleaning process. Therefore, since we know no easy way to decide whether k agents can successfully clean an instance of the *Dynamic Cooperative Cleaners* problem, producing bounds for the efficiency of any proposed cleaning protocol is important for estimating its efficiency.

Let S_t denote the size of the contaminated region F at time t, namely the number of tiles (i.e. grid points) in F.

Let us now discuss a general lower bound for the problem of dynamic cleaning. The bound is general in the sense that it bounds the problem, not a specific cleaning protocol for it. Therefore, it is applicable for any cleaning protocol which might be used by the agents. This is achieved by showing that at a specific time t, $S_t > 0$, namely — the mission could not be completed until that time (regardless of the cleaning protocol used), as the completion of the cleaning mission at time t means that $S_t = 0$.

Theorem 3.1

Using any cleaning protocol, the area of the contaminated region at time t can be recursively lower bounded, as follows:

$$S_{t+d} \geq S_t - d \cdot k + \left\lfloor 2\sqrt{2 \cdot (S_t - d \cdot k) - 1} \right\rfloor$$

Proof 3.1 Notice that a lower bound for the cleaning time can be obtained by assuming that the agents are working with maximal efficiency, meaning that at each time step every agent cleans exactly one tile. After $(d-1)$ time steps k agents will therefore clean $k \cdot (d-1)$ tiles, and thus we know that:

$$S_{d-1} \geq S_0 - (d-1) \cdot k$$

In the d-th time step, the agents clean another portion of k tiles, but the remaining contaminated tiles spread their contamination to their $4Neighbors$ and cause new tiles to become contaminated. We are interested in the *minimal* number of tiles which can become contaminated at this stage. The minimal number of $4Neighbors$ of any number of tiles is achieved when the tiles are organized in the shape of a "digital sphere" (see a detailed discussion on this topic in [11, 14] and a complete proof in Section 7 of [24]), as demonstrated in Figure 3.6. Therefore, for a region of any given area S the minimal number of its $4Neighbors$ is at least $2\sqrt{2 \cdot S_t - 1}$. As the number of tiles must be an integer value, we use $\lfloor 2\sqrt{2 \cdot S_t - 1} \rfloor$ to remain on the safe side.

After the d-th time step we thus have the following:

$$S_d \geq S_0 - d \cdot k + \left\lfloor 2\sqrt{2 \cdot (S_0 - d \cdot k) - 1} \right\rfloor$$

and the rest is implied.

Figure 3.6: The left chart shows an example of a 'sphere' in the grid. Notice that this sphere has only 16 tiles in its *4Neighbours* while the right chart, containing a shape of the same area, has 20 tiles in its *4Neighbours*.

An illustration of Theorem 3.1 is presented in Figure 3.7.

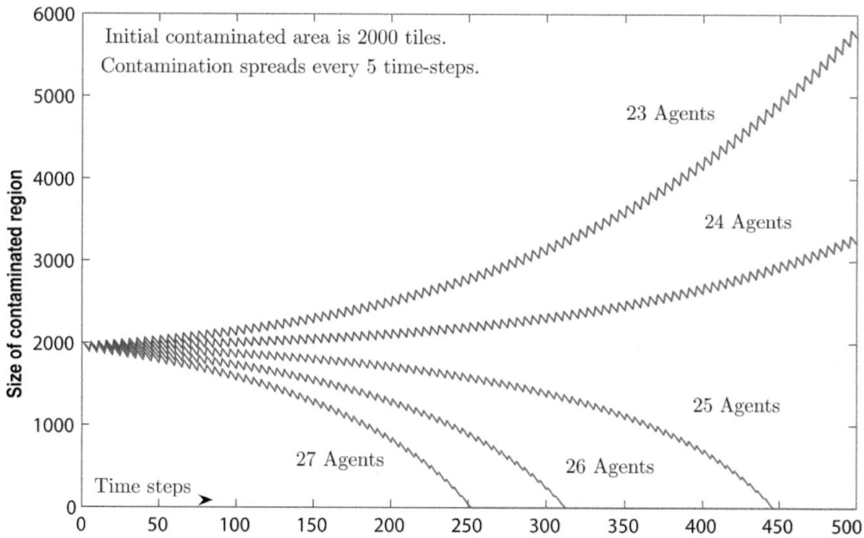

Figure 3.7: Theorem 3.1 illustrated. The graph depicts a lower bound over the size of the contaminated region as a function of t, given a various number of cleaning agents (while the initial area of the contaminated region is $S_0 = 2000$ and the contamination spreading latency is $d = 5$. The lower bound over the cleaning time can be interpreted as stating that the time for completion of the cleaning job will be higher than the point at which the graph of S_t hits 0. For example, observe how a complete cleaning is shown to be impossible for any swarm with 24 agents or less. Naturally, as Theorem 3.1 only takes into account the initial size of the contaminated region, different region's size evolutions are expected to be seen for regions of various shapes, but of the same initial size — see such examples in Section 3.7. Note that as this bound is generic, hence independent of the cleaning protocol used by the agents, the actual cleaning protocol used will have a significant effect on the actual cleaning efficiency.

The following result is a direct lower bound for the size of the contaminated region, based on the recursive bound of Theorem 3.1.

Theorem 3.2

Using any cleaning protocol, a lower bound for the area of the contaminated region at time $t = i \cdot d$ for some $i \in \mathbb{N}$ is the minimal positive S_t, which is a solution of the following equation:

$$2i = \sqrt{2(S_t - dk) - 1} - \sqrt{2(S_0 - dk) - 1} + \ln\left(\frac{\sqrt{2(S_t - dk) - 1} - \frac{dk}{2}}{\sqrt{2(S_0 - dk) - 1} - \frac{dk}{2}}\right)^{\frac{dk}{2}}$$

Proof 3.2 Observe that by denoting $y_i \triangleq S_{id}$ Theorem 3.1 can be written as:

$$y_{i+1} - y_i \geq \left\lfloor 2\sqrt{2 \cdot (y_i - d \cdot k) - 1} \right\rfloor - d \cdot k$$

Searching for the minimal area we can look at the equation:

$$y_{i+1} - y_i = \left\lfloor 2\sqrt{2 \cdot (y_i - d \cdot k) - 1} \right\rfloor - d \cdot k$$

By dividing both sides by $\Delta i = 1$ we obtain:

$$y_{i+1} - y_i \triangleq y' = \left\lfloor \sqrt{8y - 8\left(d \cdot k + \frac{1}{2}\right)} \right\rfloor - d \cdot k \tag{3.1}$$

Notice that the values of y' might be positive (stating an increase in the area), negative (stating a decrease in the area), or complex numbers (stating that the area is smaller than $d \cdot k$, and will therefore be cleaned before the next spread).

Let us denote $x^2 \triangleq 8y - 8\left(d \cdot k + \frac{1}{2}\right)$. After calculating the derivative of both sides of this expression we see that:

$$2x \cdot x' = 8y'$$

and after using the definition of y' of Equation 3.1 we see that:

$$2x \cdot \frac{dx}{di} = 2x \cdot x' = 8\left\lfloor \sqrt{8y - 8\left(d \cdot k + \frac{1}{2}\right)} \right\rfloor - 8d \cdot k \leq 8(x - d \cdot k) \tag{3.2}$$

From Equation 3.2 a definition of di can be extracted:

$$di \geq \frac{1}{8} \cdot \frac{2x}{x - d \cdot k} dx \geq \frac{1}{8} \cdot \frac{2x - 2d \cdot k + 2d \cdot k}{x - d \cdot k} dx \geq \frac{1}{4}\left(1 + \frac{d \cdot k}{x - d \cdot k}\right) dx$$

The value of x can be achieved by integrating the previous expression as follows (notice that we are interested in the equality of the two expressions):

$$\int_{i_0}^{i} di = \int_{x_0}^{x} \frac{1}{4}\left(1 + \frac{d \cdot k}{x - d \cdot k}\right) dx$$

After the integration we can see that:

$$i\Big|_{i_0}^{i} = \frac{1}{4}\left(x + d \cdot k \ln\left(x - d \cdot k\right)\right)\Big|_{x_0}^{x}$$

and after assigning $i_0 = 0$:

$$4i = x - x_0 + d \cdot k \ln \frac{x - d \cdot k}{x_0 - d \cdot k}$$

Returning back to y we get:

$$2i = \sqrt{2(y - d \cdot k) - 1} - \sqrt{2(y_0 - d \cdot k) - 1} + \ln\left(\frac{\sqrt{2(y - d \cdot k) - 1} - \frac{d \cdot k}{2}}{\sqrt{2(y_0 - d \cdot k) - 1} - \frac{d \cdot k}{2}}\right)^{\frac{d \cdot k}{2}}$$

and returning to the original size variable S_t we see that:

$$2i = \sqrt{2(S_{id} - dk) - 1} - \sqrt{2(S_0 - dk) - 1} + \ln\left(\frac{\sqrt{2(S_{id} - dk) - 1} - \frac{dk}{2}}{\sqrt{2(S_0 - dk) - 1} - \frac{dk}{2}}\right)^{\frac{dk}{2}}$$

and the rest of the Theorem is implied.

Theorem 3.2 provides an easy way of calculating the minimal possible time it takes a given number of agents to clean some portion of any given contaminated region (and specifically — to clean it entirely). A lower bound for the cleaning time of k agents for various contaminated regions based on this Theorem appears in Figures 3.8, 3.9 and 3.10. A lower bound for the minimal number of agents required to guarantee a successful cleaning of a contaminated region, as a function of the contamination spreading speed, is demonstrated in Figure 3.11, for several values of the initial contaminated area.

The influence of the initial contaminated area over the minimal number of agents required to guarantee a complete cleaning is demonstrated in Figure 3.12, in which the ratio $\frac{K_{min}(S_0,d)}{K_{min}(\frac{1}{10}S_0,d)}$ (as a function of d) is presented (suggesting that when $K_{min}(n,d)$ denotes the minimal number of agents required to enable a cleaning of a contaminated region of initial contaminated region of n tiles, for some contamination spreading d, then $K_{min}(\alpha \cdot n, d) \approx \sqrt{\alpha} \cdot K_{min}(n,d)$).

Another interesting illustration of Theorem 3.2 is presented in Figure 3.13, where the ratio $R_{S_0}(\alpha) \triangleq \frac{T_{min}(\alpha \cdot S_0,D)}{T_{min}(S_0,D)}$ (where $T_{min}(n,D)$ denotes the average lower bound for the cleaning time of $K_{min}(n,d)$ agents, where $d = [2,3,4,...,D]$) is calculated for various values of S_0 (and for $D = 20$). Based on these examples it can be conjectured that $R_{S_0}(\alpha) \approx 2\sqrt{\alpha}$ regardless of S_0 and D,

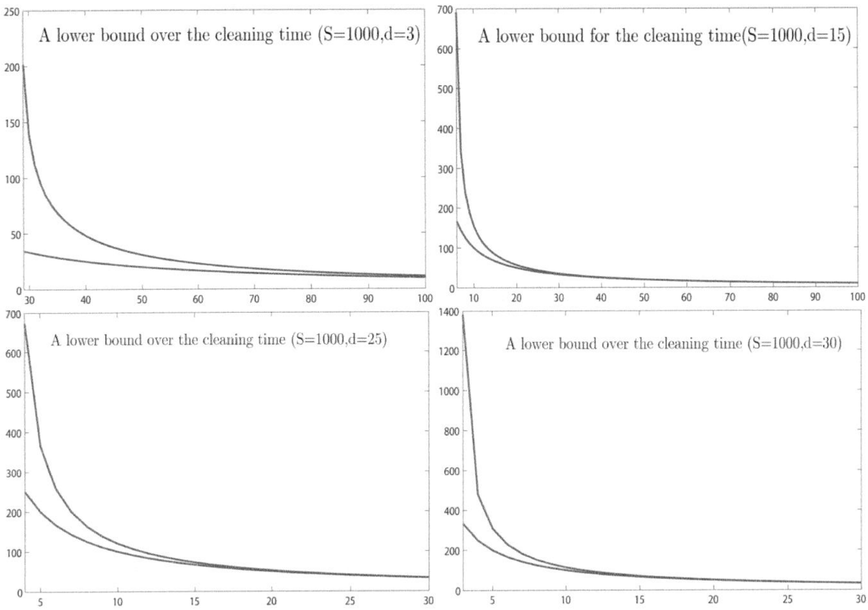

Figure 3.8: Theorem 3.2: These graphs depict lower bounds for the cleaning time of several contaminated regions (the Y axis), as a function of the number of agents (the X axis, starting from the lowest number of agents which produces a successful cleaning). A comparison to the naive $f(k) = \frac{S_0}{k}$ function, which holds for static environments, is also included (the lower curve, marked in red). Notice again that this bound is generic, and therefore will hold for any cleaning protocol used.

3.5 Impossibility Result

While designing a system intended to use cleaning agents in order to guarantee a successful completion of the cooperative cleaners problem, various cleaning methods can be used. This, combined with the dramatic changes the initial geometric properties of the contaminated region can undergo, yields a great uncertainty as to the cleaning time of a group of agents using any protocol (and to that matter, as to the possibility of guaranteeing such a successful completion, to begin with). While the theoretical lower bound shown in Section 3.4 as well as the upper bounds which are presented in the second part of this work can assist in decreasing this uncertainty, one might alternatively be concerned with the opposite question, namely — how can we guarantee that a group of k agents *will not* be able to clean a contaminated region entirely (regardless of the cleaning protocol which is being used, or the agents' sensors or communication capabilities). One solution to this problem is offered by Theorem 3.3, as follows:

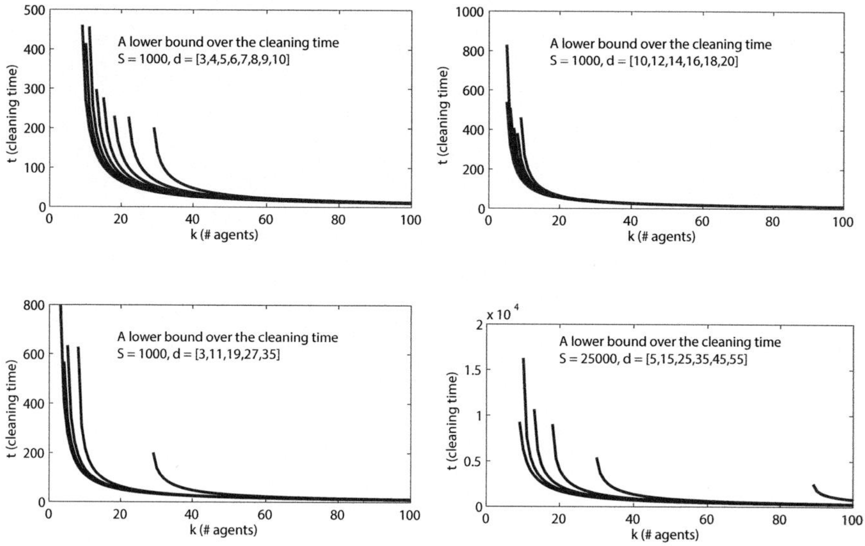

Figure 3.9: Theorem 3.2: These graphs depict lower bounds for the cleaning time of several contaminated regions (the Y axis), as a function of the number of agents (the X axis), for various values of d. Notice again that this bound is generic, and therefore will hold for any cleaning protocol used.

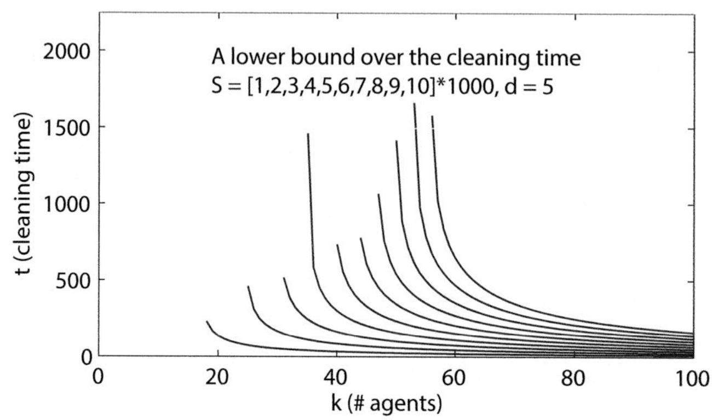

Figure 3.10: Theorem 3.2: These graphs depict lower bounds for the cleaning time of several contaminated regions (the Y axis), as a function of the number of agents (the X axis), for various values of the initial area S_0.

Theorem 3.3

Using any cleaning protocol, k agents cleaning a contaminated region of size S_0, which spreads every d time steps, will not be able to clean it if:

$$S_0 > \left\lceil \frac{1}{8}d^2k^2 + dk + \frac{1}{2} \right\rceil$$

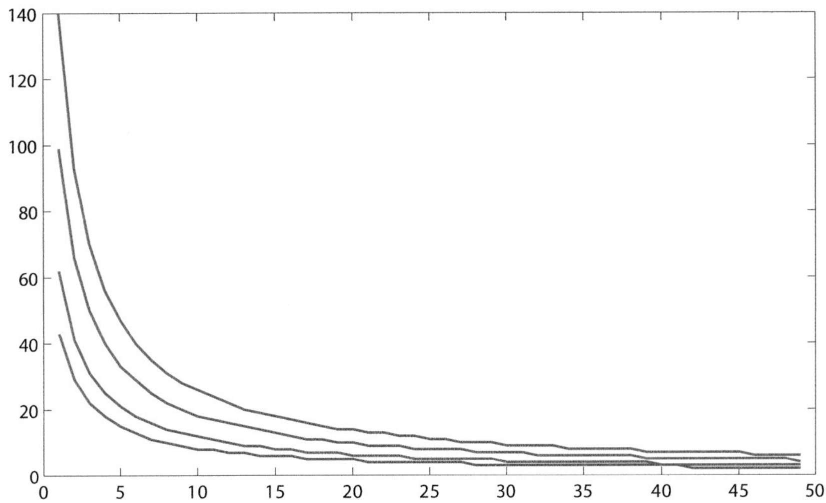

Figure 3.11: Theorem 3.2: This graph represents lower bounds for the minimal number of agents required to guarantee a successful cleaning of several contaminated regions, as a function of the contamination spreading speed d. The lower curve depicts the minimal number of agents for an initial region of $S_0 = 1000$, for spreading speeds between 2 and 50. Similarly, the other curves represent similar functions for $S_0 = 2000, 5000, 10000$, respectively.

Proof 3.3 We shall require that the size of the region increases between each two spreads (thus generating an ever expanding region, impossible to clean), or in other words, we require that:

$$S_d - S_0 > 0$$

Using Theorem 3.1 we know that regardless of the specific protocol used:

$$S_d \geq S_0 - d \cdot k + 2\sqrt{2 \cdot (S_0 - d \cdot k) - 1}$$

and therefore:

$$2\sqrt{2 \cdot (S_0 - d \cdot k) - 1} > d \cdot k$$

After some arithmetics we see that:

$$S_0 > \left\lceil \frac{1}{8} d^2 k^2 + dk + \frac{1}{2} \right\rceil$$

3.6 The Cleaning Protocol

In order to solve the *Cooperative Cleaners* problem we propose a cleaning protocol, called **SWEEP**. Generalizing an idea from computer graphics

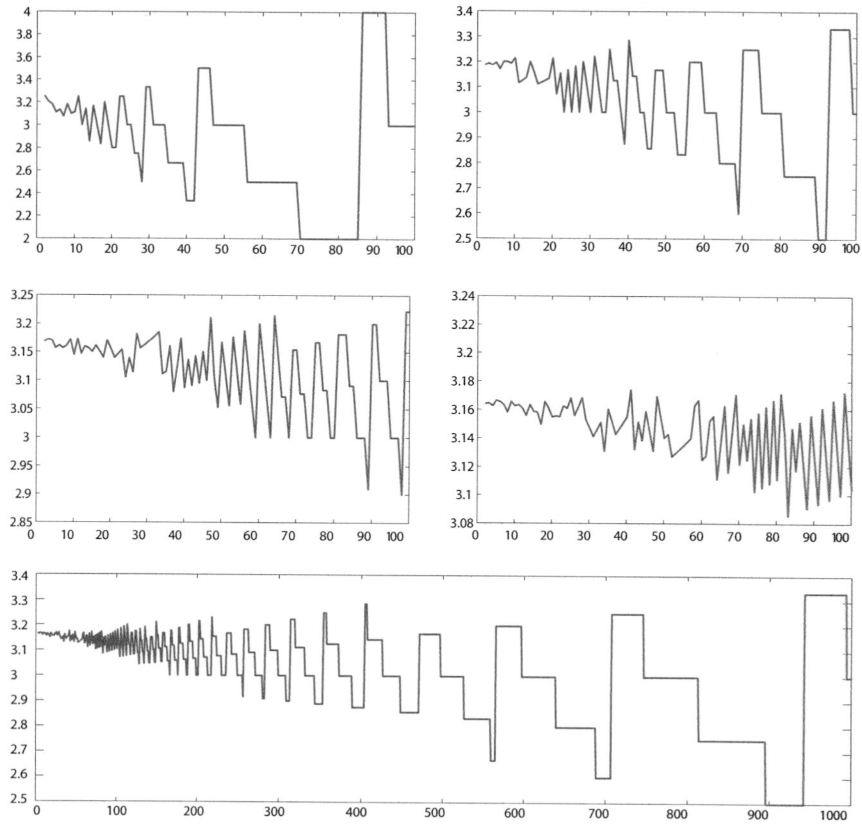

Figure 3.12: An example of Theorem 3.2. These graphs demonstrate the influence of the change in the area of the initial contaminated region over the change in the lower bound for the minimal number of agents required to guarantee a successful cleaning of the region, namely — the ratio $\frac{k(S_0,d)}{k(\frac{1}{10}S_0,d)}$ (as a function of d), when $k(S,d)$ represents the lower bound over the minimal number of agents, for a region of area S and a spreading speed d. The values of S_0 of the four upper graphs are 10000, 100000, 1000000 and 10000000, while the values of d are between 2 and 100. A continuation of the last graph (i.e. in which $S_0 = 10000000$) appears in the lower graph, in which $2 \leq d \leq 1000$. Observing these graphs it can be seen that, as expected, an increase in d results in a general decrease in the agents' ratio (disregarding the oscillatory nature of the function). The reason is that when d is increased, the problem becomes "easier", and therefore the number of agents required to complete the cleaning is smaller (regardless of the cleaning time itself). As to the effect of the initial area of the region — the ratio between the agents required for cleaning an area of S_0 tiles and the agents required for cleaning an area of $\frac{1}{10}S_0$ tiles is generally constant for smaller values of d (and surrounds $\sqrt{10}$), regardless of the ratio between S_0 and $\frac{1}{10}$.

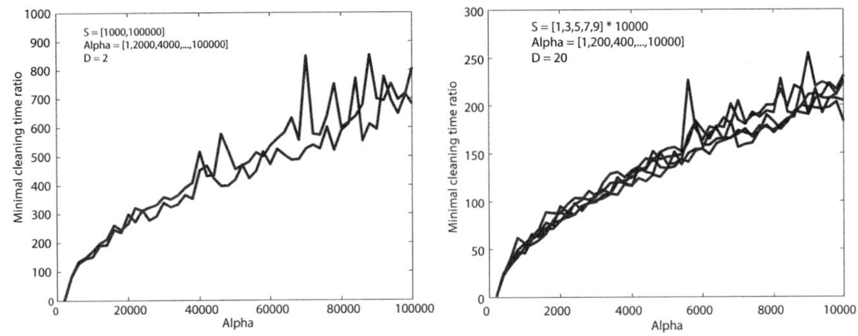

Figure 3.13: An example of Theorem 3.2, illustrating the influence of the change in the contaminated region's initial size over the lower bound of the cleaning time of the smallest group of agents which is capable of cleaning it. Namely, the graphs demonstrate $R_{S_0}(\alpha) \triangleq \frac{T_{min}(\alpha \cdot S_0, D)}{T_{min}(S_0, D)}$ as a function of α. Notice that the value of $R_{S_0}(\alpha)$ is agnostic to the selection of the initial S_0 or of D.

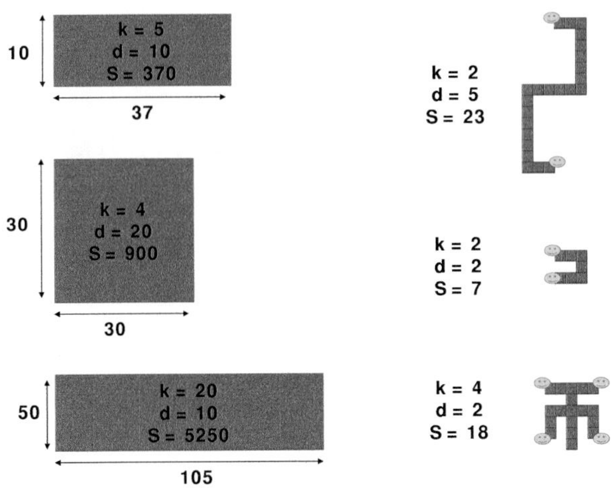

Figure 3.14: An example of Theorem 3.3, presenting several examples for initial regions, spreading latencies, and number of agents, which are guaranteed to be impossible for cleaning.

(presented in [45]), this protocol preserves the connectivity of the *dirty* region by preventing the agents from cleaning *critical points* — points which when cleaned disconnect the dirty region (see Section 3.6.1). This ensures that the agents stop only upon completing their mission.

At each time step, each agent cleans its current location (assuming it is not a critical point), and moves according to a local movement rule, creating the effect of a clockwise traversal along the boundary of the dirty region. As a result, the agents "peel" layers from the region, while preserving its connectivity, until the region is cleaned entirely. An illustration of two agents working according to the protocol can be seen in Figure 3.15.

To the basic description of the protocol given above, there are several exceptions. As the agents are equipped with very limited sensing and storage capabilities, the basic structure of the protocol must be enhanced with a set of local rules, designed for producing a pseudo-synchronization between the agents. These rules generate *resting* and *waiting* commands for the agents, capable of delaying their actions, either within the time step (until certain agents complete their cleaning process for this time step), or cause them to pause for a single time step, resuming their operation at the next time step only. A detailed description concerning the need for these additions appears in Sections 3.6.6 and 3.6.7. Note however that the rules in charge of the *resting* and *waiting* still obey the basic paradigm of this work, namely — they are local, use no prior knowledge of the agents, and can be implemented using extremely small amount of memory resources (namely — $O(\log k)$, k being the number of agents). A schematic flowchart of the protocol is presented in Figure 3.17.

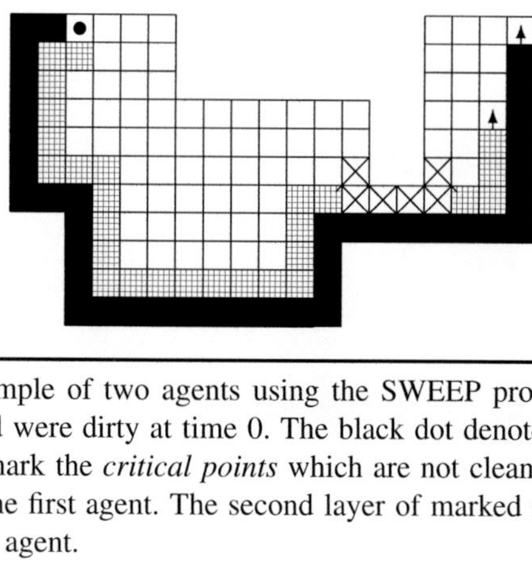

Figure 3.15: An example of two agents using the SWEEP protocol, at time step 40. All the tiles presented were dirty at time 0. The black dot denotes the starting point of the agents. The X's mark the *critical points* which are not cleaned. The black tiles are the tiles cleaned by the first agent. The second layer of marked tiles represent the tiles cleaned by the second agent.

3.6.1 Cleaning Protocol — Definitions and Requirements

Definition 3.5 Let $\tau(t) = \left(\tau_1(t), \tau_2(t), \ldots, \tau_k(t)\right)$ denote the locations of k agents at time t.

Definition 3.6 For a tile v, let $Neighborhood(v)$ denote the dirtiness states of v and its $8Neighbors$.

Let $\mathcal{M}_i$ denote some finite amount of memory contained in agent i, storing information needed for the protocol (e.g. the last moves of agent i). The requested cleaning protocol is therefore a rule f such that for every agent i:

$$f\big(\tau_i(t), Neighborhood(\tau_i(t)), \mathcal{M}_i(t)\big) \in \mathcal{D}$$

where $\mathcal{D} = \{\text{'}left', \text{'}right', \text{'}up', \text{'}down'\}$.

Definition 3.7 Let ∂F denote the boundary of F. A tile is on the boundary if and only if at least one of its $8Neighbors$ is not in F, meaning:

$$\partial F = \{v \mid v \in F \wedge 8Neighbors(v) \cap (G \setminus F) \neq \emptyset\}$$

The requested rule f should meet the following goals:

- **Successful Termination**: $(\exists t_{success}\ s.t\ F_{t_{success}} = \emptyset)$.
- **Agreement on Completion**: Within a finite time after completing the mission, all the agents must halt.
- **Efficiency**: The cleaning process should be efficient in time and in agents' memory resources.
- **Fault Tolerance**: If one or several stop working ("die") the rest of the agents will continue the cleaning process as efficiently as their number allows them.

3.6.2 The SWEEP Cleaning Protocol

The **SWEEP** protocol is implemented by each agent a_i, located at time t at $\tau_i(t) = (x, y)$. We define below several terms we use while discussing the **SWEEP** protocol. We stress the fact that this is indeed a *myopic* protocol, relying on neighborhood information, a *senile* protocol (the memory needed is constant + one counter whose size is upper bounded by $O(\log k)$, where k is the number of agents) and relies on *implicit local communication* only.

Definition 3.8 Let $\tilde{\tau}_i(t)$ denote the previous location of agent i with respect to $\tau_i(t)$, such that $\tilde{\tau}_i(t) \neq \tau_i(t)$, defined as:

$$\tilde{\tau}_i(t) \triangleq \tau_i(x)\ s.t.\ x = \max\{j \in \mathbb{N} \mid j < t\ and\ \tau_i(j) \neq \tau_i(t)\}$$

Definition 3.9 The term '*rightmost*' means:

> *Starting from $\tilde{\tau}_i(t)$ (namely, the previous boundary tile that agent a_i had been in) scan the four neighbors of $\tau_i(t)$ in a clockwise order until a boundary tile (excluding $\tilde{\tau}_i(t)$) is found. Sometimes, $\tilde{\tau}_i(t)$ might not be a boundary tile (this may occasionally happen after a contamination spread, for dirty regions of specific geometric features. In these cases,*

an agent may find itself in an internal point — a dirty tile which is not located on the boundary. From this tile the agent must move to the adjacent boundary tile, and resume its traversal along the region's boundary). In this case, if $\tau_i(t)$ *is a boundary point, then starting from* $\tilde{\tau}_i(t)$ *scan the four neighbors of* $\tau_i(t)$ *in a clockwise order until the second boundary tile is found. In case* $t = 0$ *select the tile as instructed in Figure 3.16.*

The reference to *contamination spread* used in Definition 3.9 is given with consideration to the use of this term in the next chapters.

The additional information needed for the protocol and its sub-routines is contained in $\mathcal{M}_i$ and $Neighborhood(x,y)$.

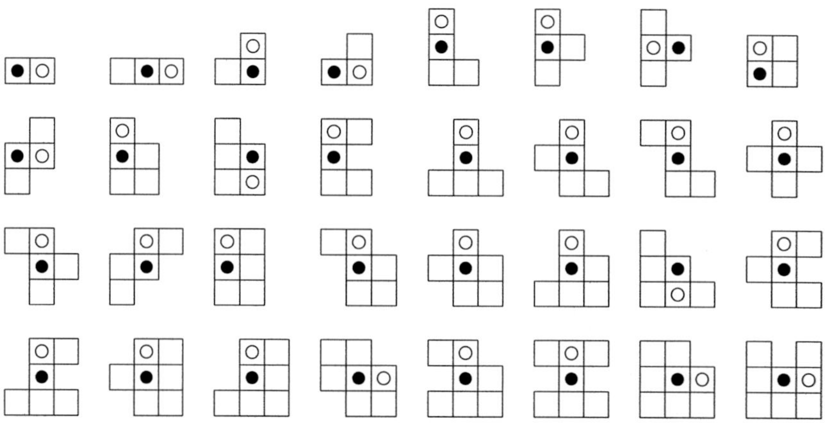

Figure 3.16: When $t = 0$ the first movement of an agent located in (x, y) should be decided according to the initial contamination status of the neighbors of (x, y), as appears in these charts — the agent's initial location is marked with a filled circle while the destination is marked with an empty one. All configurations which do not appear in these charts can be obtained by using rotations. This definition is needed in order to initialize the traversal behavior of the agents in the correct direction.

A schematic flowchart of the protocol, describing its major components and procedures is presented in Figure 3.17. The complete pseudo-code of the protocol and its sub-routines appears in Figures 3.18 and 3.19. Upon initialization of the system, the *System Initialization* procedure is called (defined in Figure 3.18). This procedure sets various initial values of the agents, and call the protocol's main procedure — *SWEEP* (defined in Figure 3.19). This procedure in turn, uses various sub-routines and functions, all defined in Figure 3.18.

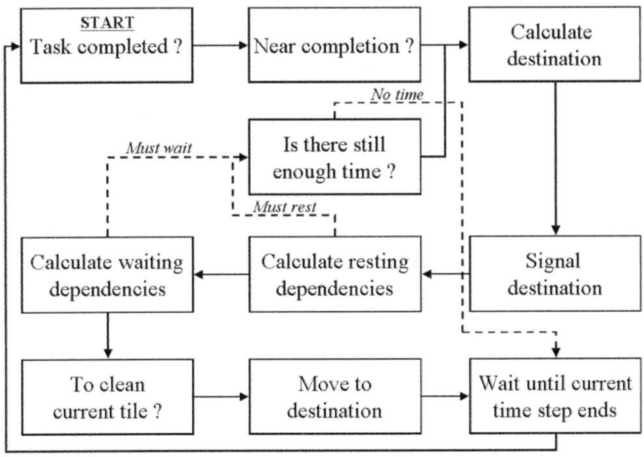

Figure 3.17: A schematic flow chart of the **SWEEP** protocol. The smooth lines represent the basic flow of the protocol while the dashed lines represent cases in which the flow is interrupted. Such interruptions occur when an agent calculates that it must not move until other agents do so (either as a result of *waiting* or *resting* dependencies — see Sections 6 and 19 of **SWEEP** for more details).

3.6.3 Pivot Point

The initial location of the agents (the *pivot point*, denoted as p_0) is artificially set to be critical during the execution, hence it is also guaranteed to be the last point cleaned. Completion of the mission can therefore be concluded from the fact that all (working) agents are back at p_0 with no contaminated neighbors to go to, thereby reporting on completion of their individual missions. Note that this artificial preservation of the criticalness of the pivot point is not necessary for the algorithm to work. Rather, it just makes life easier for the user, as one can now know where to find the agents after the cleaning has been completed. If we do not start with all agents at the pivot and force p_0 to be critical, the location of the agents upon completion will generally not be known in advance.

3.6.4 Signaling

Since by assumption the agent's sensors can detect the status of all tiles which are contained within a digital sphere of radius four placed around the current location of the agent, each agent can artificially calculate the desired destination of all the agents which are located in one of its 4*Neighbors* tiles (see Figure 3.20. Thus, the *signaling* action of each agent can be simulated by the other agents near him, and hence an explicit signaling by the agents is not actually required. However, the signaling action is kept in the description and flowchart of the protocol (in procedure 5) for the sake of simplicity and understandability.

```
System Initialization
    Arbitrarily choose a pivot point p0 in ∂F0, and mark it as critical point
    Place all the agents on p0
    For (i = 1; i ≤ k; i++) do
        Call Agent Reset for agent i
        Call SWEEP for agent i
        Wait two time steps
    End for
End procedure

Agent Reset
    resting ← false
    dest ← null /* destination */
    near completion ← false
    saturated perimeter ← false
    waiting ← ∅
End procedure

Priority
    /* Assuming the agent moved from (x0, y0) to (x1, y1) */
    priority ← 2(x1 − x0) + (y1 − y0)
End procedure

Check Completion of Mission
    If ((x, y) = p0) and (x, y) has no dirty neighbors then
        If (x, y) is dirty then
            Clean (x, y)
        STOP
End procedure

Check "Near Completion" of Mission
    /* Cases where every tile in F_t contains at least a single agent */
    near completion ← false
    If each of the dirty neighbors of (x, y) contains at least one agent then
        near completion ← true
    If each of the dirty neighbors of (x, y) has near completion = true then
        Clean (x, y) and STOP
    /* Cases where every non-critical tile in ∂F_t contains at least 2 agents */
    saturated perimeter ← false
    If ((x, y) ∈ ∂F_t) and both (x, y) and all of its non-critical neighbors in ∂F_t contain at least two agents then
        saturated perimeter ← true
    If ((x, y) ∈ ∂F_t) and both (x, y) and all of its neighbors in ∂F_t has saturated perimeter = true then
        Ignore resting commands for this time step
End procedure
```

Figure 3.18: The first part of the SWEEP cleaning protocol. The terms ∂F and $\tau_i = (x, y)$ are defined in Sections 3.6.1 and 3.6.2.

3.6.5 Connectivity Preservation

The connectivity of the region yet to be cleaned, F, is preserved by allowing the agents to clean only *non-critical* points. This guarantees both successful termination and agreement of completion (since having no dirty neighbors implies that $F = \emptyset$). Also, should several agents malfunction and halt, as long as there are still functioning agents, the mission will still be carried out, albeit slower.

Note that a tile which was originally clean, or which was cleaned by the agents, is guaranteed to remain clean. As a tile can be cleaned by the agents only when it is in ∂F, it is easy to see that the simple connectivity of F is maintained

1: **SWEEP Protocol** /* Controls agent i after **Agent Reset** */
2: **Check Completion of Mission**
3: **Check "Near Completion" of Mission**
4: *dest* $\leftarrow$ *rightmost neighbor* of (x,y) /* Calculate destination */
5: *destination signal bits* $\leftarrow$ *dest* /* Signaling the desired destination */
6: /* Calculate resting dependencies (solves agents' clustering problem) */
7: Let all agents in (x,y) except agent i be divided to the following groups :
8: A_1 : Agents signaling towards any direction different than *dest*
9: A_2 : Agents signaling towards *dest* which entered (x,y) before agent i
10: A_3 : Agents signaling towards *dest* which entered (x,y) after agent i
11: A_4 : Agents signaling towards *dest* which entered (x,y) with agent i
12: Let group A_4 be divided into the following two groups :
13: A_{4a} : Agents with lower *priority* than this of agent i
14: A_{4b} : Agents with higher *priority* than this of agent i
15: *resting* $\leftarrow$ *false*
16: **If** ($A_2 \neq \emptyset$) or ($A_{4b} \neq \emptyset$) then
17: *resting* $\leftarrow$ *true*
18: **If** (current time-step $\mathcal{T}$ did not end yet) then jump to 4 **Else** jump to 35
19: *waiting* $\leftarrow \emptyset$ /* Waiting dependencies (agents synchronization) */
20: Let *active agent* denote a *non-resting* agent which didn't move in $\mathcal{T}$ yet
21: **If** (x-1,y) $\in F_t$ contains an active agent then *waiting* $\leftarrow$ *waiting* $\cup$ $\{left\}$
22: **If** (x,y-1) $\in F_t$ contains an active agent then *waiting* $\leftarrow$ *waiting* $\cup$ $\{down\}$
23: **If** (x-1,y-1) $\in F_t$ contains an active agent then *waiting* $\leftarrow$ *waiting* $\cup$ $\{l\text{-}d\}$
24: **If** (x+1,y-1) $\in F_t$ contains an active agent then *waiting* $\leftarrow$ *waiting* $\cup$ $\{r\text{-}d\}$
25: **If** *dest* = *right* and (x+1,y) contains an active agent j, and $dest_j \neq left$, and there are no other agents delayed by agent i (i.e. (x-1,y) does not contain active agent l with $dest_l$ =*right* and no active agents in (x,y+1),(x+1,y+1), (x-1,y+1), and (x+1,y) does not contain active agent n with $dest_n = left$), then (*waiting* $\leftarrow$ *waiting* $\cup$ $\{right\}$) and $\left(waiting_j \leftarrow waiting_j \setminus \{left\}\right)$
26: **If** *dest* = *up* and (x,y+1) contains an active agent j, and $dest_j \neq down$, and there are no other agents delayed by agent i (i.e. (x,y-1) does not contain active agent l with $dest_l$ =*up* and no active agents in (x+1,y),(x+1,y+1), (x-1,y+1), and (x,y+1) does not contain active agent n with $dest_n = down$), then (*waiting* $\leftarrow$ *waiting* $\cup$ $\{up\}$) and $\left(waiting_j \leftarrow waiting_j \setminus \{down\}\right)$
27: **If** ($waiting \neq \emptyset$) then
28: **If** ($\mathcal{T}$ has not ended yet) then jump to 4 **Else** jump to 35
29: /* Decide whether or not (x,y) should be cleaned */
30: **If** $\neg$ ((x,y) $\in \partial F_t$) or ((x,y) $\equiv p_0$) or (x,y) has 2 dirty tiles in its 4*Neighbors* which are not connected via a path of dirty tiles from its 8*Neighbors* then
31: (x,y) is an *internal point* or a *critical point* and should not be cleaned
32: **Else**
33: Clean (x,y) if and only if it does not still contain other agents
34: Move to *dest*
35: Wait until $\mathcal{T}$ ends.
36: Return to 2

Figure 3.19: The **SWEEP** cleaning protocol. The term *rightmost neighbor* is defined in Section 3.6.2. *l-d* and *r-d* are left-down and right-down, respectively.

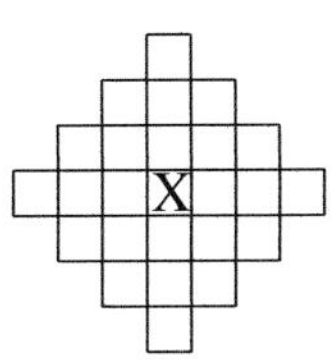

Figure 3.20: Digital sphere of diameter 7, placed around the agent.

throughout the cleaning process (as the creation of "holes" — clean tiles which are surrounded by tiles of F — is impossible).

Note that when several agents are located in the same tile, only the last one to exit cleans it (see section 33 in **SWEEP**), in order to prevent agents from being 'thrown' out of the dirty region (meaning, cleaning a tile in which another agent is located and thus preventing this agent from being able to continue the execution of the cleaning protocol). An alternative method for ensuring the agents are always capable of executing the cleaning protocol would have been the implementation of a "dirtiness seeking mechanism" (i.e. by applying methods such as suggested in [30]). Such a mechanism would have allowed an agent to clean its current location, even if other agents had also been present there. That solution, however, would have been far less elegant and would have added additional difficulties to the analysis process.

3.6.6 Agents Synchronization

Note that agents operating in the described environment must have some means to synchronize, which is mandatory in order to prevent agents from operating at the same time—risking to cut the dirty region into several connected components, as shown in Figure 3.21.

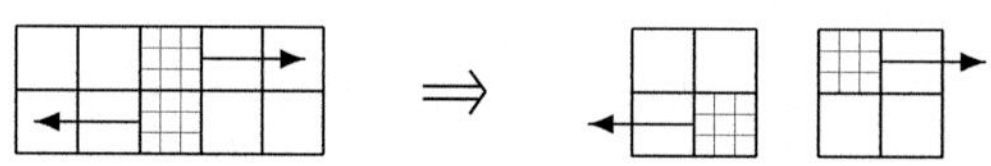

Figure 3.21: When the agents do not possess a synchronization mechanism, they may, among others, damage the region's connectivity. In this example, two agents clean their current locations, and move according to the **SWEEP** protocol. Since they are not synchronized, the tiles which they are located in are not treated as critical at the time of cleaning. However, the region's connectivity is not preserved. Had one of the agents waited for its neighbor to complete executing the protocol's steps before resuming its actions, while deciding whether to clean its current location, it would have treated this tile as critical, and therefore avoided cleaning it. In this case, the connectivity of the region would have been maintained.

To ensure such scenarios will not occur, an order between the operating agents must be implemented. Note — throughout the next paragraphs agents which are signaling a *resting* status (see Section 3.6.7 for more details) are being disregarded while calculating the dependencies of the agents' movements. The creation of the following order is implemented in Procedure 19 of **SWEEP**:

Definition 3.10 For agent i, let $P_i \subseteq \{up, down, left, right, right\text{-}down, left\text{-}down\}$ be a set of directions of tiles, in which there are currently agents, which agent i is delayed by (meaning, agent i will not start moving until the agents in these tiles move). Unless stated otherwise, $P_i = \emptyset$.

For agent i which is located in tile (x,y), if $(x-1,y)$ is a tile in F, which contains an agent, then $P_i \leftarrow P_i \cup \{left\}$. If $(x,y-1)$ is a tile in F, which contains an agent, then $P_i \leftarrow P_i \cup \{down\}$, and similarly for $(x+1,y-1)$ and *right-down* and for $(x-1,y-1)$ and *left-down*.

Definition 3.11 Let $dest_i \in \{up, down, left, right\}$ be the destination agent i might be interested in moving to, after leaving its current location.

Let each agent be equipped with a two bit flag (that may be implemented for example by two small light-bulbs). This flag states the desired destination of the agent. Alternatively, each tile can be treated as a physical tile, in which the agent can move. Thus, the agent can move towards the top side of the tile, which will be equivalent for using the flag in order to signal that it intends to move *up*.

Let each time step be divided into two phases. In phase 1, every agent "signals" the destination it intends to move towards, either by moving to the appropriate side of the tile, or by using the destination flag.

As we defined an artificial rule which states the superiority of *left* and *down* over *right* and *up* (and internally, of *down* over *left*, there are several specific scenarios in which this asymmetry should be reversed in order to ensure a proper operation of the agents. Following is such a "dependencies switching" rule: For agent i, which is located in (x,y), if $dest_i = right$ and tile $(x+1,y)$ contains an agent j, and $dest_j \neq left$, and there are no other agents which are delayed by agent i (i.e. tile $(x-1,y)$ does not contain an agent l where $dest_l = right$ and tiles $(x,y+1),(x+1,y+1),(x-1,y+1)$ do not contain any agent and tile $(x+1,y)$ does not contain an agent n where $dest_n = left$), then $P_i \leftarrow P_i \cup \{right\}$ and $P_j \leftarrow P_j \setminus \{left\}$. Also, if $dest_i = up$ and tile $(x,y+1)$ contains an agent k, and $dest_k \neq down$, and there are no other agents which are delayed by agent i (i.e. tile $(x,y-1)$ does not contain an agent m where $dest_m = up$ and tiles $(x+1,y),(x+1,y+1),(x-1,y+1)$ do not contain any agent and tile $(x,y+1)$ does not contain an agent q where $dest_q = down$), then $P_i \leftarrow P_i \cup \{up\}$ and $P_k \leftarrow P_k \setminus \{down\}$.

At phase 2 of each time step, the agents start to operate in turns, according to the order implied by P_i. This guarantees that the connectivity of the region is kept, since the simultaneous movement of two neighboring agents is prevented.

Notice that deadlocks are impossible — since the basic rule is that every agent is delayed by the agents in its *left* and *down* neighbor tiles. Therefore, at any given time, and for every possible group of agents, there exists an agent with the minimal x and y coordinates (which by definition, is not delayed by any other agent of this group). After this agent moves, all the agents which are delayed by it, can now move, and so on. As to the "dependencies switching rule" — let agent

i located in tile (x, y) have the minimal x and y values among the agents who had not moved yet, and let $dest_i = up$, and let tile $(x, y+1)$ contain an agent j such that $dest_j \neq down$. Then although agent i is located below agent j, it will be delayed by it (i.e. $(up \in P_i)$ and $\neq (down \in P_j)$) as long as agent i is not delaying any other agent (as this is the requirement of the "dependencies switching rule"). In this case, we should show that there can not be a cycle of dependencies, which starts at agent j, ends at agent i, and is closed by the dependency of agent i on agent j. Such a circle can not exist since for it to end in agent i, it means that agent i is delaying another agent k. However, this is impossible since agent i is known not to delay any agent (specifically agent k). Hence, circular dependencies are prevented and no deadlock are possible.

Note that while phase 2 is in process, F_t may change due to cleaning by the agents. As a result, the desired destinations of the agents as well as their dependencies forest, must also be dynamic. This is achieved through the repeated recalculation of these values, for every *waiting* agent. For example, assume agent i to be located in (x, y), and $dest_i = down$ without loss of generality, and let agent j located in tile $(x, y-1)$ move out of this tile and clean it. Then, $dest_i$ naturally changes (as the tile (x, y) no longer belongs to F_t, and thus is not a legitimate destination for agent i), and thus, P_i may also change. In this case, agent i should change its "destination signal" and act according to its new $dest_i$ and P_i. This is implemented in **SWEEP** by calculating the waiting agents' destinations and dependencies lists repeatedly, until either all agents have moved or until the time step has ended (meaning that some agents had to change their status to *resting*, and pause until the next time step — see Section 3.6.7 for more details). Note that every *waiting* agent is guaranteed either to complete its movement in the current time step, or to be forced to wait for the next time step, by switching its status to *resting*.

Notice that the "dependencies switching rule" is not required in order to ensure a proper completion of the mission, but rather to improve the agents' performance, by preventing a bug demonstrated in Figure 3.22.

3.6.7 Clustering Problem

Since we are interested in preventing the agents from moving together and functioning as a single agent (due to a resulting decrease in the system's performance), we would like to impose a "resting policy" which will ensure that the agents do not group into clusters which move together. This is done by artificially delaying agents in their current locations, should several agents placed in the same tile wish to move to the same destination. However, we would like the resting time of the agents to be minimal, for performance and analysis reasons.

The following resting policy is implemented by Section 6 of **SWEEP**. Using its sensors, an agent intending to move into tile v is aware of other agents which

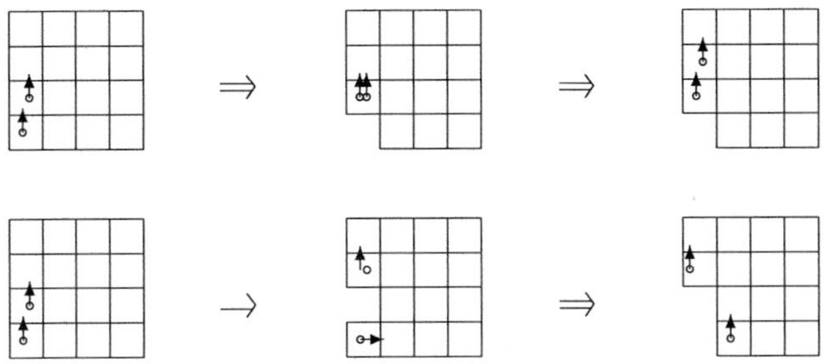

Figure 3.22: The upper charts demonstrate a performance bug which may be caused due to local dependencies. The agents are advancing according to the **SWEEP** protocol, but their cleaning performance is decreased. The lower charts demonstrate the cleaning operation after adding the dependencies switching mechanism.

intend to move into v as well. Thus, when an agent enters a tile which already contains other agents, it knows whether these agents entered this tile at the same time step it did, or whether they had occupied this tile before the current time step had started.

Note that in phase 1 of each time step all the agents are signaling their desired destinations. Thus, an agent can identify which ones of the agents located in its current tile intend to move to the same destination. Only those agents may cause this agent to delay its actions and rest.

From the agents which intend to move in the same direction as agent i, the agent can distinguish between three groups of agents — the agents which entered this tile before the time step agent i did (group A_2), those which entered the tile in the same time step as i (group A_4), and those who entered it after agent i did (group A_3). If $A_2 \neq \emptyset$, agent i waits, changes its status to *resting agent*, signals this status to the other agents in its vicinity, and does not move. As this rule is kept by every other agent, agent i is guaranteed that the agents of group A_3, in their turn, will wait for agent i to move before being free to do so as well.

As to the agents of A_4, the problem is solved by induction over the time steps, namely — that at any given time, two agents can not leave the same tile in the same direction at the same time step. The base of this induction holds for small values of t as the agents are periodically released by the initialization procedure of **SWEEP**, while no two agents are released at the same time step. Later, as we are assured that all the agents of group A_4 arrived to v from different tiles (which are also different than the tile agent i had entered v from), then since all the agents in group A_4 know each other's previous locations, a consensus over a local ordering of A_4 is established (note that no explicit communication is needed to form this ordering). An example for such an order is the **Priority** function of the **SWEEP** protocol. As a result, the agents are able to exit the tile they are currently located in, in an orderly fashion, according to a well defined order. Thus, the following invariant holds : "*at any given time t, for any two tiles v, u,* *there can only be a single agent which moves from v to u at time step t*". Thus, the clustering problem is solved.

Notice that there is a single exception to this mechanism, in which the resting commands are overruled. This happens when all non-critical perimeter tiles contain at least two agents. In this scenario, following the resting commands would have created a situation in which no tile can be cleaned, as for every agent which leaves a certain tile, there are still "resting" agents located in the same tile. Therefore, once this scenario is detected by the *check near completion of mission* procedure of **SWEEP**, the agents ignore their resting commands momentarily, in order to be able to clean the non-critical perimeter tiles. The order and internal prioritization of the agents are maintained, for calculating the new *resting* commands in the following time step.

3.6.8 Mission Termination

The termination of the protocol is done in one of two cases — either an agent finds itself in the pivot point, while all of its neighbors are clean (which means that this is the last dirty tile and therefore should be cleaned), or when each dirty tile contains at least one agent (which is a generalization of the previous scenario). The second case is implemented by allowing the agents to signal to their four neighbors whether all of their dirty neighbors contain at least a single agent.

3.7 Experimental Results vs. Lower Bound

Figures 3.23, 3.24 and 3.25 depict the cleaning process of several contaminated regions — for each one, the cleaning time for various number of agents is presented. As the agents' cleaning time is influenced by their initial location a multitude of simulations were performed for every number of agents. Specifically, one simulation was conducted for each possible location (namely the total number of simulations for any number of agents equaled $|\partial F_0|$). The maximal and minimal cleaning times for each number of agents is presented, as well as the average cleaning time (calculated for all the initial locations). Figure 3.26 demonstrates how placing the agents at different starting points allows them to successfully clean regions which were impossible to clean when the agents were concentrated at a single starting point. Figure 3.27 demonstrates the dynamic nature of the agents' cleaning efficiency, by comparing the contaminated region's size to the whereabouts of the agents at various times. Figure 3.28 demonstrates the changes in the geometric features of a region while it is being cleaned, as well as the influence the number of agents has on the efficiency of the cleaning.

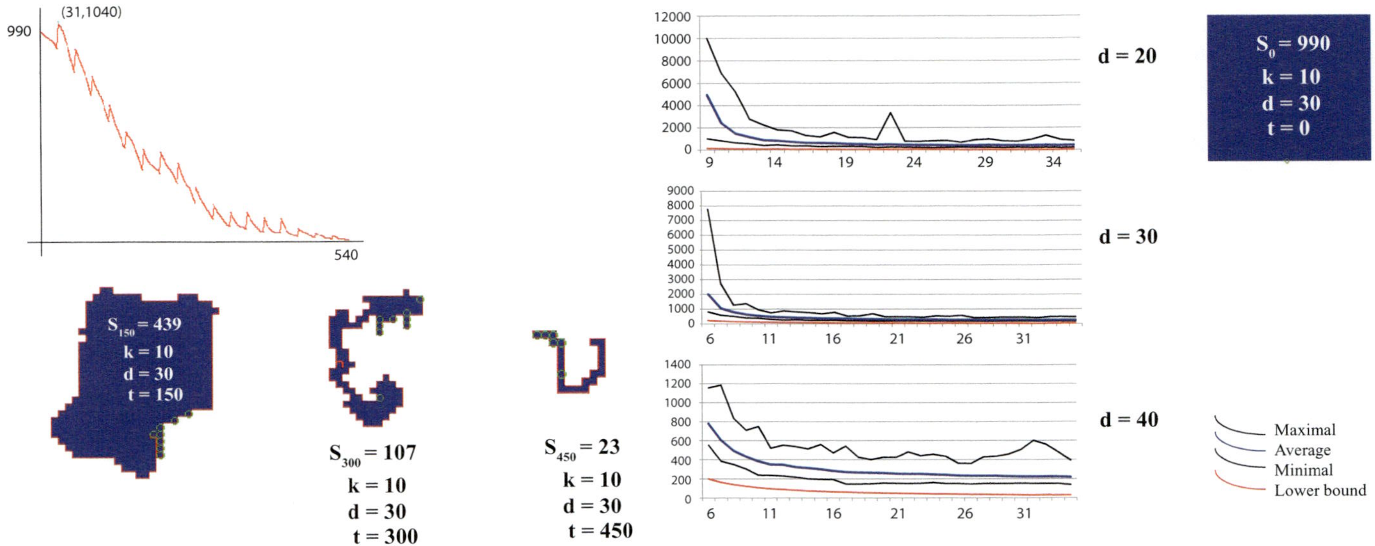

Figure 3.23: An example of a cleaning mission by agents using the **SWEEP** cleaning protocol, trying to clean a contaminated square (see the original contaminated shape at the top, on the right). A graph of the size of the region as a function of the time, for 10 agents and spreading speed of $d = 30$ appears on the left top corner, as well as size and shape of the region at several times during the cleaning. The three graphs at the bottom depict the cleaning time (maximal, minimal and average, for 25 different starting points of the agents) for various numbers of agents, and for three values of spreading speeds. In addition, the cleaning times are compared to the lower bound, as presented in Theorem 3.2.

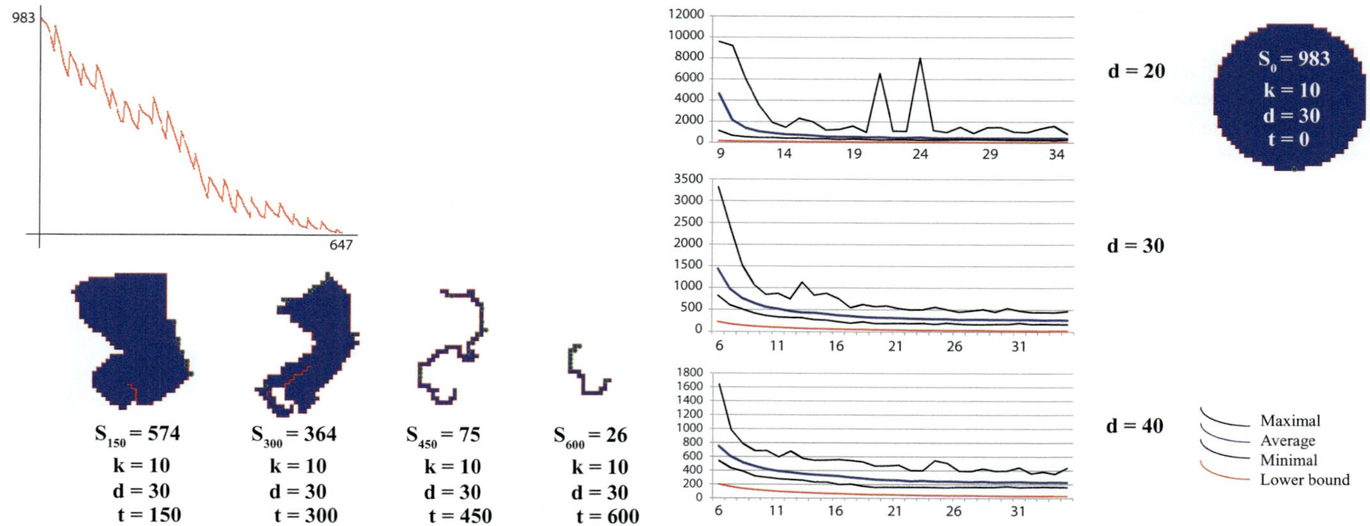

Figure 3.24: An example of a cleaning mission by agents using the **SWEEP** cleaning protocol, trying to clean a contaminated circle (see the original contaminated shape at the top, on the right). A graph of the size of the region as a function of the time, for 10 agents and spreading speed of $d = 30$ appears on the left top corner, as well as size and shape of the region at several times during the cleaning. The three graphs at the bottom depict the cleaning time (maximal, minimal and average, for 25 different starting points of the agents) for various numbers of agents, and for three values of spreading speeds. In addition, the cleaning times are compared to the lower bound, as presented in Theorem 3.2.

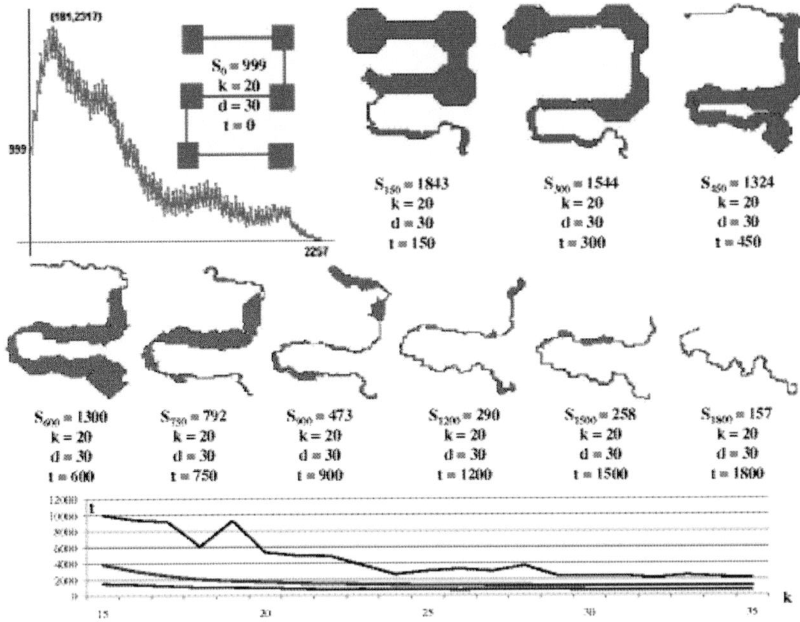

Figure 3.25: An example of a cleaning mission by agents using the **SWEEP** cleaning protocol, trying to clean a contaminated region in the shape of several rooms, connected by narrow corridors (see the original contaminated shape at the top, on the right). A graph of the size of the region as a function of the time, for 20 agents and spreading speed of $d = 30$ appears on the left top corner, as well as size and shape of the region at several times during the cleaning. The graph at the bottom depicts the cleaning time (maximal, minimal and average, for 25 different starting points of the agents) for various numbers of agents, and for a spreading speed of d=30.

3.8 Upper Bound for Static Domains ($d = \infty$)

The introduction of the notion of critical points makes time-complexity analysis of the **SWEEP** protocol significantly harder since a critical point may be visited several times before it is cleaned. We conjecture that the protocol proposed is efficient, and additional agents will speed up the cleaning process only up to a limit. In order to give this argument a rigorous form, some definitions are required.

3.8.1 Definitions

Definition 3.12 Let S_t denote the size of the dirty region F at time t, namely the number of grid points (or tiles) in F_t.

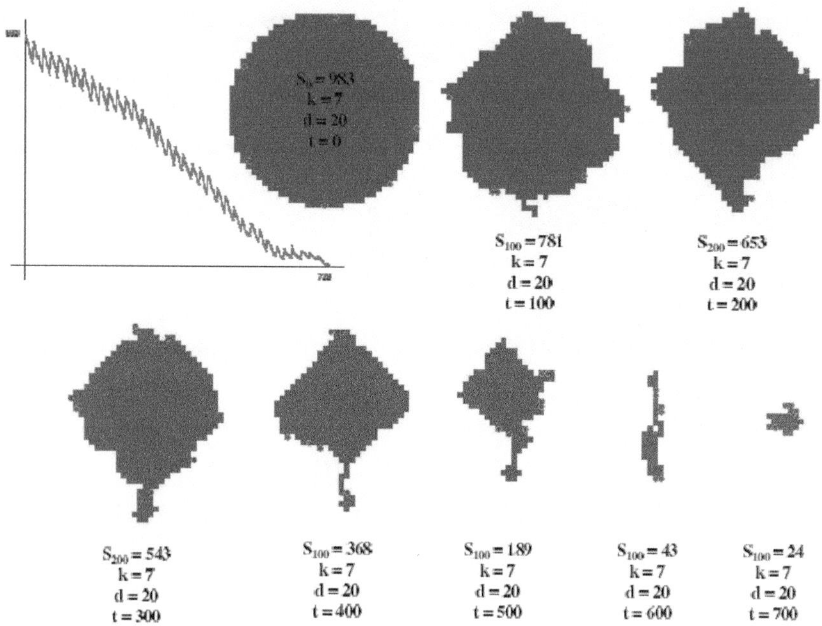

Figure 3.26: The graph presents the cleaning process of a (relatively large) digital sphere. Observe how the size of the region increases at various times. Examining the appearance of the region at those times reveals that the majority of agents are being located in the vicinity of the initial location of the agents (a point which is artificially kept critical, according to the **SWEEP** protocol) — see snapshots at the bottom. After several traversals of the region, this part of the region is composed mainly of critical points, which are not cleaned by the agents. Hence, when traveling through this part, the cleaning efficiency of the agents diminishes temporarily, and as the region continues to spread, its global size increases.

Actually, F defines a dichotomy of $\mathbf{Z}^2$ into F and $\overline{F} = \mathbf{Z}^2 \setminus F$.

Recalling Definition 3.7, the boundary of the dirty region F is denoted as ∂F and defined as:

$$\partial F = \{(x,y) \mid (x,y) \in F \land (x,y) \text{ has an } 8Neighbor \text{ in } (G \setminus F)\}$$

Definition 3.13 A *path* in F is defined to be a sequence $(v_0, v_1, \ldots, v_n)$ of tiles in F such that every two consecutive tiles are 4 *connected* (the *Manhattan* distance between them is 1). The *length* of a *path* is defined to be the number of tiles in it.

Definition 3.14 Let tile v be called a *critical point* if there exist $v_1, v_2 \in 4Neighbors(v)$ for which all paths connecting v_1 and v_2, included in $8Neighbors(v)$, necessarily pass through v (where v, v_1, v_2 and all said paths are in F).

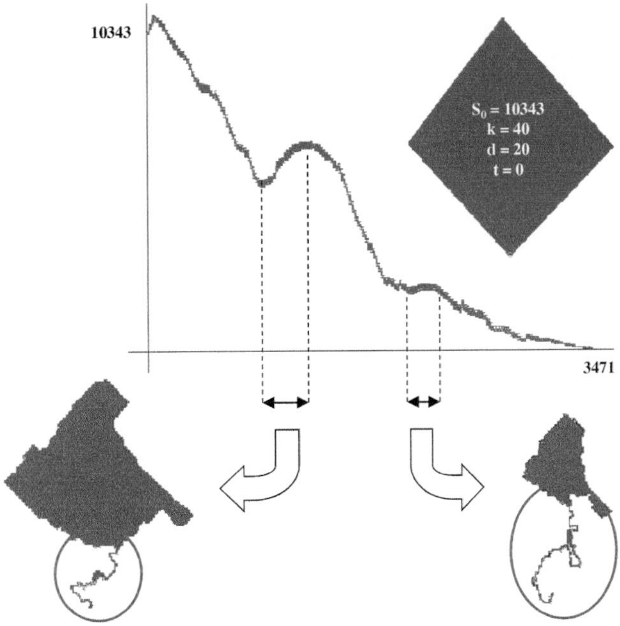

Figure 3.27: The graph presents the cleaning process of a (relatively large) "digital sphere". Observe how the size of the region increases at various times. Examining the appearance of the region at those times reveals that the majority of agents are being located in the vicinity of the initial location of the agents (a point which is artificially kept critical, according to the **SWEEP** protocol) — see snapshots at the bottom. After several traversals of the region, this part of the region is composed mainly of critical points, which are not cleaned by the agents. Hence, when traveling through this part, the cleaning efficiency of the agents diminishes temporarily, and as the region continues to spread, its global size increases.

Definition 3.15 We shall denote by c_t the circumference of F at time t, defined as follows: let v_0 and v_n be two 4 *connected* tiles in ∂F_t, and let $C_t = (v_0, v_1, \ldots, v_n)$ be the shortest *path* connecting v_0 and v_n, which contains all the tiles of ∂F_t and only such tiles. Notice that C_t may contain several instances of the same tile, if this tile is a *critical point* (meaning that C_t is an ordering of the tiles of ∂F_t, in which multiple instances of tiles that are *critical points* are allowed). c_t will be defined as the length of C_t. An example appears in Figure 3.29.

Definition 3.16 For some $v \in F_t$ let $Strings_t(v)$ denote the set of all *paths* in F_t that begin in v and end at any *non-critical point* in ∂F_t, and let $w(F_t, v)$ denote the *depth* of v — the length of the shortest *path* in $Strings_t(v)$ (unless v is a *critical point* in which case its *depth* is defined to be zero).

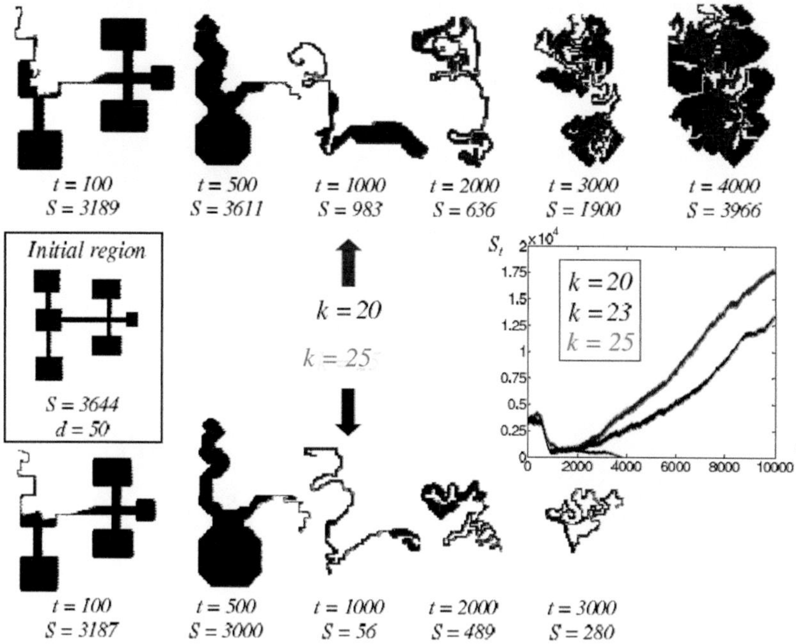

Figure 3.28: This example demonstrates the change in the geometric properties of the contaminated region as a result of the cleaning process.

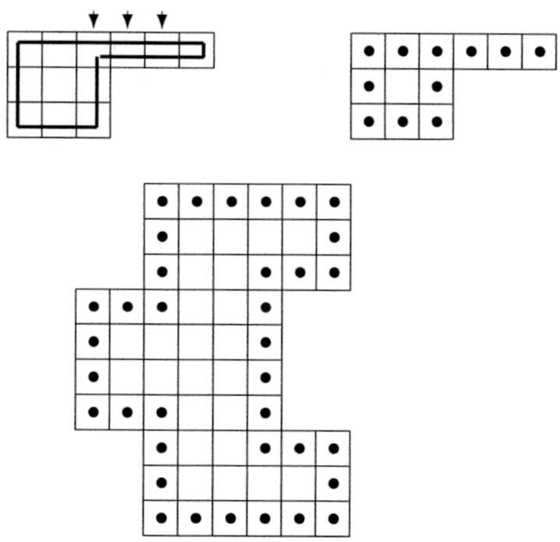

Figure 3.29: The line in the left upper chart goes through the tiles of C_t where the 3 arrows denote tiles that are included twice. The circles in the right upper chart denote the tiles of ∂F_t. Note that while ∂F_t contains 11 tiles, C_t contains 14. In the lower chart, there are no *critical points* in ∂F_t and therefore $c_t = |\partial F_t| = |C_t| = 36$.

Definition 3.17 Let $W(F_t)$ denote the *width* of F_t, defined as the maximal depth of all the tiles in F_t, i.e.:

$$W(F_t) = \max\{w(F_t, v) \mid v \in F_t\}$$

An example appears in Figure 3.30.

Definition 3.18 Let F_t' denote the region one could get, having one agent traverse F_t once, using the **SWEEP** protocol (i.e. when $k = 1$, $F_t' = F_{t+c_t}$). For the sake of simplicity, F_t'' will be used instead of $(F_t')'$, and so on.

Definition 3.19 The longest in-region distance between p_0 and any other point in F_t will be referred to as $l(F_t)$, the *length* of F_t:

$$l(F_t) = \max_{v \in F_t} \{d_{F_t}(p_0, v)\}$$

where $d_{F_t}(x, y)$ is the distance (shortest path within F_t) between x and y.

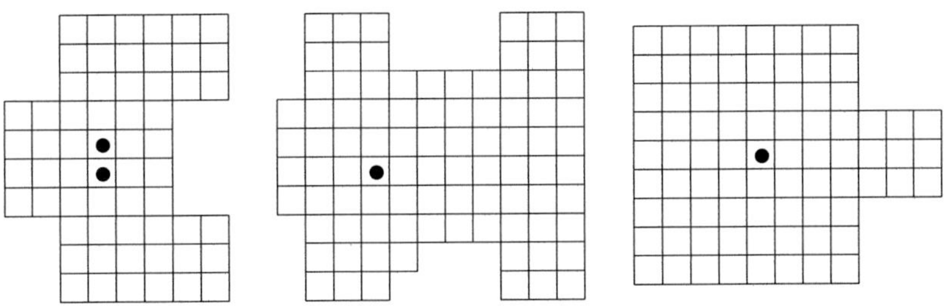

Figure 3.30: The chart contains three regions. The black circles denote the deepest tiles within the regions. Notice that while in the left region there are two tiles whose depth is the maximal, in the other regions there is only one tile of maximal depth. The widths of the regions equal the depths of the deepest tiles, and are (from left to right) 2, 3, and 4.

3.8.2 Correctness

Lemma 3.1
If $S_t > 0$ then $W(F_t) \geq 1$, that is: any finite simple region has at least two non-critical points on its boundary.

Proof 3.4 The boundary ∂F_t is a connected graph, hence it has a spanning tree. In such a tree there are at least two vertices with degree 1. These vertices are necessarily not critical in F_t, since any boundary point which is critical in F_t is also critical in ∂F_t.

Lemma 3.2
During the course of the cleaning process, if all the agents are using the ***SWEEP***

protocol the agents are never going outside the dirty region. Namely, the only time an agent is located on a clean tile is when $F = \emptyset$.

Proof 3.5 The only movements allowed by the **SWEEP** protocol are towards a dirty tile. The only times an agent is allowed to clean a tile are when it is the last agent leaving this tile, or when cleaning this tile means that the mission is completed (for example, when this is the only dirty tile left). In addition, the various preemptions which are in charge of agents synchronization prevent an agent to move to a dirty tile, which is getting cleaned during its movement. As a result, an agent is guaranteed to be located in dirty tiles throughout the entire cleaning process.

Theorem 3.4
A group of agents executing the ***SWEEP*** *protocol will eventually clean a simply connected region and stop together at the pivot point p_0.*

Proof 3.6 While F_t has not yet been cleaned, $S_t > 0$ and hence, by Lemma 3.1 there is a non-critical point on ∂F_t. Since the agents obey the **SWEEP** protocol, while $F_t \neq \emptyset$, there is at least a single agent still traversing F_t. As this agent does not change the direction of its traversal, it is guaranteed to arrive to (and clean) a non-critical point at least once during the course of its traversal (thus decreasing S_t by at least 1). As $c_t \leq 4S_t$ (since during a traversal around F_t any tile can be entered to from each one of its 4 neighbors at most once), the maximal length of this traversal is $4S_t$. As agents may indeed delay one another, it is shown in Lemma 3.5 that the actual time required for an agent to complete its traversal around F_t is at most four times the length of this traversal, namely — $16S_t$. As $\forall t, S_t \leq S_0$, it can be seen that after no more than $16s_0^2$ time steps we shall have $S_t = 0$. The fact that all agents will meet at the same point is implied by the following two rules that are implemented in the **SWEEP** protocol:

- **Rule 1:** F_t is always being kept connected (we never clean a tile that has no clean neighbor and never clean a critical tile either).
- **Rule 2:** The pivot p_0 is cleaned only at the last time step of the cleaning process.

3.8.3 Detailed Analysis

In this section, we shall discuss an upper bound for the cleaning time of k agents employing the **SWEEP** cleaning protocol, for some dirty region F_0. From an upper bound for the cleaning time of a swarm comprising k agents, an upper bound for the number of agents needed to ensure successful cleaning of a given region in a given number of time-steps can be derived.

The following Lemma states the change in the cardinality of a region's circumference after being traversed by an agent using the **SWEEP** protocol.

Lemma 3.3
The cardinality of the region's circumference always decreases after being traversed by an agent applying the **SWEEP** *protocol (assuming no spread occurs), namely:*

$$|\partial F'| \leq \max\{1, |\partial F| - 8\}$$

Proof 3.7 For any region F such that the number of tiles in F' is at least 2, note that a traversal around F is a closed, simple, rectilinear polygon. $\partial F'$ was obtained after deleting all the *non-critical* points of ∂F. Traversing such a polygon, an agent either goes straight, makes an internal turn ("left" turn, if we assume a clockwise movement), or makes an external turn ("right" turn). Suppose without loss of generality that an agent is moving up and making an internal turn left. Assume that there are no critical points. The path around this turn, along the newly created boundary tiles is now longer by two, as it contains an additional movement up, and another one towards the left. Similarly, an external turn, creates a new path which is shorter by two tiles (as a single horizontal movement and a single vertical movement are no longer necessary). Since ∂F is a simple rectilinear polygon, it always has four "right" turns more than "left" ones[1]. When *critical* points are met, they are not being cleaned, which means that they exist in both ∂F and $\partial F'$. Re-visiting these points, however, does not change the overall size of the set of tiles (see an example in Figure 3.31).

This proof, though, does not always hold for regions that, after being traversed once, produce either $\emptyset$ or a single tile. The reason is that the proof of the Lemma that appears in [73] relies on the fact that for each "turn" in F, there is a corresponding "turn" in F'. However, the concept of "turn" requires the existence of a possible movement of an agent from one tile to another time. This obviously cannot be done for $F' = \emptyset$ or for $F' = \{v\}$. For such regions it can easily be seen the Lemma holds, as $|\partial\emptyset| = 0$ and as $|\partial\{v\}| = 1$.

While producing a bound for the cleaning time of the agents we shall demonstrate that while being traversed by the agents using the **SWEEP** protocol, the width of the initial region F_0 decreases monotonically. The main component of this proof is presented in the following Lemma.

Lemma 3.4
Every time a region is traversed by an agent using the **SWEEP** *protocol, its width is decreased by at least one, namely:*

$$W(F_t') \leq W(F_t) - 1$$

[1] This is a simple consequence of the "rotation index" Theorem (see e.g. [39] pp. 396): If $\alpha : [0,1] \rightarrow R^2$ is a plane, regular, simple, closed curve, then $\int_0^1 k(s)ds = 2\pi$, where $k(s)$ is the curvature of $\alpha(s)$ and the curve is traversed in the positive direction.

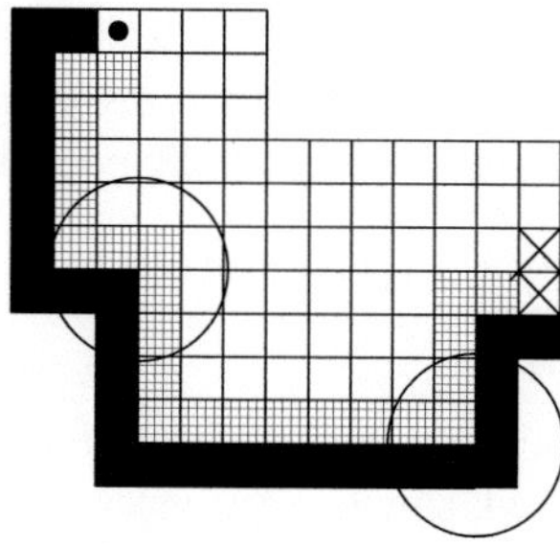

Figure 3.31: An illustration of Lemma 3.3 and of the **SWEEP** protocol. The black dot denotes the starting point of the agents cleaning F (which is the entire region presented). The X's mark the *critical points* which are not cleaned. The black tiles are some of the tiles of ∂F, cleaned by the first agent. The second layer of marked tiles represent some of the tiles of $\partial F'$, cleaned by the second agent. The effect of corners on the traversal cardinality is either an increase (marked by the left circle) or a decrease (marked by the right circle).

A corollary is that the number of tours around a region F that k agents must accomplish before $F = \emptyset$ is at most $(\frac{W(F)}{k} + 1)$.

Proof 3.8 By the definition of the depth of a tile, it is the shortest path to a *non-critical* tile in ∂F. According to the **SWEEP** protocol, after an agent had traversed F, all of the *non-critical points* in ∂F were cleaned. This is true as the agent is instructed to clean all the *non-critical points* it is going through. The only exception are tiles which an agent does not clean upon exit, as there are other agents currently located in it. However, in this case, this tile will be cleaned by the last agent leaving it, at the following time steps. Therefore, after an agent traverses the region, the depth of every internal tile was decreased by (at least) one, meaning that the total width of the region was decreased by (at least) one. As a result, as all agents are identical, when k agents complete a traversal of F, the width of F is decreased by at least k.

When examining the cleaning time of the agents, one must remember that merely calculating the length of the paths the agents must move along is not enough. Since in various scenarios the **SWEEP** protocol may direct an agent to stop and wait (for example, when several agents are located in the same tile), a way to bound the actual time it takes an agent to move along a specific path must be found. The following Lemma establishes the relation between the length of a path and the maximal time it takes an agent using the **SWEEP** protocol to move along it.

Lemma 3.5

The time it takes an agent which uses the ***SWEEP*** *protocol to move along a path* *of length c_t (including delays caused by other agents located in the same tiles) is at most $4 \cdot c_t$.*

Proof 3.9 According to the problem's specifications, each agent moves along the dirty region F at a pace of one tile per time step. The only exception may occur when several agents are delayed after entering the same tile.

According to the **SWEEP** protocol, although several agents can enter the same tile at the same time step, agents located in the same tile can leave it at the same time step only if they do so towards different directions. As a result, there arises the question whether more and more agents can enter a certain tile without leaving it, causing a decrease in the swarm's performance.

Let v be an empty tile. Let us assume without loss of generality that at some time step t, 4 agents enter v (this is of course the maximal number of agents which can enter v at the same time step, due to the invariant stating that no two agents can exit u to v at the same time step, for u, some neighbor of v, and since v has at most 4 neighbors). Thus, in time step $t-1$ v had exactly 4 neighbors in ∂F. Let us assume that in time step $t+1$ v still has 4 neighbors in ∂F. Thus, all the 4 agents which entered v intends to move to 4 different directions (since according to the **SWEEP** protocol, each of them has different *rightmost* neighbor). Thus, none of them will be delayed. Alternatively, let us assume that at time step $t+1$ v has less than 4 neighbors in ∂F. Let α denote the number of neighbors of v in ∂F. Then, since the 4 agents which entered v in time t did so from different directions, according to the **SWEEP** protocol, α of them will be able to leave v towards the α neighbors of v, and $(4-\alpha)$ will stay in v. However, in this time step, only α new agents can enter v, since v has only α neighbors in ∂F. Thus, in time step $t+2$ there are at most 4 agents in v. This is true of course for each time steps $T > t+2$. Since the number of agents in each tile is at most 4, and since each tile with agents in it has at least one neighbor of ∂F (since the agents had to enter it from a neighbor in ∂F, then there are at least $\frac{k}{4}$ agents which are able to move. Thus, even if the agents collide with each other continuously, the time it takes k agents to traverse a region of circumference c_t is at most $4 \cdot k \cdot c_t$ (and the average traversal time of an agent, amortized for the entire group of agents, is at most 4 times its minimal traversal time).

Notice that this still holds for the rare occasions in which the *check near completion of mission* procedure detects that every non-critical perimeter tile of F_t contains at least two agents. In his case, the *resting* commands are indeed overruled, allowing temporarily more than 4 agents to be located into the same tile, however — as the internal prioritization of the agents is maintained, this is corrected at the following time steps, as the "redundant" agents will wait (sometimes for more than 4 time steps). Notice though that this "long resting" was already compensated by the "untimely movement" those agents had performed while entering this tile.

It should be stated that in reality, the agents' traversal time is only slightly more than c_t, although an analytic proof is yet to be found.

When an agent using the **SWEEP** protocol is traversing F_t it does so by moving along a path which contains all the tiles of ∂F_t. Since by doing that, the agent may pass through the same tile more than once, the length of this path may be larger than the number of tiles in ∂F_t. A bound for the ratio between the two is described in the following Lemma.

Lemma 3.6
c_t, the length of the circumference of F_t never exceeds twice its cardinality, namely:

$$c_t \leq 2 \cdot |\partial F_t| - 2$$

Proof 3.10 ∂F_t is a connected graph and thus it has spanning trees. An algorithm for constructing a spanning tree for ∂F_t and a *Depth First Search* (DFS) scan of it was constructed, such that the path generated by using the DFS algorithm contains a traversal around F_t (meaning that the DFS scan generates a path which after having several of its tiles removed, equals a path which traverses F_t).

This recursive spanning tree construction algorithm appears in Figure 3.32. The algorithm receives p_s, a *non-critical* tile as its starting point and an empty list of tiles, L.[2] The algorithm constructs a spanning tree of ∂F_t as well as a DFS scan for this tree, by adding tiles to the list, while optionally marking some of them as tiles which should later be deleted. After the deletion of these tiles the remaining tiles form a path which traverses F_t. An example of the above appears in Figure 3.33.

A DFS scan of a tree can go through each edge at most twice. A tree of $|V|$ vertices contains exactly $(|V| - 1)$ edges. Thus, a DFS scan of a spanning tree of ∂F_t contains at most $2 \cdot (|\partial F_t| - 1)$ transitions of edges. Since there exists such a spanning tree that contains a 'tour' around F_t then: $c_t \leq 2 \cdot |\partial F_t| - 2$.

Definition 3.20 Let $W_{REMOVED}(t)$ denote the decrease in F's width due to the agents cleaning activity until time t.

Lemma 3.7

$$\text{If } \left\lfloor \sum_{i=1}^{t} \frac{k}{4 \cdot c_i} \right\rfloor \geq W(F_0) \text{ then } W_{REMOVED}(t) \geq W(F_0)$$

Proof 3.11 $W_{REMOVED}(t)$ can be defined as the number of completed traversals around the dirty region, multiplied by the number of the agents k. Note that at time step t, each agent completes $\frac{1}{c_t}$ of the circumference. Since we know the value of c_t for every t, and since we know that the time it takes an agent to move along a path is at most 4 times the length of this path (see Lemma 3.5), then the decrease in the

[2]The existence of a *non-critical* point is guaranteed since ∂F_t is a connected graph and thus has a spanning tree, in which at least two tiles has a degree of 1, which makes them *non-critical* tiles.

```
Add (x,y) to L and mark it as OLD
Let v denote the rightmost neighbor of (x,y)

If v ≡ p_s then
    Delete all the marked tiles from L
    STOP
End if

If (v is marked as OLD) and (no circle was formed) then
    /* The traversal repeats this tile as well */
    Call SPAN-DFS for v
    Upon return — STOP
End if

If (v is marked as OLD) and (a circle was formed) then
    /* Continue, and clean the redundant tiles */
    Add to L the sequence of tiles from L, starting with (x,y) and backwards to v
    Mark these tiles (excluding (x,y)) to be deleted
    Call SPAN-DFS for v
    Upon return — STOP
End if

/* The rightmost neighbor was not OLD */
Call SPAN-DFS for v
Upon return — STOP
```

Figure 3.32: The recursive spanning tree construction algorithm. The term *rightmost* has the same meaning as in the **SWEEP** protocol. By checking L we can find out whether continuing to an **OLD** neighbor is a "clean return" (e.g. (A,B,C,B)) or whether it completes a circle (e.g. (A,B,C,A)).

original width of the region caused by the cleaning process of the agents is bounded as described.

Note that as the expression above does not define between a multitude of agents working for a short while, and a single agent which is working for a long period of time, it may not hold for smaller values of t (in which the accumulated partial peelings is smaller than $W(F_0)$). Therefore, while merely writing $W_{REMOVED}(t) \geq \left\lfloor \sum_{i=1}^{t} \frac{k}{4 \cdot c_i} \right\rfloor$ might not always hold, the expression as appears above does.

A bound over the cleaning time of a given region F_0 can then be produced as follows:

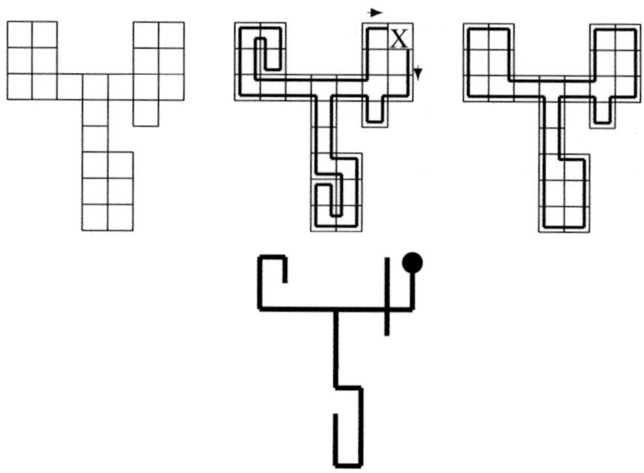

Figure 3.33: An example of the spanning tree protocol. The left chart represents the dirty region F_t. The middle chart shows the DFS scan of ∂F_t according to the **SPAN-DFS** protocol, when the big X marks p_s, the *non-critical* starting point. The right chart shows the traversal path which is contained in this DFS scan. The drawing on the bottom shows the spanning tree that is created by the protocol and is searched by the DFS algorithm.

Theorem 3.5

*Assuming that k agents start cleaning a simple connected region F_0 at some boundary point p_0 and work according to the **SWEEP** protocol, and denote by $t_{success}(k)$ the time needed for this group to clean F_0. Then it holds that:*

$$\max\left\{2k, \left\lceil \frac{S_0}{k} \right\rceil, 2l(F_0)\right\} \leq t_{success}(k) \leq \frac{8(|\partial F_0|-1)\cdot(W(F_0)+k)}{k} + 2k$$

Proof 3.12 The lower bound is quite obvious — the left term $2k$ is the time needed to release the k agents from the pivot point, $\frac{S_0}{k}$ is a lower bound on the time necessary to cover the region if the agents were optimally located at the beginning. The right term, $2l(F_0)$, comes from the observation that at least one agent should visit (and return from) any point of F_0, including the one farthest from p_0, to which the distance is $l(F_0)$.

Considering the upper bound, in order for F_0 to be cleaned, there should exist a time $t_{success}$ in which the width of the region will be 0. Note that once the dirty region was reduced to a 'skeleton' comprises only of tiles of depth zero, an additional traversal must be done in order to clean it entirely.

By combining Lemma 3.3 and 3.6 we know that for $c_j(i)$ and $\partial F_j(i)$ be the length and cardinality respectively of the i-th traversal of agent j:

$$c_j(i) \leq 2(|\partial F_j(i)|-1) \leq 2(|\partial F_0|-1)$$

Using Lemma 3.7 we see that:

$$W_{REMOVED}(t) \geq \left\lfloor \sum_{i=1}^{t} \frac{k}{4 \cdot c_i} \right\rfloor \geq \left\lfloor \frac{k \cdot t}{8(|\partial F_0| - 1)} \right\rfloor \tag{3.3}$$

Thus, we are interested in:

$$\left\lfloor \frac{k \cdot t_{success}(k)}{8(|\partial F_0| - 1)} \right\rfloor \geq W(F_0) + k \tag{3.4}$$

adding the "$+k$" for the additional traversal needed for cleaning the dirty 'skeleton'.

Since releasing the agents requires an additional $2k$ time steps, the final bound for this case is:

$$t_{success}(k) \geq \frac{8(|\partial F_0| - 1) \cdot (W(F_0) + k)}{k} + 2k$$

Using the above Theorem we can bound the *efficiency ratio*, defined as $\frac{t_{success}(k)}{S_0}$ which expresses the benefit of using k agents for a cleaning mission:

Corollary 3.1

$$\max \left\{ \frac{2k}{S_0}, \frac{1}{k}, \frac{2l(F_0)}{S_0} \right\} \leq \frac{t_{success}(k)}{S_0} \leq \frac{8(|\partial F_0| - 1) \cdot (W(F_0) + k)}{k \cdot S_0} + \frac{2k}{S_0} \tag{3.5}$$

An interesting result of Corollary 3.1 is that when $W(F_0) \ll k \ll S_0$, i.e. the number of agents is large relative to the width but small compared to the area, then the efficiency ratio is bounded below by twice the ratio of the length and area, and bounded above by $8\frac{|\partial F_0|}{S_0}$. Note here the similarity to the ratio $\frac{c_0}{\sqrt{S_0}}$, known as the *shape-factor*.

Another conclusion is that when we scale up the region by a factor of n, the area increases as n^2 but the width, length and circumference all increase as n so we get the following:

Corollary 3.2

$$as\ n \to \infty, \quad L_\infty \leq \frac{t_{success}(k)}{S} \leq U_\infty$$

where

$$L_\infty = \frac{1}{k}, \quad U_\infty = 8\frac{\partial F \cdot W}{k \cdot S} = 8\frac{|\partial F_0| \cdot W(F_0)}{k \cdot S_0}$$

and $S = S_0 n^2, \partial F = |\partial F_0| n, W = W(F_0) n$ *are the scaled area, circumference cardinality and width, respectively.*

3.8.4 Robustness

The issue of an algorithm's robustness and fault-tolerance is a major aspect of robotics systems, as the environments in which the agents operate may often include unpredicted changed from the original specification of the system, such as various kinds of noises, robots' malfunctions, etc.

While analyzing the performance of the cleaning protocol, the robotic agents were assumed to work perfectly. In real robotic system, however, this is not always the case. In simulation, however, it was quite easy to examine what happens when one or more agents disappear. The disappearance of an agent which dies is necessary in our framework in order for the system to work properly.

As discussed in Section 3.6.1, one of the requirements of the **SWEEP** protocol is a complete fault-tolerance to malfunctions in the agents, meaning — even if one or several agents stop working ("die" and disappear) the rest of the agents will continue the cleaning process as efficiently as their number allows them. The basic idea enabling this robustness is the complete independence between the agents, due to the full decentralized nature of the system (as the agents do not share the cleaning mission between them, nor do they rely on one another in the performance of their cleaning process).

A different aspect of the algorithm's robustness is the performance of the agents in noisy environments. In general, random noise can appear and influence various aspects of the system — in sensing whether a tile should be cleaned, in sensing the presence of other agents in an agent's vicinity, in the agent's movement, etc. As in almost any system, it is easy to see that severe noise may significantly affect the algorithm's efficiency, and may even prevent the agents from completion the mission altogether. For example, difficulties in sensing whether a tile should be cleaned or not, may cause an agent to clean a critical point, thus separating the region into several connected components. In addition, noises in the agents' movement system may cause the agents to detach from the contaminated region, thus delaying, or even preventing the completion of the cleaning process. Whereas a situation in which the noise level is kept at a reasonable level (i.e. correct identification of clean and contaminated tiles, and movement noises), it is the authors' opinion that the completion of the algorithm may still be achieved, although the cleaning will increase. When the agent has difficulties in the sensors in charge of detecting other agents in its vicinity, and interpreting their signaling (see relevant sections for details about signaling and synchronization) several problematic scenarios (which these signaling mechanisms were designed to avoid), may occur. However, simulations show that in such cases often only the cleaning time is affected, while the completion of the cleaning mission is preserved. Nevertheless, as this issue was not fully investigated yet, a perfect neighbors detection is still one of the algorithm's requirements.

One example for the above described problem is presented Figure 3.21, in which a failure in the sensing mechanism may damage the simple-connectivity of the contaminated region. Note that in such scenarios, there is least a single agent in each of the new contaminated components. Therefore, each component is an instance of the cooperative cleaners problem, for which the upper bound for the cleaning time may be applied.

Another example is demonstrated in Figure 3.22, where a bug in the *waiting* mechanism (see Section 3.6.6 for more details) may prevent some of the agents from cleaning contaminated tiles. The same holds for possible bugs in the *resting* mechanism (see Section 3.6.7 for more details) which may cause several agents to move together, acting as a single cleaning agent.

It is also interesting to examine the scenario in which all the robots are picked up, and re-placed on some arbitrary positions onboard the dirty region. Thanks to the definition of the SWEEP protocol, the robots will navigate to the boundary of the region, and resume their cleaning. Naturally, the overall cleaning time might be affected, however — the completion of the cleaning is still guaranteed.

3.8.5 Comparison to Existing Work

Looking at the existing work that has been done with respect to multi robotic systems designed for a cooperative cleaning/coverage/etc., several analytic results concerning the completion time of the mission can be found.

An interesting result is presented in [69]. In this work, a group of simple robots is used for periodically covering an unknown area, using the nodes counting method. Based on a more general result for undirected domains shown in [54], the following bound is given:

> The cover time of teams of ant robots (of a given size) that use node counting on strongly connected undirected graphs can be exponential in the square root of the number of vertices.

Namely:

$$f(k) = O(2^{\sqrt{S_0}}) \tag{3.6}$$

denoting the covering time of k robots by $f(k)$.

Recalling Theorem 3.5 in which:

$$t_{success}(k) \leq \frac{8(|\partial F_0| - 1) \cdot (W(F_0) + k)}{k} + 2k$$

as $|\partial F_0| = O(S_0)$ and as $W(F_0) = O(\sqrt{S_0})$ we see that:

$$t_{success}(k) = O\left(\frac{1}{k} S_0^{1.5} + S_0 + k\right)$$

As for practical reasons we assume that $k < S_0$ we can see that:

$$t_{success}(k) = O\left(\frac{1}{k}S_0^{1.5} + S_0\right)$$

and when the number of robots is independent in the size of the region, the time complexity can be reduced to:

$$t_{success}(k) = O\left(S_0^{1.5}\right)$$

Namely:

Corollary 3.3

$$\left(\frac{8(|\partial F_0|-1)\cdot(W(F_0)+k)}{k} + 2k < d\right) \Rightarrow \left(t_{success}(k) = O\left(S_0^{1.5}\right)\right)$$

Comparing this to the bound of Equation 3.6, we see that the time complexity of the **SWEEP** protocol is much smaller than this of the nodes counting technique.

It should be mentioned though, that in [69] the authors clearly state that it is their belief that the coverage time of robots using nodes counting in grids is much smaller. This estimation is also demonstrated experimentally. However, it should also be pointed out that the methods described in [69] require the robots to have some sort of *marking* capability (namely, leaving trails along the region's tiles). Such a requirement is not necessary for the **SWEEP** protocol.

3.9 Experimental Results vs. Upper Bound (Static)

A computer simulation, implementing the **SWEEP** protocol, was constructed. The protocol was run on several shapes of regions and for numbers of agents varying from 1 to 20. Figure 3.34 illustrates the evolution of the regions' layout with time. The gray level of each pixel designates the index of the agent that actually cleaned this point. The right side of Figure 3.34 shows the same region and number of agents as in the left side of this figure, but with randomly chosen initial locations at the corners. It can be seen that the dirty region is cleaned in a similar way. It should be said here that all the theory we developed in the previous sections (up to a small additive constant) applies to the case where the agents are initially located in randomly selected points on the boundary of F_0 (rather than starting from p_0).

Figure 3.35 depicts the timing results (as produced by computer simulations) describing the cleaning time as a function of the number of agents, compared to the theoretical bounds presented in previous sections (the cleaning time is normalized by the initial area of the dirty region).

Additional result are presented in Figure 3.36, comparing the various efficiency ratios obtained for initial dirty regions of various sizes. This is examined even further by Figure 3.37, which shows the behavior of $\frac{t_{success}(50)}{t_{success}(1)}$ namely, the ratio between the cleaning time of 50 agents and the cleaning time required for a single agent. The change in this ratio can be seen, hinting the fact that as the initial region is larger, the addition of more agents can more efficiently decrease the cleaning time of the agents.

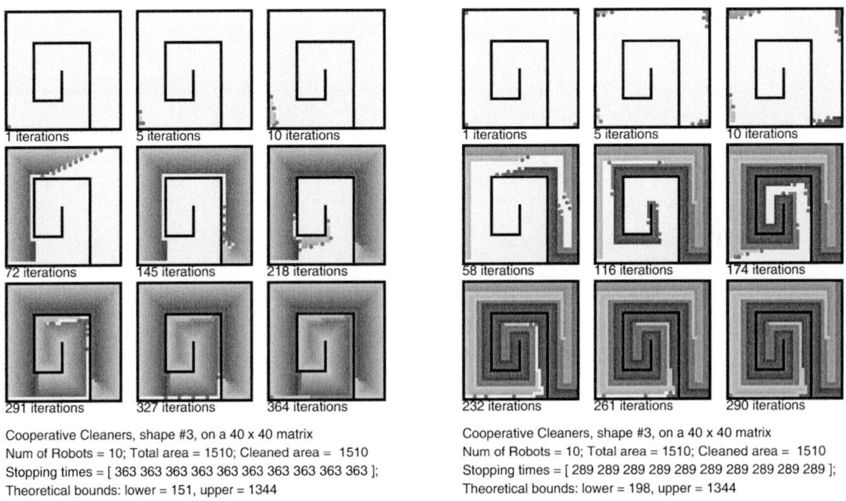

Figure 3.34: Cooperative cleaners: maze, 10 agents (left), and 10 agents with random initial locations on the boundary (right).

3.10 Upper Bound for Expanding Domains

3.10.1 Recursive Upper Bound

Due to the preservation of the *critical points*, such points can be visited many times by the agents before being cleaned. Furthermore, due to the dynamic nature of the problem, the shape of the contaminated region can change dramatically during the cleaning process.

Lemma 3.8

*Every time a region is traversed by an agent using the **SWEEP** protocol, its width is decreased by at least one (assuming no spread occurs), namely:*

$$W(F_t') \leq W(F_t) - 1$$

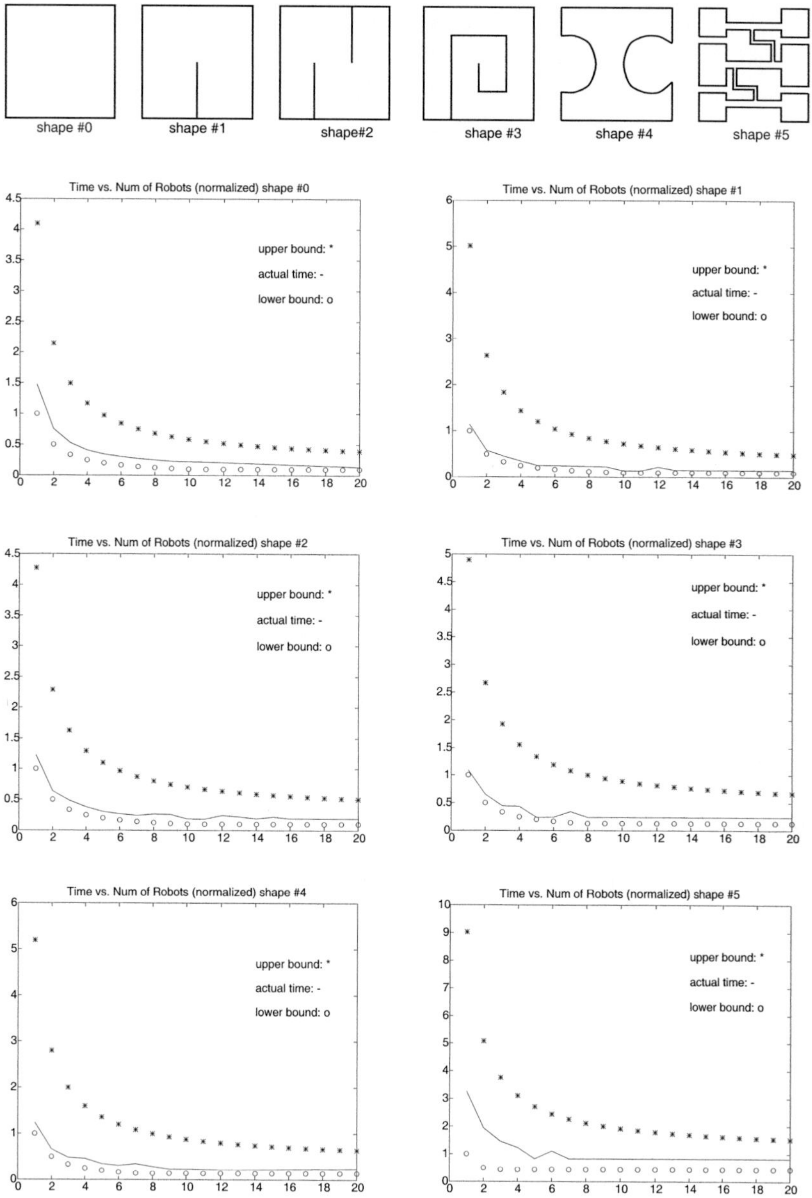

Figure 3.35: Various shapes tested in **SWEEP** simulations and their normalized cleaning time $t(k) = \frac{t_{success}(k)}{S_0}$ as a function of the number of agents, k, for each region, together with the lower and upper bounds according to Theorem 3.5. For a confidence level of 95%, based on at least 10 simulation instances per each dirty region, the appropriate confidence interval is less than $\pm 6\%$.

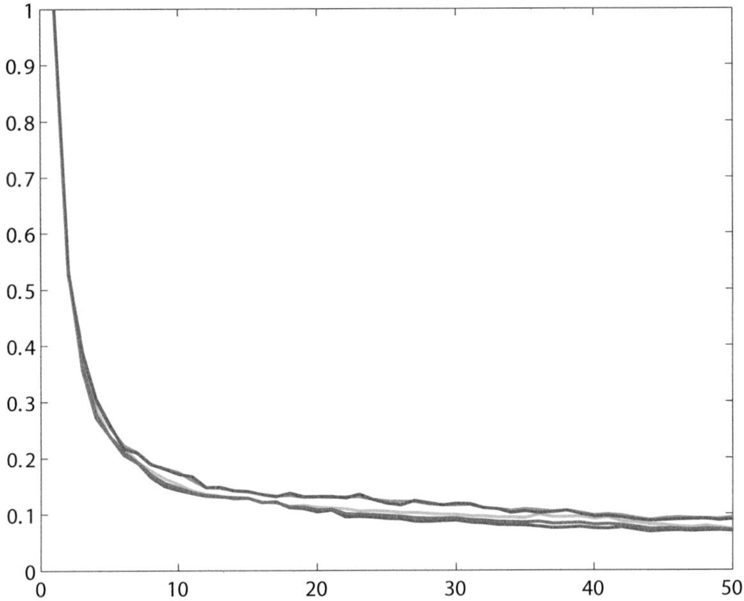

Figure 3.36: Cooperative cleaners: from 1 to 50 agents, dirty regions of sizes 600 to 1500 tiles, 10 simulations per region's size, cleaning time normalized by the cleaning time of a single agent. Note that as the initial dirty area is larger, the ratio between $t_{success}(k)$ and $t_{success}(1)$ increases. In fact, comparing the time ratios of regions of sizes 600 and 1500 tiles, the difference reaches 30%. For a confidence level of 95% the confidence interval is less than $\pm 15\%$.

Proof 3.13 By the definition of the depth of a tile, it is the shortest path to a *non-critical* tile in ∂F. According to the **SWEEP** protocol, after an agent had traversed F, all of the *non-critical points* in ∂F were cleaned. This is true as the agent is instructed to clean all the *non-critical points* it is going through. The only exception are tiles which an agent does not clean upon exit, as there are other agents currently located in it. However, in this case, this tile will be cleaned by the last agent leaving it, at the following time steps. Therefore, after an agent traverses the region, the depth of every internal tile was decreased by (at least) one, meaning that the total width of the region was decreased by (at least) one. As a result, as all agents are identical, when k agents complete a traversal of F, the width of F is decreased by at least k.

An important feature of a region F_t is its size — a bound over the contaminated region's size will later be used for producing an upper bound over the cleaning time of the agents. The following Lemma presents an upper bound over the size of the contaminated region at time t.

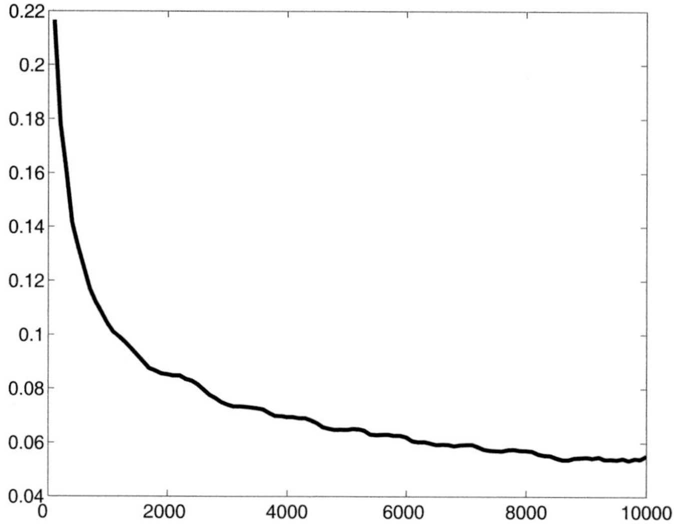

Figure 3.37: Cooperative cleaners: graph demonstrates the behavior of $\frac{t_{success}(50)}{t_{success}(1)}$ for dirty regions of sizes 100 to 10000 tiles. 10 simulations were done per region's size. Note how as the initial dirty area gets larger $\frac{t_{success}(50)}{t_{success}(1)}$ decreases. The minimal value possible is of course $\frac{1}{50} = 0.02$. For a confidence level of 90% the confidence interval is less than 10%.

Lemma 3.9
At any given time, the size of the contaminated region (i.e. $S_t = |F_t|$) can be bounded as follows:

$$S_t \leq S_0 + \eta \cdot c_0 + 2(\eta^2 + \eta) \qquad \textit{where} \qquad \eta \triangleq \left\lfloor \frac{t}{d} \right\rfloor$$

Proof 3.14 The only time clean tiles can become contaminated is when the contamination spreads. When producing an upper bound for S_t, we can therefore disregard the cleaning done by the agents, as it can never result in the contamination of clean tiles. The maximal number of new contaminated tiles is achieved when the already contaminated tiles have the maximal boundary area possible. It is easy to see that this happens when the tiles are arranged in the form of a straight line. Since F_t is connected it has at least $S_t - 1$ edges, meaning that the sum of the degrees of the tiles is at least $2(S_t - 1)$. Thus, the maximal boundary area is at most $4S_t - 2(S_t - 1)$ (four possible neighbors per contaminated tile minus the edges already connecting the contaminated tiles). In a straight line all tiles but two have two clean neighbors while two tiles have three clean neighbors. This sums up to $2(S_t - 2) + 6$ which is exactly the maximal boundary area possible.

In such a case, when the contamination spreads for the i-th time, the area of the region is increased by $(c_0 + 2 + 2\iota_i)$ when ι_i is the i-th element of the series

$(1,3,5,7,\ldots)$. Let S_{s_i} denote the area of F_t after the i-th spread. Then for any region F_t, this is bounded as follows:

$$S_{s_i} \leq S_0 + i(c_0 + 2) + 2\big(1 + 3 + 5 + \ldots + (2i - 1)\big)$$

The last can be simplified to:

$$S_{s_i} \leq S_0 + i \cdot c_0 + 2(i^2 + i) \tag{3.7}$$

Since a spread occurs every d time steps, t can be written in a form of: $t = i \cdot d + \alpha$ where i is the number of spreads and α — the remainder. Since we are only interested in the number of spreads (i.e. i), we can use:

$$i = \left\lfloor \frac{t}{d} \right\rfloor$$

By applying this to Equation (3.7) we get the requested result.

It is obvious that the number of tiles in the boundary of F_t is equal or smaller than its total size, namely:

$$\partial F_t \leq S_t$$

After combining this with Lemma 3.6 and Lemma 3.9 and with the "intra-spreads $|\partial F_t|$ preservation" property of Lemma 3.3, we produce the following bound for c_t.

Lemma 3.10
At any given time, the circumference of the contaminated region F_t is bounded as follows:

$$c_{t+1} \leq 2\Big[\Lambda_t + \eta \cdot c_0 + 2(\eta^2 + \eta) - 1\Big]$$

where

$$\eta \triangleq \left\lfloor \frac{t}{d} \right\rfloor \qquad \textit{and} \qquad \Lambda_t = \begin{cases} |\partial F_0|, & t \leq d \\ S_0, & t > d \end{cases}$$

Proof 3.15 According to Lemma 3.6, between two spreads, c_t is bounded by twice its original value (from just after the last spread). After a spread occurs, the value of c_t is being reset, according to the bound for the total size of F_t (Lemma 3.9).

It is easy to see that the width of F_0 at any time is bounded by its initial width plus the number of contamination spreads. This notion is expressed in the following Lemma.

Lemma 3.11

$$\begin{aligned} W('F_t) &\leq W(F_t)+1 \\ W(F_{t+1}) &\leq W(F_0)+\eta \qquad \textit{where} \qquad \eta \triangleq \left\lfloor \frac{t}{d} \right\rfloor \end{aligned}$$

With $W_{REMOVED}(t)$ denoting the decrease in F's width due to the agents cleaning activity until time t, we can now have the following:

Lemma 3.12

$$\text{If} \quad \left\lfloor \sum_{i=1}^{t} \frac{k}{4 \cdot c_i} \right\rfloor \geq \overline{W}(F_{t+1}) \quad \text{then} \quad W_{REMOVED}(t) \geq \overline{W}(F_{t+1})$$

where $\overline{W}(F_{t+1})$ denotes the width of F_0 at time $t+1$, assuming that no cleaning took place by the agents thus far (namely, the width predicted by Lemma 3.11).

Proof 3.16 $W_{REMOVED}(t)$ can be defined as the number of completed traversals around the contaminated region, multiplied by the number of the agents k. Note that at time step t, each agent completes $\frac{1}{c_t}$ of the circumference. Since we know the value of c_t for every t, and since we know that the time it takes an agent to move along a path is at most 4 times the length of this path (see Lemma 3.5), then the decrease in the original width of the region caused by the cleaning process of the agents is bounded as described.

Note that as the expression which appears in the Lemma does not distinguish between a multitude of agents working for a short while, and a single agent which is working for a long period of time, it may not hold for smaller values of t (in which the accumulated partial peelings is smaller than $\overline{W}(F_{t+1})$). Therefore, while writing $W_{REMOVED}(t) \geq \left\lfloor \sum_{i=1}^{t} \frac{k}{4 \cdot c_i} \right\rfloor$ might not always hold, the expression as appears above does.

Note that this holds also when taking into account the effect the spreading contamination has on the region's width. Let $p = (v_1, v_2, \ldots, v_n)$ denote a "tour", traversing a *contaminated* region F, and let us assume that a spread occurs at time $t = t_s$. Let $p = p_1 \cdot p_2$, and let p_1 denote the part of the 'tour' which took place before the contamination spread.

The width of the region F is defined as the depth of the deepest tile in F. Let us assume without loss of generality that there is only one tile of maximal depth, u (if there is more than a single deepest tile, the same observation can be applied to all deepest tiles, separately). Therefore, $W(F_t)$ is the length of the shortest *path* (or paths) from u to a *non-critical* tile in ∂F_t. Let us assume without loss of generality that there is only one such shortest path, from u to a *non-critical* boundary tile u_{END} (if there are more than a single shortest path, the same observation can be applied to all, separately). u_{END} is a boundary tile, and therefore must be included at least once in p (either in p_1 or in p_2, or in both). If $(u_{END} \in p_1 \wedge u_{END} \in p_2)$ then u_{END} is a *critical point*(and we already know this is not the case).

If $(u_{END} \in p_1)$ then it was already cleaned prior to t_s, decreasing the width of the region by 1. After the contamination spread, u_{END} might become *contaminated* once again, restoring the region's width to its original value.

If $\neg(u_{END} \in p_1)$ it therefore means that $(u_{END} \in p_2)$. As u_{END} is a boundary tile of F it has at least one $8-Neighbor$ which is clean prior to t_s. In order for the contamination spread to increase the width of F, it means that all the $8-Neighbors$ of u_{END} which were clean have become contaminated after the spread. However, as those neighbors are perimeter tiles, they will be visited by the agent. As at least one of them can be cleaned (there does not exist a tile whose $8-Neighbors$ are all critical points), u_{END} will become a perimeter tile again, restoring the original value of the width of F.

Spreads can naturally occur more than once in the middle of a 'tour'. In this case, the same principal can be applied recursively. An example of the above appears in Figure 3.38.

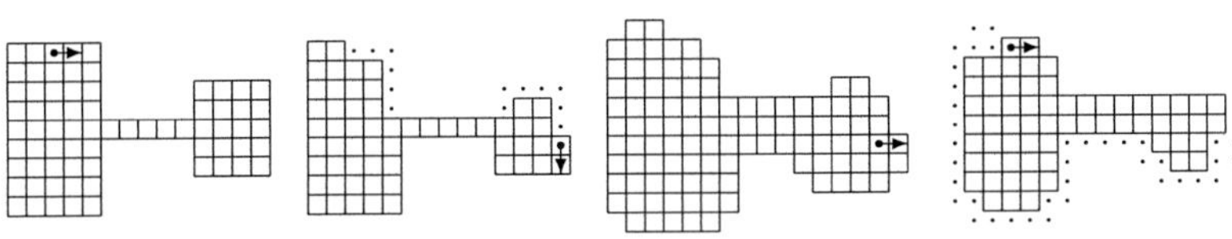

Figure 3.38: An example of an "interrupting spread" — a spread that occurs in a middle of a 'tour'. The first chart on the left is the original contaminated region, whose width is 3 (the black circle is the agent and the arrow marks its next movement). The second chart is the region just before the spread (the dots are the cleaned tiles). The third chart is the region just after the spread (notice that the width is now 4 and that the 'tour' was changed). The fourth chart is the region after the agent completes its (new) 'tour'. Notice that the width is 3 again.

An upper bound over the cleaning time of a contaminated region F_0 can now be constructed, as follows:

Theorem 3.6
*Assume that k agents start cleaning a simply connected region F_0 at some boundary point p_0 and work according to the **SWEEP** protocol, and denote by $t_{success}(k)$ the time needed for this group to clean F_0. Then it holds that:*

1. *If $(t = \frac{8(|\partial F_0|-1)\cdot(W(F_0)+k)}{k} + 2k)$ is not greater than d, then $t_{SUCCESS} = \lceil t \rceil$. This also holds for static environments, since in such cases $d \to \infty$.*

2. *Otherwise* $(t > d)$: $t_{SUCCESS}$ *is the minimal (integer)* t *for which:*

$$\sum_{i=d+1}^{t} \frac{1}{S_0 - 1 + 2\lfloor \frac{i}{d} \rfloor^2 + (c_0 + 2)\lfloor \frac{i}{d} \rfloor} \geq \gamma + \frac{8}{k} \cdot \left\lfloor \frac{t}{d} \right\rfloor$$

where

$$\gamma \triangleq \frac{8(k + W(F_0))}{k} - \frac{d - 2k}{|\partial F_0| - 1}$$

Proof 3.17 In order for F_0 to be cleaned, there should exist a time $t_{SUCCESS}$ in which the width of the region will be 0. Remembering that $W(F_t)$ is monotonically increasing and disregards the cleaning performed by the agents, and that $W_{REMOVED}(t)$ denotes the decrease in the width of F_0 due to the cleaning of the agents, we are interested in:

$$W(F_{t_{SUCCESS}}) + k \leq W_{REMOVED}(t_{SUCCESS})$$

The purpose of the "$+k$" is to guarantee the cleaning of the 'skeleton' that remains when the width of the region decreases to zero. In such a time, the agents function as a single agent, and we must guarantee that an additional traversal will be performed. We now apply Lemma 3.11 and Lemma 3.12.

Note that either $t_{SUCCESS} \leq d$ or $t_{SUCCESS} > d$. We first examine the case of $t_{SUCCESS} \leq d$. Regarding $W(F_t)$, c_t and $W_{REMOVED}(t)$ this means that:

$$\forall i \leq t \text{ , } W(F_i) \leq W(F_0)$$

$$\forall i \leq t \text{ , } c_{i+1} \leq 2(|\partial F_0| - 1)$$

$$W_{REMOVED}(t) \geq \left\lfloor \sum_{i=1}^{t} \frac{k}{4 \cdot c_i} \right\rfloor \geq \left\lfloor \frac{k \cdot t}{8|\partial F_0| - 8} \right\rfloor$$

Thus, we are interested in:

$$\left\lfloor \frac{k \cdot t_{SUCCESS}}{8(|\partial F_0| - 1)} \right\rfloor \geq W(F_0) + k$$

Since releasing the agents requires an additional $2k$ time steps, the final bound for this case is:

$$t_{SUCCESS} \geq \frac{8(|\partial F_0| - 1) \cdot (W(F_0) + k)}{k} + 2k$$

If the $t_{SUCCESS}$ received is greater than d, we continue to the next step. Lemma 3.10 provides us with c_t. As for $W_{REMOVED}(t_{SUCCESS})$, remembering that the actual cleaning begins at $t = 2 \cdot k$ (the time it takes to release the agents) we get:

$$W_{REMOVED}(t_{SUCCESS}) \geq$$

$$\geq \left\lfloor \sum_{i=2k}^{t_{SUCCESS}} \frac{k}{4 \cdot c_i} \right\rfloor \geq \left\lfloor \sum_{i=2k}^{d} \frac{k}{4 \cdot c_i} \right\rfloor + \left\lfloor \sum_{i=d+1}^{t_{SUCCESS}} \frac{k}{4 \cdot c_i} \right\rfloor \geq$$

$$\geq \left\lfloor \frac{(d-2k)\cdot k}{8(|\partial F_0|-1)} \right\rfloor + \left\lfloor \sum_{i=d+1}^{t_{SUCCESS}} \frac{k}{8\cdot(S_0+\lfloor\frac{i}{d}\rfloor\cdot c_0+2(\lfloor\frac{i}{d}\rfloor^2+\lfloor\frac{i}{d}\rfloor)-1)} \right\rfloor$$

As we already know that $W(F_{i+1}) \leq W(F_0)+\lfloor\frac{i}{d}\rfloor$ using the previous expression for $W_{REMOVED}$ yields:

$$\left\lfloor \sum_{i=2k}^{d} \frac{k}{4\cdot c_i} \right\rfloor + \left\lfloor \sum_{i=d+1}^{t} \frac{k}{4\cdot c_i} \right\rfloor \geq W(F_0)+k+\left\lfloor\frac{t}{d}\right\rfloor$$

$$\implies \quad \left\lfloor \sum_{i=d+1}^{t} \frac{k}{4\cdot c_i} \right\rfloor \geq W(F_0)+k+\left\lfloor\frac{t}{d}\right\rfloor - \frac{(d-2k)\cdot k}{8(|\partial F_0|-1)}$$

$$\implies \quad \sum_{i=d+1}^{t} \frac{1}{S_0-1+2\lfloor\frac{i}{d}\rfloor^2+(c_0+2)\lfloor\frac{i}{d}\rfloor} \geq$$

$$\geq \left(\frac{8(k+W(F_0))}{k} - \frac{d-2k}{|\partial F_0|-1}\right) + \frac{8}{k}\cdot\left\lfloor\frac{t}{d}\right\rfloor$$

In order to find such t which satisfies this expression, one should add elements to the sum iteratively, recalculating the value of the right expression. The minimal t for which the above holds is $t_{SUCCESS}$. Performing the above takes $O(t_{SUCCESS})$ time (since every iteration requires a single comparison).

Notice that the expression $t_{NoSpread}(k) \triangleq \frac{8(|\partial F_0|-1)\cdot(W(F_0)+k)}{k} + 2k$ is not monotonic. Moreover, we can see that:

$$\lim_{k\to\infty} t_{NoSpread}(k) = \infty$$

Note that given k agents, using merely a portion of them is naturally a valid strategy, should better results are to be obtained by such action. Therefore, we can state that:

Corollary 3.4

*Given a contaminated region F_0 and k cleaning agents using the **SWEEP** protocol. Let us define:*

$$\hat{k} \triangleq \sqrt{4\cdot W(F_0)\cdot(|\partial F_0|-1)}$$

and let $\hat{t}_{NoSpread}$ be defined as:

$$\min\left\{ \frac{8(|\partial F_0|-1)\cdot(W(F_0)+\lceil\hat{k}\rceil)}{\lceil\hat{k}\rceil} + 2\lceil\hat{k}\rceil, \frac{8(|\partial F_0|-1)\cdot(W(F_0)+\lfloor\hat{k}\rfloor)}{\lfloor\hat{k}\rfloor} + 2\lfloor\hat{k}\rfloor \right\}$$

If $\hat{k} < k$ and $\hat{t}_{NoSpread} \leq d$ then the completion of the cleaning mission before a contamination spread is guaranteed, and $t_{success} = \lceil\hat{t}_{NoSpread}\rceil$.

Proof 3.18 We are interested in the "effective" number of agents $\hat{k}$, which minimizes the value of $t_{NoSpread}$. Namely, a value $\hat{k}$ for which $\frac{dt_{NoSpread}}{dk}(\hat{k}) = 0$, provided that $\hat{k} \leq k$.

Theorem 3.6 is illustrated in Figures 3.39 and 3.40.

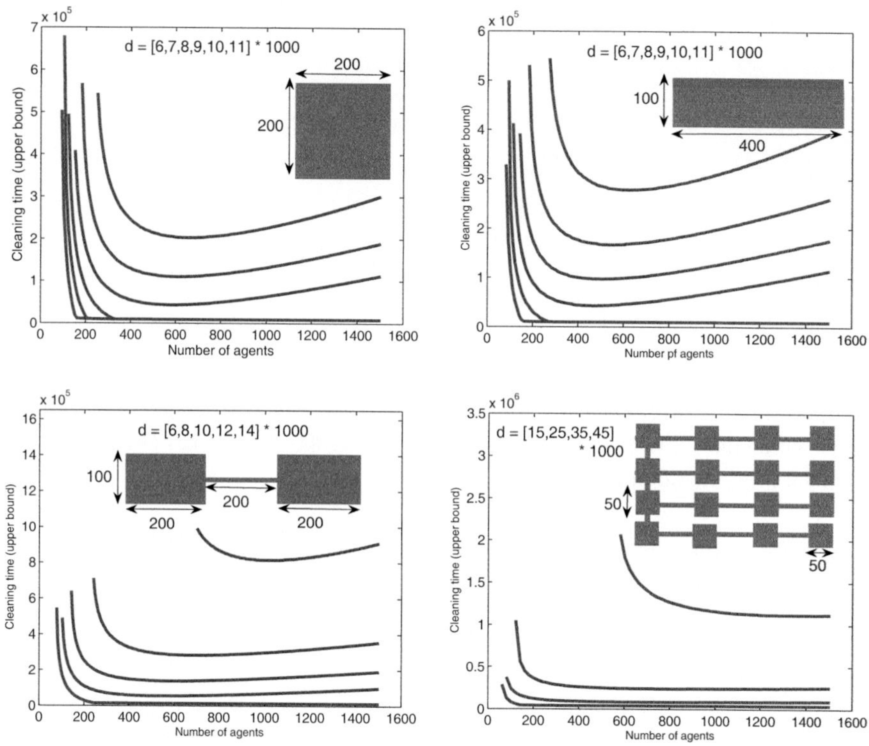

Figure 3.39: Theorem 3.6 illustrated. Upper bound over the cleaning time is shown as a function of the number of cleaning agents using the **SWEEP** protocol. All four contaminated regions shown are of the same initial size.

3.10.2 Direct Upper Bounds

The bound in Theorem 3.6 contains a recursive expression and requires an iterative process in order to calculate the minimal time $t_{SUCCESS}$ in which it is guaranteed that the agents will complete the cleaning mission. The following Theorem presents an upper bound for the cleaning time given in the form of a direct expression, whose solution is the requested value for $t_{SUCCESS}$.

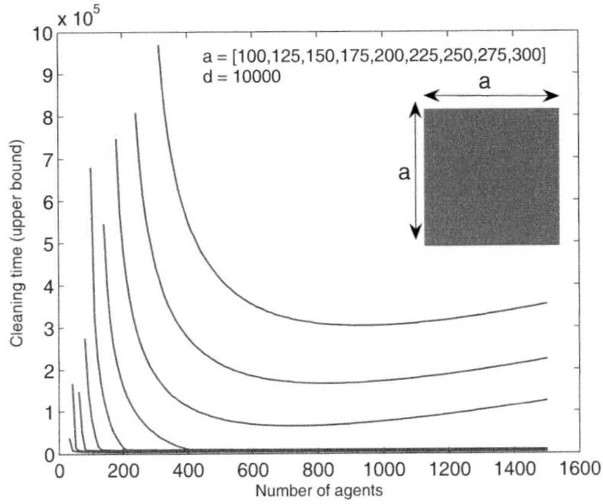

Figure 3.40: Theorem 3.6 illustrated. Upper bound over the cleaning time is shown as a function of the number of cleaning agents using the **SWEEP** protocol, for $d = 10000$ and size of initial region varying between $S_0 = 10000$ and $S_0 = 90000$ (all other geometric properties remaining the same).

Theorem 3.7

If $(t = \frac{8(|\partial F_0|-1)\cdot(W(F_0)+k)}{k} + 2k)$ *is not greater than d, then* $t_{SUCCESS} = \lceil t \rceil$. *This also holds for static environments, since in such cases* $d \to \infty$.

Otherwise $(t > d)$*: find the minimal* μ *for which:*

$$\psi\left(\mu + \frac{c_0+2-\gamma_2}{4}\right) - \psi\left(\mu + \frac{c_0+2+\gamma_2}{4}\right) + \gamma_1 - \frac{\gamma_2 \cdot \gamma}{d} - \frac{8\cdot\gamma_2}{d\cdot k}\cdot\mu \geq 0$$

where

$$\gamma_1 \triangleq \psi\left(1+\frac{c_0+2+\gamma_2}{4}\right) - \psi\left(1+\frac{c_0+2-\gamma_2}{4}\right)$$

and

$$\gamma_2 \triangleq \sqrt{(c_0+2)^2 - 8\cdot(S_0-1)} \qquad \textit{and} \qquad \gamma \triangleq \frac{8(k+W(F_0))}{k} - \frac{d-2k}{|\partial F_0|-1}$$

where $\psi()$ *is the* Digamma *function (studied in [1]) — the logarithmic derivative of the Gamma* function, *defined as:*

$$\psi(x) = \frac{d}{dx}\ln\Gamma(x) = \frac{\Gamma'(x)}{\Gamma(x)}$$

or as:

$$\psi(x) = \int_0^\infty \left(\frac{e^{-t}}{t} - \frac{e^{-xt}}{1-e^{-t}}\right)dt$$

Then $t_{SUCCESS} = \lceil \mu \cdot d \rceil$.

Proof 3.19 The first part of the Theorem is equivalent to the first part of Theorem 3.6.

For the second part, recall that in the second part of Theorem 3.6 we were interested in finding the minimal t for which the following inequality holds:

$$\sum_{i=d+1}^{t} \frac{1}{S_0 - 1 + 2\lfloor \frac{i}{d} \rfloor^2 + (c_0+2)\lfloor \frac{i}{d} \rfloor} \geq \gamma + \frac{8}{k} \cdot \left\lfloor \frac{t}{d} \right\rfloor$$

where

$$\gamma \triangleq \frac{8(k + W(F_0))}{k} - \frac{d - 2k}{|\partial F_0| - 1}$$

Let us denote $\mu \triangleq \lfloor \frac{t}{d} \rfloor$. Therefore we can now search for the minimal μ for which:

$$\sum_{i=1}^{\mu} \frac{d}{S_0 - 1 + 2 \cdot i^2 + (c_0+2) \cdot i} \geq \gamma + \frac{8}{k} \cdot \mu$$

After extracting d we get:

$$d \cdot \sum_{i=1}^{\mu} \frac{1}{S_0 - 1 + 2 \cdot i^2 + (c_0+2) \cdot i} \geq \gamma + \frac{8}{k} \cdot \mu$$

Observe that the left side of the inequality is in the form,

$$d \cdot \sum_{i=1}^{\mu} \frac{1}{A + B \cdot i + C \cdot i^2} \tag{3.8}$$

where $A = (S_0 - 1) \quad B = (c_0 + 2) \quad C = 2$.

This expression can be integrated[3] into the form of:

$$d \cdot \frac{\psi(\frac{2 \cdot C \cdot x + B - \sqrt{B^2 - 4 \cdot C \cdot A}}{2 \cdot C}) - \psi(\frac{2 \cdot C \cdot x + B + \sqrt{B^2 - 4 \cdot C \cdot A}}{2 \cdot C})}{\sqrt{B^2 - 4 \cdot C \cdot A}} \Bigg|_1^{\mu} \tag{3.9}$$

After assigning the proper values of A, B and C, we receive the following formula, where as the minimal μ which satisfies it, will be denoted as $\mu_{SUCCESS}$:

$$\frac{d}{\gamma_2} \cdot \left(\psi(\mu + \frac{c_0 + 2 - \gamma_2}{4}) - \psi(\mu + \frac{c_0 + 2 + \gamma_2}{4}) \right) + \frac{d}{\gamma_2} \cdot \gamma_1 \geq \gamma + \frac{8}{k} \cdot \mu$$

where:

$$\gamma_1 \triangleq \psi\left(1 + \frac{c_0 + 2 + \gamma_2}{4}\right) - \psi\left(1 + \frac{c_0 + 2 - \gamma_2}{4}\right)$$

and:

$$\gamma_2 \triangleq \sqrt{(c_0+2)^2 - 8 \cdot (S_0 - 1)}$$

[3]The integral was produced using MATLAB [46].

By rewriting the above, we get the following formula:

$$\psi\left(\mu+\frac{c_0+2-\gamma_2}{4}\right)-\psi\left(\mu+\frac{c_0+2+\gamma_2}{4}\right)+\gamma_1-\frac{\gamma_2\cdot\gamma}{d}-\frac{8\cdot\gamma_2}{d\cdot k}\cdot\mu\geq 0$$

where:

$$\gamma_1 \triangleq \left(\psi(1+\frac{c_0+2+\gamma_2}{4})-\psi(1+\frac{c_0+2-\gamma_2}{4})\right)$$

and:

$$\gamma_2 \triangleq \sqrt{(c_0+2)^2-8\cdot(S_0-1)}$$

and:

$$\gamma \triangleq \left(\frac{8(k+W(F_0))}{k}-\frac{d-2k}{|\partial F_0|-1}\right)$$

Since $\mu=\lfloor\frac{t}{d}\rfloor$ then $\mu>\frac{t}{d}+1$. Thus, $t_{SUCCESS}<(\mu_{SUCCESS}-1)\cdot d$.

Figure 3.41 shows an example of applying Theorem 3.7 for a cross shaped region.

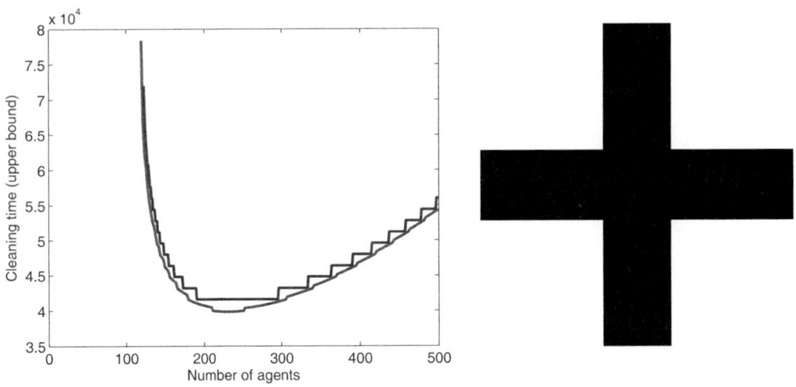

Figure 3.41: An example of Theorem 3.7. The lower curve represents a bound over the cleaning time, as produced by Theorem 3.6, for the cross region which appears on the right (for which $S_0=2960$, $c_0=326$, d = 1600). The upper curve is produced by using Theorem 3.7.

3.10.3 *Simplifying Theorem 3.7*

Although Theorem 3.7 presents a direct way of calculating an upper bound over the cleaning time of k agents using the **SWEEP** protocol, this expression is still implicit. Theorem 3.8 presents an explicit approximation of the result of Theorem 3.7.

Theorem 3.8

Let F_0 be a contaminated region for which Theorem 3.7 predicts that the contamination will spread before being cleaned. Then let $\mu_{SUCCESS} \triangleq \min\{x \in \{\mu_1, \mu_2\} | x > 0\}$ where μ_1 and μ_2 are defined as:

$$\mu_1, \mu_2 = \frac{-(A_1A_3 - A_4) \pm \sqrt{(A_1A_3 - A_4)^2 - 4A_3(A_2 - A_1 - A_1A_4)}}{2A_3}$$

where:

$$A_1 = \frac{c_0 + 2 - \gamma_2}{4}$$

$$A_2 = \frac{c_0 + 2 + \gamma_2}{4}$$

$$A_3 = \frac{8 \cdot \gamma_2}{d \cdot k}$$

$$A_4 = \gamma_1 - \frac{\gamma_2 \cdot \gamma}{d}$$

and where:

$$\gamma_1 \triangleq \psi\left(1 + \frac{c_0 + 2 + \gamma_2}{4}\right) - \psi\left(1 + \frac{c_0 + 2 - \gamma_2}{4}\right)$$

and:

$$\gamma_2 \triangleq \sqrt{(c_0 + 2)^2 - 8 \cdot (S_0 - 1)}$$

and:

$$\gamma \triangleq \frac{8(k + W(F_0))}{k} - \frac{d - 2k}{|\partial F_0| - 1}$$

If such $\mu_{SUCCESS}$ exists then:

$$t_{SUCCESS} = (\mu_{SUCCESS} - 1) \cdot d$$

Proof 3.20 From Theorem 3.7 we know that if $\mu_{SUCCESS}$ is the minimal μ for which:

$$\psi\left(\mu + \frac{c_0 + 2 - \gamma_2}{4}\right) - \psi\left(\mu + \frac{c_0 + 2 + \gamma_2}{4}\right) + \gamma_1 - \frac{\gamma_2 \cdot \gamma}{d} - \frac{8 \cdot \gamma_2}{d \cdot k} \cdot \mu \geq 0$$

(for some constant γ, γ_1 and γ_2) then:

$$t_{SUCCESS} = (\mu_{SUCCESS} - 1) \cdot d$$

Note that this expression is in the form of:

$$\psi\left(\mu + A_1\right) - \psi\left(\mu + A_2\right) - A_3 \cdot \mu + A_4 \geq 0 \tag{3.10}$$

As to the *digamma* function, we know that (Eq. 6.3.5 in [1]):

$$\psi(z + 1) = \psi(z) + \frac{1}{z}$$

Thus, we can see that:

$$\frac{A_1 - A_2}{\mu + A_1} \leq \psi\left(\mu + A_1\right) - \psi\left(\mu + A_2\right) \leq \frac{A_1 - A_2}{\mu + A_2} \tag{3.11}$$

Combining (3.10) and (3.11) we can see that we must find the minimal μ for which:

$$\frac{A_1 - A_2}{\mu + A_1} \geq A_3 \cdot \mu - A_4 \tag{3.12}$$

Meaning that we must find the minimal μ for which:

$$A_3 \cdot \mu^2 + (A_1 A_3 - A_4) \cdot \mu + (A_2 - A_1 - A_1 A_4) \leq 0 \tag{3.13}$$

Note that $A_3 > 0$ and that $\mu \in (0..\infty)$. Thus, both roots are of the same sign (where for negative roots there is no valid solution, and for positive roots the minimal of them is the solution). Roots of opposites signs are not possible, since it implies that the solution is $\mu = 0$, which in turn is impossible as $F_0 \neq \emptyset$, and as at least one contamination spread took place. μ_1 and μ_2 are defined as:

$$\mu_1, \mu_2 = \frac{-(A_1 A_3 - A_4) \pm \sqrt{(A_1 A_3 - A_4)^2 - 4A_3(A_2 - A_1 - A_1 A_4)}}{2A_3}$$

After some arithmetics, and by assigning the values of A_1, A_2, A_3 and A_4 we get the requested result.

3.10.4 An Upper Bound for k

Using the previous upper bound for the time it takes k agents to clean an initial contaminated region of known geometric properties, a similar bound for the number of agents required to guarantee a successful cleaning of the region (in a finite amount of time) can be produced.

Theorem 3.9
Given a contaminated region of properties S_0, $|\partial F_0|$ and $W(F_0)$, spreading every d time steps, then in order to guarantee a successful cleaning of the region before the contamination is able to spread even once, the minimal number of cleaning agents required is upper bounded as follows:

$$(k_1 \leq k \leq k_2) \wedge (k > 0)$$

where:

$$k_{1,2} = 2\left(-|\partial F_0| + 1 + \frac{d}{8}\right) \pm 2\sqrt{\left(|\partial F_0| - 1 - \frac{d}{8}\right)^2 - (|\partial F_0| - 1)W(F_0)}$$

Proof 3.21 Let us use the first part of Theorem 3.7:

$$\frac{8(|\partial F_0| - 1) \cdot (W(F_0) + k)}{k} + 2k \leq d$$

After some arithmetics we see that:

$$2k^2 + (8|\partial F_0| - 8 - d)k + (8|\partial F_0| - 8)W(F_0) \leq 0$$

After extracting the value of k we get:

$$k_{1,2} = \frac{-(8|\partial F_0| - 8 - d) \pm \sqrt{(8|\partial F_0| - 8 - d)^2 - 8(8|\partial F_0| - 8)W(F_0)}}{4}$$

and the rest is implied.

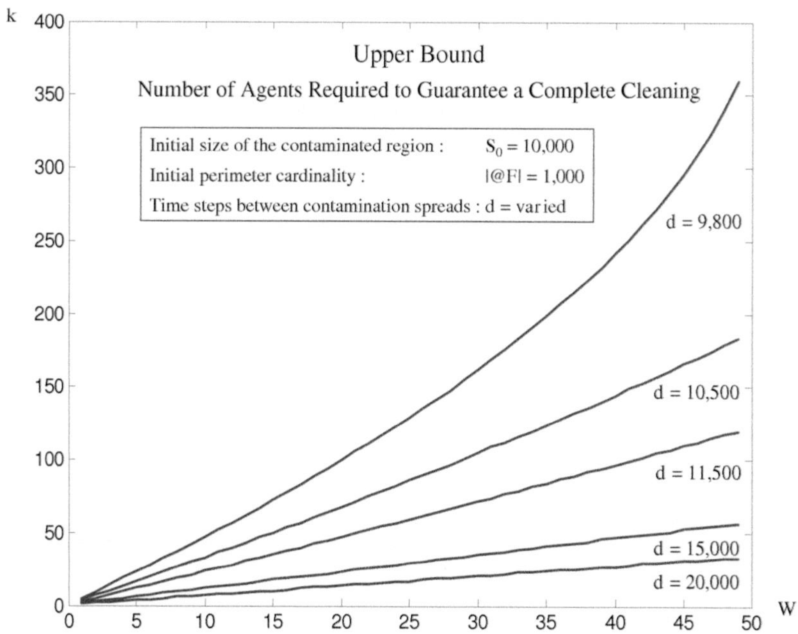

Figure 3.42: Theorem 3.9 illustrated. The graph presents an upper bound over the number of agents needed to guarantee a successful cleaning of a region of initial size of 10,000 tiles and a perimeter cardinality of 1,000 tiles, as a function of the width of the region (namely — its "bulkiness"). Notice how the effect this geometric feature has on the number of agents required increases dramatically as the time intervals between contamination spreads decreases.

3.11 Complexity Analysis

We note again that when discussing the coverage of regions on the grid, either static or expanding, it is enough to show that the region can be cleaned by the

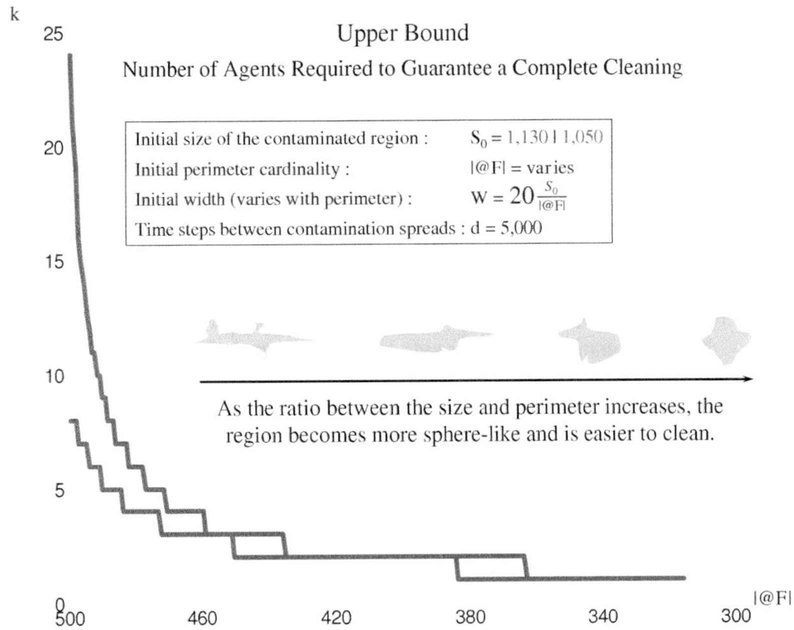

Figure 3.43: Theorem 3.9 illustrated. Observe how as $\frac{S_0}{\partial F_0}$ increases (namely, the initial region is "bulkier"), it becomes easier to guarantee the proper cleaning of the region with a smaller number of agents. The width of the initial region was selected to reflect a reasonable proportion considering the changing cardinality value of the perimeter ∂F_0.

team of robots, as clearly the cleaned sites are always a subset of the visited ones. We first present the cover time of a group of robots operating in non-expanding domains, using the **SWEEP** algorithm.

Theorem 3.10
Given a connected region of $S_0 = n$ grid tiles and perimeter c_0, that should be covered by a team of k ant-like robots, the robots can cover it using $O\left(\frac{1}{k}S_0^{1.5} + S_0\right)$ time.

Proof 3.22 Since $|\partial F_0| = \Theta(c_0)$, and $W(F_0) = O(\sqrt{S_0})$, recalling Theorem 3.5 we can see that:

$$t_k(n) = t_{static}(k) = O\left(\frac{1}{k}\sqrt{S_0} \cdot c_0 + c_0 + k\right)$$

As $c_0 = O(S_0)$ and as for practical reasons we assume that $k < n$ this would equal in the worst case to:

$$t_k(n) = t_{static}(k) = O\left(\frac{1}{k}S_0^{1.5} + S_0\right)$$

As $c_0 \geq \sqrt{S_0}$ then $\frac{c_0^2}{dk^2} \geq \frac{c_0\sqrt{S_0}}{dk^2}$. In addition, $1 \geq \frac{1}{d}$ and also $1 \geq \frac{\ln S_0}{c_0}$ and $1 \geq \frac{\ln S_0}{k}$ (as Eq. 3.19 shows that $k \geq \ln S_0$). In order to have also $1 \geq \frac{c_0}{dk}$ we must have:

$$d \cdot k = \Omega(c_0) \tag{3.17}$$

In addition, we should also require that the result μ would be positive (as it denotes the coverage time). Namely, that:

$$A_4 + \sqrt{(A_1A_3 - A_4)^2 - 4A_3(A_2 - A_1 - A_1A_4)} > A_1A_3$$

For this to hold we shall merely require that:

$$A_2 - A_1 - A_1A_4 \leq 0$$

(as A_3 is known to be positive). Assigning the values of A_1, A_2, A_4, this translates to:

$$c_0 + c_0^2 \frac{\gamma}{d} \leq O(c_0\gamma_1)$$

Dividing by c_0 we can write:

$$1 + c_0 \frac{1 + \frac{\sqrt{S_0}}{k} - \frac{d}{c_0} + \frac{k}{c_0}}{d} \leq O(\gamma_1)$$

Namely:

$$c_0 + \frac{c_0\sqrt{S_0}}{k} + k \leq dO(\gamma_1)$$

As c_0 is the dominant element of the left side of the inequation, we see that:

$$d = \Omega\left(\frac{c_0}{\gamma_1}\right) \tag{3.18}$$

Assigning this lower bound for d we can now see that:

$$\Omega(\sqrt{S_0}) \leq k \leq O(c_0) \tag{3.19}$$

Therefore, we shall select the value of k such that:

$$k = \Theta(\sqrt{S_0})$$

This also satisfies Equation 3.17.

Theorem 3.14

*The time it takes a group of $k = \Theta(\sqrt{S_0})$ robots using the **SWEEP** algorithm to cover a connected region of size S_0 tiles, that expands every $d = \Omega(\frac{c_0}{\gamma_1})$ time steps, is upper bounded as follows:*

$$t_{SUCCESS} = O\left(S_0^2 \ln S_0\right)$$

where γ_1 shifts from $O(1)$ to $O(\ln S_0)$ as c_0 grows from $O(\sqrt{S_0})$ to $O(S_0)$, defined as:

$$\gamma_1 = \psi\left(1+\frac{c_0+2+\gamma_2}{4}\right) - \psi\left(1+\frac{c_0+2-\gamma_2}{4}\right)$$

$$\gamma_2 = \sqrt{(c_0+2)^2 - 8S_0 + 8}$$

Proof 3.28 Recalling Theorem 3.5 we know that if:

$$\frac{8(|\partial F_0|-1)\cdot(W(F_0)+k)}{k} + 2k < d$$

then the robots could clean the region before it expands even once. In this case, the cleaning time would be $O(\frac{1}{k}\sqrt{S_0}\cdot c_0 + c_0)$ as was shown in Theorem 3.10. Therefore, we shall assume that:

$$\frac{8(|\partial F_0|-1)\cdot(W(F_0)+k)}{k} + 2k \geq d \tag{3.20}$$

Observing Theorem 3.8 we see that:

$t_{SUCCESS} =$

$$d\cdot O\left(A_1 + \frac{|A_4|}{A_3} + \sqrt{A_1^2 + \frac{|A_1A_4|+A_1+A_2}{A_3} + \frac{A_4^2}{A_3^2}}\right) \leq$$

$$d\cdot O\left(A_1 + \frac{|A_4|}{A_3} + \frac{\sqrt{A_3}\sqrt{|A_1A_4|+A_1+A_2}}{A_3} + \frac{|A_4|}{A_3}\right) \leq$$

$$d\cdot O\left(A_1 + \frac{|A_4|}{A_3} + \frac{\sqrt{|A_1A_4|}+\sqrt{A_1}+\sqrt{A_2}}{\sqrt{A_3}}\right) =$$

$$d\cdot O\left(\begin{array}{c} c_0+\gamma_2+dk\frac{\gamma_1}{\gamma_2}+k\gamma+ \\ \sqrt{k}\frac{\sqrt{c_0+\gamma_2}\sqrt{d\gamma_1+\gamma_2\cdot\gamma}}{\sqrt{\gamma_2}} + \sqrt{\frac{kd}{\gamma_2}}\sqrt{c_0+\gamma_2} \end{array}\right)$$

Using the fact that $\gamma_2 = \Theta(c_0)$ (Lemma 3.13) we can rewrite this expression as:

$$d\cdot O\left(c_0 + dk\frac{\gamma_1}{c_0} + k\gamma + \sqrt{k}\sqrt{d\gamma_1 + c_0\gamma} + \sqrt{kd}\right) \tag{3.21}$$

Recalling Equation 3.20, and as $W(F_0) = O(\sqrt{S_0})$, we can see that:

$$d = O\left(\frac{\sqrt{S_0}\cdot c_0}{k} + c_0 + k\right)$$

Therefore, $|\gamma|$ can now be written as:

$$|\gamma| = O\left(\frac{\sqrt{S_0}}{k} + \sqrt{S_0} + \frac{k}{\sqrt{S_0}} + 1\right)$$

In addition, remembering that $O(\sqrt{S_0}) \leq c_0 \leq O(S_0)$ we can rewrite the expression of Equation 3.21 as follows:

$$d \cdot O\left(\frac{c_0 + dk\frac{\gamma_1}{c_0} + k\sqrt{S_0} + \frac{k^2}{\sqrt{S_0}} +}{\sqrt{kc_0}\sqrt{\frac{d}{c_0}\gamma_1 + \frac{\sqrt{S_0}}{k} + \sqrt{S_0} + \frac{k}{\sqrt{S_0}}} + \sqrt{kd}}\right) =$$

$$d \cdot O\left(\frac{c_0 + dk\frac{\gamma_1}{c_0} + k\sqrt{S_0} + \frac{k^2}{\sqrt{S_0}} +}{\sqrt{kc_0}\left(\sqrt{\gamma_1} + \sqrt[4]{S_0}\sqrt{\frac{\gamma_1}{k}} + \sqrt[4]{S_0} + \frac{\sqrt{k}}{\sqrt[4]{S_0}}\right) + \sqrt{kd}}\right)$$

Using Lemma 3.14 we see that:

$$d \cdot O\left(\frac{c_0 + dk\frac{\ln S_0}{c_0} + k\sqrt{S_0} + \frac{k^2}{\sqrt{S_0}} +}{\sqrt{c_0 \ln S_0 \sqrt{S_0}} + \sqrt{c_0 k \sqrt{S_0}} + k\sqrt[4]{S_0} + \sqrt{kd}}\right) =$$

$$d \cdot O\left(\frac{c_0 + k\sqrt{S_0}\ln S_0 + \frac{k^2}{\sqrt{S_0}} +}{\sqrt{c_0 \ln S_0 \sqrt{S_0}} + \sqrt{c_0 k \sqrt{S_0}}}\right) \tag{3.22}$$

Assuming that $k > O(\ln S_0)$ and as $c_0 = O(S_0)$ we can now write:

$$O\left(\begin{array}{c}\frac{S_0^{2.5}}{k} + S_0^2 \ln S_0 + S_0^{1.5} k \ln S_0 + k^2\sqrt{S_0}\ln S_0 + \\ \frac{k^3}{\sqrt{S_0}} + \frac{S_0^{2.25}}{\sqrt{k}} + S_0^{1.75}\sqrt{k} + S_0^{0.75}k^{1.5}\end{array}\right) \tag{3.23}$$

Using Equation 3.19 we can see that this translates to:

$$O\left(S_0^{1.5} k \ln S_0\right) \tag{3.24}$$

3.12 Conclusion

In this chapter we have discussed the problem of cleaning or covering a connected region on the grid using a collaborate team of simple, finite-state-automata robotic agents. This problem has several interesting applications, some of which are for example an autonomous patrolling of a pre-defined area using a swarm of collaborative drones [12, 13, 17], or an implementation of a distributed anti-virus system for computer networks [16,50]. Additional applications are distributed search engines [33], as well as the coordination of fire-fighting units (see [38] for details).

We have shown that when the regions are static, this can be done in $O(\frac{1}{k}\sqrt{n} \cdot c_0 + c_0 + k)$ time which equals $O(\frac{1}{k}n^{1.5} + n)$ time in the worst case, thus improving the previous results for this problem. In addition, we have shown that when the region is expanding at a constant rate (which is "slow enough"), a team of $\Theta(\sqrt{n})$ robots can still be guaranteed to clean or cover it, in $O(n^2 \ln n)$ time.

In addition, we have shown that teams of less than $\Omega(\sqrt{n})$ robots can *never* cover a region that expands every $O(1)$ time steps, regardless of their sensing capabilities, communications and memory resources employed, or the algorithm used. As to regions that expand slower than every $O(1)$ time steps, we have shown the following impossibility results. First, a region of n tiles that expands every d time steps cannot be covered by a group of k agents if $dk \leq O(\sqrt{n})$. Using Theorem 3.13 we can guarantee a coverage when $dk = \Omega(\frac{n^{1.5}}{\ln n})$, or even for $dk = \Omega(n))$ (when the region's perimeter c_0 equals $O(n)$).

Second, a spreading region cannot be covered when d^2k is smaller than the order of the perimeter of the bounding rectangle of the region (which equals $O(n)$ in the worst case and $O(\sqrt{n})$ for shapes with small perimeters). Using Theorem 3.13 we can guarantee a coverage when $d^2k = \Omega(\frac{n^{2.5}}{\ln^2 n})$, or for $d^2k = \Omega(n^{1.5})$ (when the region's perimeter c_0 equals $O(n)$).

We believe that these results can be easily applied to various other problems that can be reduced or modeled as a swarm of autonomous agents that are required to operate in an expanding grid domain. For example, this result can show that a team of $\Theta(\sqrt{n})$ cops can always catch a robber (or for that matter — a group of robbers), moving (slowly) in an unbounded grid. Alternatively, robbers can be guaranteed to escape a team of less than $O(\sqrt{n})$ cops, if the area they are located in is unbounded.

As in the case of the static variant of the problem, the question of how fast can k drones scan a pre-defined expanding region, is *NP-hard* (for any number of drones, as this can be reduced to finding a Hamilton path in a non-simple grid-graph [49]). Various effects on the completion time as a function of the region's geometric properties are analyzed in [7, 8, 29], and upper bounds are discussed in [10]. A stochastic variant of this problem, involving a probabilistic analysis of the search problem in domains that spread nondeterministically is discussed in [18].

References

[1] M. Abramowitz and I.A. Stegun. *Handbook of Mathematical Functions*, page 55. Applied Mathematics Series. National Bureau of Standards, 1964.

[2] E.U. Acar, Y. Zhang, H. Choset, M. Schervish, A.G. Costa, et al. Path planning for robotic demining and development of a test platform. In: *International Conference on Field and Service Robotics*, pages 161–168, 2001.

[3] M. Ahmadi and P. Stone. A multi-robot system for continuous area sweeping tasks. In: *IEEE International Conference on Robotics and Automation, 2006 (ICRA 2006)*, 2006.

[4] S. Alpern and S. Gal. *The Theory of Search Games and Rendezvous*. Kluwer Academic Publishers, 2003.

[5] Tal Altshuler, Yaniv Altshuler, Rachel Katoshevski, and Yoram Shiftan. Modeling and prediction of ride-sharing utilization dynamics. *Journal of Advanced Transportation*, 2019, 2019.

[6] Y. Altshuler, A.M. Bruckstein and I.A. Wagner. Swarm robotics for a dynamic cleaning problem. In: *IEEE Swarm Intelligence Symposium*, pages 209–216, 2005.

[7] Y. Altshuler, I.A. Wagner and A.M. Bruckstein. Shape factor‘s effect on a dynamic cleaners swarm. In: *Third International Conference on Informatics in Control, Automation and Robotics (ICINCO), the Second International Workshop on Multi-Agent Robotic Systems (MARS)*, pages 13–21, 2006.

[8] Y. Altshuler, I.A. Wagner and A.M. Bruckstein. On swarm optimality in dynamic and symmetric environments. Volume 7, page 11, 2008.

[9] Y. Altshuler, I.A. Wagner and A.M. Bruckstein. Collaborative exploration in grid domains. In: *Sixth International Conference on Informatics in Control, Automation and Robotics (ICINCO)*, 2009.

[10] Y. Altshuler, I.A. Wagner, V. Yanovski and A.M. Bruckstein. Multi-agent cooperative cleaning of expanding domains. *International Journal of Robotics Research*, 30: 1037–1071, 2010.

[11] Y. Altshuler, V. Yanovski, D. Vainsencher, I.A. Wagner and A.M. Bruckstein. On minimal perimeter polyminoes. In: *The 13th International Conference on Discrete Geometry for Computer Imagery (DGCI2006)*, pages 17–28, 2006.

[12] Y. Altshuler, V. Yanovski, I.A. Wagner and A.M. Bruckstein. The cooperative hunters - efficient cooperative search for smart targets using uav swarms. In: *Second International Conference on Informatics in Control, Automation and Robotics (ICINCO), the First International Workshop on Multi-Agent Robotic Systems (MARS)*, pages 165–170, 2005.

[13] Y. Altshuler, V. Yanovsky, A.M. Bruckstein and I.A. Wagner. Efficient cooperative search of smart targets using uav swarms. *ROBOTICA*, 26: 551–557, 2008.

[14] Y Altshuler and A Bruckstein. On short cuts-or-fencing in rectangular strips. *arXiv preprint arXiv:1011.5920*, 2010.

[15] Y Altshuler and A.M. Bruckstein. Static and expanding grid coverage with ant robots: Complexity results. *Theoretical Computer Science*, 412(35): 4661–4674, 2011.

[16] Y Altshuler, S Dolev, Y Elovici and N Aharony. Ttled random walks for collaborative monitoring. In *NetSciCom 2010 (Second International Workshop on Network Science for Communication Networks)*, San Diego, CA, USA, 3 2010.

[17] Y. Altshuler, A. Pentland, S. Bekhor, Y. Shiftan and A. Bruckstein. Optimal dynamic coverage infrastructure for large-scale fleets of reconnaissance uavs. *arXiv preprint arXiv: 1611.05735*, 2016.

[18] Y. Altshuler, A. Pentland and A.M. Bruckstein. Collaborative patrolling swarms in stochastically expanding environments. In: *Swarms and Network Intelligence in Search*, pages 155–185. Springer, 2018.

[19] Y. Altshuler, A. Pentland and A.M. Bruckstein. The cooperative hunters–efficient and scalable drones swarm for multiple targets detection. In: *Swarms and Network Intelligence in Search*, pages 187–205. Springer, 2018.

[20] Y. Altshuler, A. Pentland and A.M. Bruckstein. Cooperative "swarm cleaning" of stationary domains. In: *Swarms and Network Intelligence in Search*, pages 15–49. Springer, 2018.

[21] Y. Altshuler, A. Pentland and A.M. Bruckstein. Introduction to swarm search. In: *Swarms and Network Intelligence in Search*, pages 1–14. Springer, 2018.

[22] Y. Altshuler, A. Pentland and A.M. Bruckstein. Optimal dynamic coverage infrastructure for large-scale fleets of reconnaissance uavs. In: *Swarms and Network Intelligence in Search*, pages 207–238. Springer, 2018.

[23] Y. Altshuler, A. Pentland and A.M. Bruckstein. The search complexity of collaborative swarms in expanding z^2 grid regions. In: *Swarms and Network Intelligence in Search*, pages 129–153. Springer, 2018.

[24] Y. Altshuler, A. Pentland and A.M. Bruckstein. Swarm search of expanding regions in grids: Lower bounds. In: *Swarms and Network Intelligence in Search*, pages 51–89. Springer, 2018.

[25] Y. Altshuler, A. Pentland and A.M. Bruckstein. Swarm search of expanding regions in grids: Upper bounds. In: *Swarms and Network Intelligence in Search*, pages 91–127. Springer, 2018.

[26] Y. Altshuler, A. Pentland and A.M. Bruckstein. *Swarms and Network Intelligence in Search.* Springer, 2018.

[27] Y. Altshuler, R. Puzis, Y. Elovici, S. Bekhor and A. Sandy Pentland. On the rationality and optimality of transportation networks defense: a network centrality approach. *Securing Transportation Systems*, pages 35–63.

[28] Y. Altshuler, E. Shmueli, G. Zyskind, O. Lederman, N. Oliver, et al. Campaign optimization through behavioral modeling and mobile network analysis. *Computational Social Systems, IEEE Transactions on*, 1(2):121–134, 2014.

[29] Y. Altshuler, I.AWagner and A.M. Bruckstein. On swarm optimality in dynamic and symmetric environments. *economics*, 7:11, 2008.

[30] R.B. Yates and R. Schott. Parallel searching in the plane. *Computational Geometry*, 5: 143–154, 1995.

[31] M.A. Batalin and G.S. Sukhatme. Spreading out: A local approach to multi-robot coverage. In: *6th International Symposium on Distributed Autonomous Robotics Systems*, 2002.

[32] R. Bejar, B. Krishnamachari, C. Gomes and B. Selman. Distributed constraint satisfaction in a wireless sensor tracking system. In: *Proceedings of the IJCAI-01 Workshop on Distributed Constraint Reasoning*, 2001.

[33] M. Bender, S. Michel, P. Triantafillou, G. Weikum and C. Zimmer. Minerva: collaborative p2p search. In: *Proceedings of the 31st international conference on Very large data bases*, pages 1263–1266, 2005.

[34] R. Borie, C. Tovey and S. Koenig. Algorithms and complexity results for pursuit-evasion problems. In: *Proceedings of the International Joint Conference on Artificial Itelligence (IJCAI)*, pages 59–66, 2009.

[35] V. Braitenberg. *Vehicles.* MIT Press, 1984.

[36] A.Z. Broder, A.R. Karlin, P. Raghavan and E. Upfal. Trading space for time in undirected $s-t$ connectivity. In: *ACM Symposium on Theory of Computing (STOC)*, pages 543–549, 1989.

[37] Z. Butler, A. Rizzi and R. Hollis. Distributed coverage of rectilinear environments. In: *Proceedings of the Workshop on the Algorithmic Foundations of Robotics*, 2001.

[38] D.W. Casbeer, D.B. Kingston, R.W. Beard and T.W. McLain. Cooperative forest fire surveillance using a team of small unmanned air vehicles. *International Journal of Systems Science*, 37(6): 351–360, 2006.

[39] M.P. Do-Carmo. Differential geometry of curves and surfaces. In: *Prentice-Hall, New-Jersey*, 1976.

[40] G. Dudek, M. Jenkin, E. Milios and D. Wilkes. Robotic exploration as graph construction. *IEEE Transactions on Robotics and Automation*, 7: 859–865, 1991.

[41] Y. Elovici and Y. Altshuler. Introduction to security and privacy in social networks. *Security and Privacy in Social Networks*, pages 1–6.

[42] J.O. Flynn. Lion and man: the general case. *SIAM Journal of Control*, 12: 581–597, 1974.

[43] Y. Gabriely and E. Rimon. Competitive on-line coverage of grid environments by a mobile robot. *Computational Geometry*, 24: 197–224, 2003.

[44] A.S. Goldstein and E.M. Reingold. The complexity of pursuit on a graph. *Theoretical Computer Science*, 143: 93–112, 1995.

[45] D. Henrich. Space efficient region filling in raster graphics. *The Visual Computer*, 10: 205–215, 1994.

[46] The MathWorks Inc. Matlab—the language of technical computing, ver. 6.5, 2002.

[47] R.P. Isaacs. *Differential Games: A Mathematical Theory with Applications to Warfare and Pursuit, Control and Optimization*. John Wiley and Sons, Reprinted by Dover Publications 1999, 1965.

[48] V. Isler, S. Kannan and S. Khanna. Randomized pursuit-evasion with local visibility. *SIAM Journal of Discrete Mathematics*, 20: 26–41, 2006.

[49] A. Itai, C.H. Papadimitriou and J.L. Szwarefiter. Hamilton paths in grid graphs. *SIAM Journal on Computing*, 11: 676–686, 1982.

[50] R. Janakiraman, M. Waldvogel and Q. Zhang. Indra: A peer-to-peer approach to network intrusion detection and prevention. In: *Proceedings of IEEE WETICE*, 2003.

[51] Johan Jonasson and Oded Schramm. On the cover time of planar graphs. *Electron. Comm. Probab.*, 5: 85–90 (electronic), 2000.

[52] W. Kerr and D. Spears. Robotic simulation of gases for a surveillance task. In: *Intelligent Robots and Systems (IROS 2005)*, pages 2905–2910, 2005.

[53] S. Koenig and Y. Liu. Terrain coverage with ant robots: A simulation study. In: *AGENTS'01*, May 2001.

[54] S. Koenig, B. Szymanski and Y. Liu. Efficient and inefficient ant coverage methods. *Annals of Mathematics and Artificial Intelligence*, 31: 41–76, 2001.

[55] S. Koenig, M. Likhachev and X. Sun. Speeding up moving-target search. In: *AAMAS '07: Proceedings of the 6th international joint conference on Autonomous agents and multiagent systems*, pages 1–8, New York, NY, USA, 2007. ACM.

[56] B. Koopman. *Search and Screening: General Principles with Historical Applications*. Pergamon Press, 1980.

[57] B.O. Koopman. The theory of search ii, target detection. *Operations Research*, 4(5): 503–531, 1956.

[58] R. Korf. Real-time heuristic search. *Artificial Intelligence*, 42: 189–211, 1990.

[59] D. Latimer, S. Srinivasa, V. Lee-Shue, S. Sonne, H. Choset, et al. Toward sensor based coverage with robot teams. In: *Proceedings of the 2002 IEEE International Conference on Robotics and Automation*, 2002.

[60] L. Lovasz. Random walks on graphs: A survey. *Combinatorics, Paul Erdos is Eighty*, 2: 353—-398, 1996.

[61] T.W. Min and H.K. Yin. A decentralized approach for cooperative sweeping by multiple mobile robots. In: *IEEE/RSJ International Conference on Intelligent Robots and Systems*, pages 380–385, 1998.

[62] P.M. Morse and G.E. Kimball. *Methods of operations research*. MIT Press and New York: Wiley, 1951.

[63] K. Passino, M. Polycarpou, D. Jacques, M. Pachter, Y. Liu, et al. *Cooperative Control for Autonomous Air Vehicles*, chapter Cooperative Control and Optimization. Kluwer Academic, Boston, 2002.

[64] M. Polycarpou, Y. Yang and K. Passino. A cooperative search framework for distributed agents. In *IEEE International Symposium on Intelligent Control*, pages 1–6, September 2001.

[65] R. Puzis, Y. Altshuler, Y. Elovici, S. Bekhor, Y. Shiftan, et al. Augmented betweenness centrality for environmentally-aware traffic monitoring in transportation networks. *Journal of Intelligent Transportation Systems*, 17(0):91—105, 2013.

[66] I. Rekleitis, V. Lee-Shuey, A. Peng Newz and H. Choset. Limited communication, multi-robot team based coverage. In: *IEEE International Conference on Robotics and Automation*, April 2004.

[67] E. Shmueli, I. Mazeh, L. Radaelli, A.S. Pentland and Y. Altshuler. Ride sharing: A network perspective. In: *Social Computing, Behavioral-Cultural Modeling, and Prediction*, pages 434–439. Springer, 2015.

[68] L.D. Stone. *Theory of Optimal Search*. Academic Press, New York, 1975.

[69] J. Svennebring and S. Koenig. Building terrain-covering ant robots: A feasibility study. *Autonomous Robots*, 16(3): 313–332, 2004.

[70] B. Szymanski and S. Koenig. The complexity of node counting on undirected graphs. *Technical Report, Computer Science Department, Rensselaer Polytechnic Institute, Troy, NY*, 1998.

[71] A. Thorndike. Summary of antisubmarine warfare operations in world war ii. Summary report, NDRC Summary Report, 1946.

[72] Sebastian B. Thrun. Efficient exploration in reinforcement learning—technical report cmu-cs-92-102. Technical report, Carnegie Mellon University, 1992.

[73] I.A. Wagner, Y. Altshuler, V. Yanovski and A.M. Bruckstein. Cooperative cleaners: A study in ant robotics. *The International Journal of Robotics Research (IJRR)*, 27(1): 127–151, 2008.

[74] I.A. Wagner, M. Lindenbaum and A.M. Bruckstein. Efficiently searching a graph by a smell-oriented vertex process. *Annals of Mathematics and Artificial Intelligence*, 24: 211–223, 1998.

[75] X. Zheng and S. Koenig. Robot coverage of terrain with non-uniform traversability. In: *Proceedings of the IEEE International Conference on Intelligent Robots and Systems (IROS)*, pages 3757–3764, 2007.

[76] X. Zheng, S. Jain, S. Koenig, and D. Kempe. Multi-robot forest coverage. In: *In Proc. IROS*, pages 3852–3857. press, 2005.

Chapter 4

Defending Large-Scale Critical Infrastructures Using a Swarm of Drones

Yaniv Altshuler[1], **Alex Pentland**[2], **Alfred Bruckstein**[3]
[1]Massachusetts Institute of Technology (yanival@mit.edu)
[2]Massachusetts Institute of Technology (pentland@mit.edu)
[3]Alfred Bruckstein, Israeli Institute of Technology (freddy@cs.technion.ac.il)

4.1 Introduction

The last decade has seen a paradigmatic change in the operational doctrines of modern militaries aerial forces, as designers and commanders have been gradually shifting their focus towards Unmanned Aerial Vehicles (UAVs) [46, 67]. Indeed, over 50% of the aerial vehicles in use by the US Military today are unmanned, with this trend expected to further increase in the coming years, and be mimicked by other defense forces [48, 61, 95]. Israel, as a leader in drone technology and its military adoption [50, 86] is currently heavily relying on the use of UAVs in its military force, with such vehicles becoming an increasingly dominant element of its intelligence platforms [71]. This has been the case in visual intelligent (VISINT), and recently also in tactical signal intelligence (SIGINT), which is traditionally in charge of 80% of the information gathered by the intelligence corps [2, 37].*[0]

Therefore, as the use of UAVs as an integral component of ongoing intelligence gathering in wartime, as well as during the "battle within the wars" increases, so grows the importance of the need to base this use on an efficient infrastructure. In other words, an innovative small scale dedicated UAV squadron designed for special missions may function perfectly with high redundancy and inefficient use of its resources, but a regular large-scale information gathering functional system that is based on unmanned vehicles operating in swarms, cannot. Furthermore, the lack of an efficient infrastructure that assumes control of the low-level resource utilization tasks means that these tasks must ultimately be taken care of by the human operators (as is being done today) —- dramatically reducing the number of tasks these can engage, increasing the time it takes them to do so, as well as the overall cost of this process, and ultimately significantly limiting the vehicles' operational potential.

As the complexity of such problems increases so does the impact of optimizing capabilities on the overall resources required in order to guarantee a pre-defined level of performance. A successful use of large scale swarms of drones as a combat and intelligence gathering tool necessitates the development of an efficient mechanism for their optimal utilization, specifically in the design and maintenance of their patrolling routes.

This work proposes an efficient and robust analytic infrastructure for the deployment of collaborative drone swarms, focusing on its application for tactic intelligence gathering. Specifically, we present an analytic model for devising an optimal reconnaissance strategy for any given threat scenario, defined as: (a) The correlation function between the cost of a single drone and its detection performance; (b) The deployment method used (represented as a monotonically increasing function that models the coverage percentage as a function of the number of units); and (c) The estimated expected cost of an undetected threat.

In other words, for any deployment method of the drones swarm (three of which are discussed in this paper) and an answer to the question "*how much detection performance do I get by using a more expensive unit?*" our model generates an optimal deployment strategy that is threat-specific, namely—that is economically optimizing the type and quantity of drones with respect to the expected cost of the damage incurred by an undetected threat.

The proposed technique enables swarms of semi-autonomous vehicles to perform an efficient ongoing dynamic patrolling and scan of the entire roads and transportation infrastructure in a pre-defined "search region" in a robust and near-optimal way, while guaranteeing detection of targets that are traveling in that region. We demonstrate the applicability of our model using a

[0]This work was previously published in [27].

comprehensive roads network of the Israeli roads and highways system, containing over 15,000 directed links.

The rest of the paper is organized as follows: related work is presented in Section 4.2. The problem and the proposed patrolling model is presented in Section 4.3. An analysis is presented in Sections 4.4 and 4.5, containing among others an analytic estimation of the required number of drones for a given threat scenario (*Theorem 4.1*), as well as an analytical estimation of the optimal types of drones to be used, if several drones of different costs and performances are available (*Theorem 4.2*). Section 4.6 presents a theoretical analysis of a case-study of the proposed monitoring method. Section 4.7 presents a second case study demonstrating the use of a real world transportation network dataset in order to illustrate several deployment strategies. The data used for this analysis is presented and discussed in Section 4.8. Concluding remarks appear in Section 4.9. Readers interested in further expansion on the mathematical analysis of this problem and similar ones are invited to review [32, 34] which also partially overlap with this work. Further discussion on the concepts of *Swarm Intelligence* can be found in [31] with a quick intro in [26] and a detailed discussion of its applications for search in [24, 25, 28, 29, 30].

4.2 Related Work

The need to efficiently monitor physical network layouts has been the topic of many studies throughout the recent decades, ranging from determining the optimal deployment of air quality monitoring units [72] through finding an efficient deployment of security monitors [84], studying submarine ecology [99] and even mapping the boundaries between tectonic plates [109]. In land transportation, many works had focused on monitoring the transportation of hazardous materials. For example, infectious disease outbreaks pose a critical threat to public health and national security [41, 51]. Utilizing today's expanded trade and travel routes, infectious agents can be distributed easily within and across country borders as part of a biological terror attack, resulting in potentially significant loss of life, major economic crises, and political instability. Such threats stress even more the importance of a reliable and efficient transportation monitoring infrastructure. A survey of homeland security related threats and risks regarding transportation infrastructure can be found in [47]. Alternatively, analysing the traffic dynamics along a transportation infrastructure was shown to enable new kind of collaborative travel, or 'ridesharing' algorithms, that can operate with greater efficiency [4, 97].

In [80] the trade-off between accuracy and coverage for given limited resources of sensor devices is discussed. Lam and Lo [78] proposed a heuristic approach to select locations for traffic volume count sensors in a roadway network. [116] proposed a sensor deployment framework to maximize such utilities. This framework has been extended to accommodate turning traffic information [42], existing installations and O-D information content [55], screen line problems [115], time-varying network flows [59] and unobserved link flow estimates [70]. The ability to utilize data from physical events manifested in social networks for the detection of potential emergencies was demonstrated in [20, 21]. Similar approaches for data derived from financial networks were also discussed in [100, 101, 120].

In its military aspects and its relevant scientific literature it is interesting to mention the roots of this field, dating back to World War II [83, 106]. The first planar search problem considered is patrolling a corridor between parallel borders separated by a constant width. This problem aimed at determining optimal patrol strategies for aircraft searching for ships in a channel [75], with a more generic theory of optimal scanning that later proposed in [85].

In recent years this problem had gained popularity with the emphasis on large scale swarms of drones [11, 15, 16, 17, 39, 73, 76, 85, 98, 107, 108, 110], using a variety of methods ranging from the analysis of the geometrical properties of the search space [9, 10], the use of multi-agent robotics approaches [68, 102, 111] or [36, 66, 82] for biology inspired designs (behavior based control models, flocking and dispersing models and predator-prey approaches, respectively), network theory oriented architectures [22, 32, 87] or economics inspired designs [64, 92, 105, 113] or physics inspired approaches [49, 74] (see [3] for a survey of search and evasion strategies).

In general, most of the techniques used for the task of a distributed monitoring use some sort of cellular decomposition. For example, in [93] the area to be monitored is divided between "monitoring agents" based on their relative locations. In [45] a different decomposition method is being used, which is analytically shown to guarantee a complete coverage of the area. Another interesting work is presented in [1], discussing two methods for cooperative monitoring (one probabilistic and the other based on an exact cellular decomposition). Another work called *Exploration as Graph Construction*, provides a coverage of degree bounded graphs in $O(n^2)$ time, is described in [54]. Here a group of ant robots with a limited capability explores an unknown graph using special "markers". Another mechanism of using "pheromone markers" for the purpose of intra-swarm synchronization is discussed in [60], however, a work that discusses a stochastic collaborative exploration of a virtual network for malicious cyber threats is discussed in [18].

While in most works the targets of the search mission were assumed to be idle, recent works considered dynamic targets, meaning — targets which after being detected by the searching robots, respond by performing various evasive maneuvers intended to prevent their interception. Some of these works can be seen in [7, 12, 14], however, a variant that emphasizes the stochastic aspects of the problem can be found in [23].

4.3 Patrolling System Optimizing—Problem Definitions

This work analyzes the design of efficient monitoring systems through its economic perspective. Namely, when a multitude of threat scenarios are available, the main challenge is mitigating these scenarios in a cost-effective way. That is, resolving the real-time uncertainty regarding the exact kinds of threats through the optimization of its overall costs (both the direct costs of the system, as well as the indirect costs of undetected threats). It is therefore crucial to provide planners and operators with a monitoring system that can be reconfigured in real-time, having drones deployed/activated gradually as they are needed (due to new information regarding the unfolding situation, and taking into account other operational requirements or budget constraints).

Our proposed model does not assume any constraint on the deployment scheme itself, as the latter can be highly influenced by a large variety of considerations. Rather, it enables planners to model any given deployment scheme they desire, providing the optimal number and types of drones for each one. Our proposed model also offers operators high flexibility regarding the types of drones to be used—modeled using a trade off between the number of monitoring drones and their detection quality with respect to the number of potential targets each drone can monitor simultaneously (to be denoted as the drones' "*Sampling*" quality, a number ranging between 0 and 1). In this work we assume that a higher sampling quality implies a higher cost per drone. Therefore, the overall cost of the monitoring system can be modeled as:

$$Overalll\ Cost = Cost\ per\ Unit \times Number\ of\ Units$$

whereas the overall monitoring performance of the system can be modeled as:

$$Overalll\ Performance = Monitoring\ Coverage\ Percentage \times Sampling$$

We note that the quality of monitoring units (i.e. their sampling rate) has in fact a *non linear* effect on the overall monitoring quality. For example, assuming the sampling rate of each monitor is $\frac{1}{2}$, it will be able to detect slightly more than half of the vehicles traveling in the network, because some of them might pass by more than one monitor and thus have a chance of being detected by either one. Specifically, such vehicles will have a probability of $1-(0.5)^2 = 0.75$ to be detected. Since drones' deployments is likely be done is much smaller numbers compared to the overall possible locations, it is unlikely that many potential targets would pass by two drones during a single trip. We can therefore use a linear approximation.

Definition 4.1 Let $C_{ATTACK}(\chi)$ denote the expected cost associated with the damage caused by an attack χ, used to model an undetected target.

The estimation of the damage costs are beyond the scope of this paper.

Definition 4.2 Let $M(x) \in [0,1]$ be a monotonically increasing function denoting the percentage of potential targets moving on the roads network, that are detected and analyzed using x monitoring units, using some given monitoring deployment scheme.

A monitoring system of n units would therefore be cost-effective as long as:

$$C_{ATTACK} \geq \frac{n}{M(n)} \cdot Cost\ per\ Unit \tag{4.1}$$

For modeling the function $M(x)$ we propose to use the well-known *Gompertz function* [65], whose general form is:

$$y(t) = ae^{be^{ct}}$$

The *Gompertz function* is widely used for modeling a great variety of processes, (due to the flexible way it can be controlled using the parameters a, b and c), such as mobile phone uptake [94], population in a confined space [56], or growth of tumors [53] (see an illustration in Figure 4.1). Its ability to model the progress of the optimization process as a function of the available resources can also be seen in [5, 6, 8, 19].

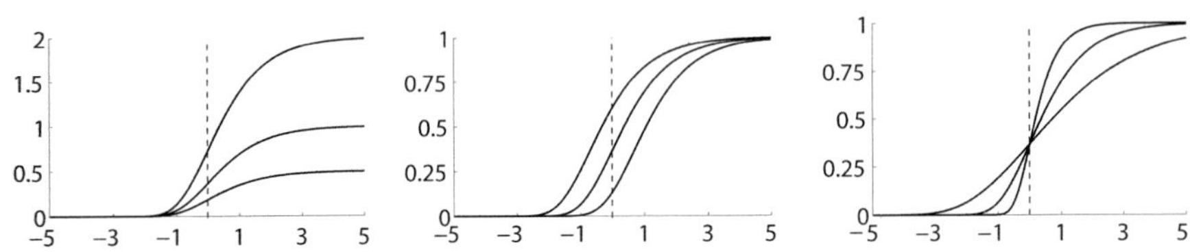

Figure 4.1: An illustration of the *Gompertz function*. The charts represent the following functions (from left to right): $y = x\,e^{-e^t}$, $y = e^{-xe^t}$ and $y = e^{e^{-xt}}$, for $x = \frac{1}{2}$, $x = 1$ and $x = 2$.

The function $M(x)$ can be extrapolated using simulations, as demonstrated in Figure 4.12. Note that $M(x)$ is domain dependant and may significantly change for different networks.

In the following sections we show how this model can be used in order to calculate the optimal number and types of the monitoring drones.

4.4 Finding the Optimal Number of Drones

Definition 4.3 The *Normalized Benefit* of a monitoring system of n units is defined as the expected monetary saving from preventing attacks from which the fixed cost of the monitoring units is subtracted:

$$\omega \triangleq C_{ATTACK} \cdot M(n) - n \cdot Cost\ per\ Unit$$

Note that this function can refer also to non-monetary aspects of a successful attack. It also resembles the “Net Present Value” calculation, but without explicitly accounting for discount rates.

Theorem 4.1
For values of C_{ATTACK}, a, b, c and Cost per Unit*, the optimal number of monitoring units that would maximize the Normalized Benefit of a monitoring system is:*

$$n_1 = \frac{\ln\left(\frac{1}{a \cdot b \cdot c} \cdot \frac{\text{Cost per Unit}}{C_{ATTACK}}\right) - W\left(\frac{1}{a \cdot c} \cdot \frac{\text{Cost per Unit}}{C_{ATTACK}}\right)}{c}$$

$$n_2 = \frac{\ln\left(\frac{1}{a \cdot b \cdot c} \cdot \frac{\text{Cost per Unit}}{C_{ATTACK}}\right) - W_{-1}\left(\frac{1}{a \cdot c} \cdot \frac{\text{Cost per Unit}}{C_{ATTACK}}\right)}{c}$$

where $W(x)$ is the Lambert product log*, that can be calculated using the series:*

$$W(x) = \sum_{n=1}^{\infty} \frac{(-1)^{n-1} n^{n-2}}{(n-1)!} x^n$$

Proof 4.1 The optimal value of the Normalized Benefit is received for the number of monitoring units that nullifies the derivative $\frac{\partial \omega}{\partial n}$:

$$\frac{\partial \omega}{\partial n} = C_{ATTACK} \cdot \frac{\partial M(n)}{\partial n} - Cost\ per\ Unit$$

Namely:

$$\frac{\partial M(n)}{\partial n} = \frac{Cost\ per\ Unit}{C_{ATTACK}} \tag{4.2}$$

(notice that we disregard interest rates, as both cost and damage can assumed to be subject to a similar change along time).

In that case, using Equation 4.2 we obtain:

$$a \cdot b \cdot c \cdot e^{cn} \cdot e^{be^{cn}} = \frac{Cost\ per\ Unit}{C_{ATTACK}}$$

which in turn implies:

$$be^{cn} + cn - \ln \frac{Cost\ per\ Unit}{a \cdot b \cdot c \cdot C_{ATTACK}} = 0 \tag{4.3}$$

We note that $a > 0$ whereas $b, c < 0$. Analyzing Equation 4.3 we can then see that in cases where:

$$\frac{Cost\ per\ Unit}{C_{ATTACK}} \leq -\frac{a \cdot c}{e} \tag{4.4}$$

the optimal value of n would equal:

$$\begin{aligned} n_1 &= \frac{\ln\left(\frac{1}{a \cdot b \cdot c} \cdot \frac{Cost\ per\ Unit}{C_{ATTACK}}\right) - W\left(\frac{1}{a \cdot c} \cdot \frac{Cost\ per\ Unit}{C_{ATTACK}}\right)}{c} \\ n_2 &= \frac{\ln\left(\frac{1}{a \cdot b \cdot c} \cdot \frac{Cost\ per\ Unit}{C_{ATTACK}}\right) - W_{-1}\left(\frac{1}{a \cdot c} \cdot \frac{Cost\ per\ Unit}{C_{ATTACK}}\right)}{c} \end{aligned} \tag{4.5}$$

where $W(x)$ is the *Lambert product log*, and where $W_k(x)$ is its analytic continuation over the complex plane (the values of the functions $W(x)$ and $W_{-1}(x)$ in the segment implied by the constraint of Equation 4.4 are illustrated in Figure 4.2).

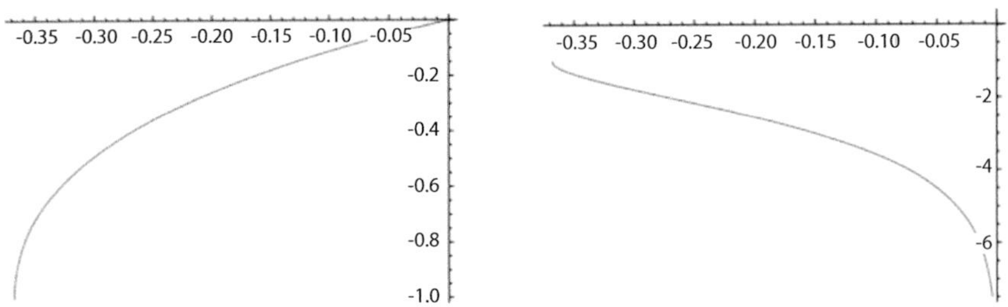

Figure 4.2: The left and right charts depict the values of the *Lambert* functions $W(x)$ and $W_{-1}(x)$ in the segment $[-\frac{1}{e}, 0]$, respectively. The segment is implied by the constraint of Equation 4.4.

Returning to the Normalized Benefit of the system, we can now assign the values of the optimal number of monitoring units that appear in Equation 4.5 into the definition of ω, as follows:

$$\omega_1 = C_{ATTACK} \cdot ae^{be^{cn_1}} - n_1 \cdot Cost\ per\ Unit$$

$$\omega_2 = C_{ATTACK} \cdot ae^{be^{cn_2}} - n_2 \cdot Cost\ per\ Unit$$

and using the properties of the W function, simplify it into the following form:

$$\omega_{max} = \max\{\omega_1, \omega_2\} \qquad where: \tag{4.6}$$

$$\begin{aligned} \omega_1 &= a \cdot b \cdot C_{ATTACK} \cdot \gamma \cdot \left(W(b \cdot \gamma) + \frac{1}{W(b \cdot \gamma)} - \ln(\gamma)\right) \\ \omega_2 &= a \cdot b \cdot C_{ATTACK} \cdot \gamma \cdot \left(W_{-1}(b \cdot \gamma) + \frac{1}{W_{-1}(b \cdot \gamma)} - \ln(\gamma)\right) \end{aligned}$$

where the *Monitoring Benefit Factor* γ is defined as:

$$\gamma = \frac{1}{a \cdot b \cdot c} \cdot \frac{Cost\ per\ Unit}{C_{ATTACK}}$$

From Equation 4.6 we can see that the Normalized Benefit of a drones monitoring swarm is a function of the *Monitoring Benefit Factor* γ, which takes into account both the parameters of the threat (through the overall potential damages and the costs of the monitoring units required to detect it) as well as the properties of the monitoring method (characterized by the values of a, b and c). Notice that the value of *Cost per Unit* does affect the values of a, b and c—as they are solely derived from the coverage efficiency of the network's traffic. However, we also note that the actual detection of the drones swarm is affected by their sampling quality, which in turn is monotonously affected by their cost.

4.5 Selecting Which Type of Drones to Use

In this section we address the issue of finding the optimal type of monitoring drones that should be deployed. This is done by optimizing Equation 4.6 with respect to the cost per single unit.

Definition 4.4 Let $Cost_{Base}$ denote the cost of the "most expensive" drone model. That is, the monitoring drone model with the highest level of detection available.

There is a wide range of monitoring drone models, where in most cases the cheapest ones are expected to have the poorest performance. We define the correlation between the sampling (or detection) quality and the cost of a single unit using a generic function, as follows:

$$Sampling = f_S\left(\frac{Cost\ Per\ Unit}{Cost_{Base}}\right) \tag{4.7}$$

Theorem 4.2
For values of C_{ATTACK}, a, b, c, Cost per Unit, $\text{Cost}_{\text{Base}}$ *and for every function* $f_S : [0,1] \to [0,1]$, *the optimal cost for a monitoring unit that would maximize the Normalized Benefit of a monitoring system is a value that satisfies at least one of the following expressions:*

$$W(b\cdot\gamma) - \ln(\gamma) + \frac{1 - \frac{\text{Cost Per Unit}}{\text{Cost}_{\text{Base}}} \cdot \frac{\partial f_S}{\partial \text{Cost Per Unit}}\left[\frac{\text{Cost Per Unit}}{\text{Cost}_{\text{Base}}}\right] \cdot \left(f_S\left[\frac{\text{Cost Per Unit}}{\text{Cost}_{\text{Base}}}\right]\right)^{-1}}{W(b\cdot\gamma)} = 0$$

$$W_{-1}(b\cdot\gamma) - \ln(\gamma) + \frac{1 - \frac{\text{Cost Per Unit}}{\text{Cost}_{\text{Base}}} \cdot \frac{\partial f_S}{\partial \text{Cost Per Unit}}\left[\frac{\text{Cost Per Unit}}{\text{Cost}_{\text{Base}}}\right] \cdot \left(f_S\left[\frac{\text{Cost Per Unit}}{\text{Cost}_{\text{Base}}}\right]\right)^{-1}}{W_{-1}(b\cdot\gamma)} = 0$$

where:

$$\gamma = \frac{1}{a \cdot b \cdot c} \cdot \frac{\text{Cost per Unit}}{C_{ATTACK}} \cdot \left(fs \left[\frac{\text{Cost Per Unit}}{\text{Cost}_{\text{Base}}} \right] \right)^{-1}$$

Proof 4.2 We first revise Equation 4.6 in order to take into account the different types of monitoring units. as follows:

$$\omega_{max} = \max\{\omega_1, \omega_2\} \tag{4.8}$$

where ω_1 and ω_2 equal:

$$\begin{aligned}
\omega_1 &= a \cdot b \cdot C_{ATTACK} \cdot Sampling \cdot \gamma \cdot \left(W(b \cdot \gamma) + \frac{1}{W(b \cdot \gamma)} - \ln(\gamma) \right) \\
\omega_2 &= a \cdot b \cdot C_{ATTACK} \cdot Sampling \cdot \gamma \cdot \left(W_{-1}(b \cdot \gamma) + \frac{1}{W_{-1}(b \cdot \gamma)} - \ln(\gamma) \right)
\end{aligned}$$

and where:

$$\gamma = \frac{1}{a \cdot b \cdot c} \cdot \frac{Cost\ per\ Unit}{C_{ATTACK} \cdot Sampling}$$

We now proceed to the maximizing the financial merits of the monitoring system (namely, $\max\{\omega_1, \omega_2\}$). For this, we shall calculate the partial derivatives $\frac{\partial \omega_1}{\partial Cost\ Per\ Unit}$ and $\frac{\partial \omega_2}{\partial Cost\ Per\ Unit}$:

$$\begin{aligned}
\frac{\partial \omega_1}{\partial Cost\ Per\ Unit} &= \frac{1}{c} \cdot \left(W(b \cdot \gamma) - \ln(\gamma) + \frac{1 - \frac{\partial fs}{\partial Cost\ Per\ Unit} \cdot \frac{Cost\ Per\ Unit}{Cost_{Base} \cdot fs}}{W(b \cdot \gamma)} \right) \qquad (4.9) \\
\frac{\partial \omega_2}{\partial Cost\ Per\ Unit} &= \frac{1}{c} \cdot \left(W_{-1}(b \cdot \gamma) - \ln(\gamma) + \frac{1 - \frac{\partial fs}{\partial Cost\ Per\ Unit} \cdot \frac{Cost\ Per\ Unit}{Cost_{Base} \cdot fs}}{W_{-1}(b \cdot \gamma)} \right)
\end{aligned}$$

and the rest is implied.

Theorem 4.2 can now be used in order to calculate the *optimal aspired cost* of a single drone, for every correlation between its cost and its detection quality, and for every deployment scheme and estimated threat's potential.

4.6 Case Study I—Theoretical Analysis

In this section we demonstrate how the proposed model can be used in order to produce the optimal number and types of monitoring drones, for a given threat scenario and drones deployment scheme, selected by the system's operators.

For the sake of simplicity we assume in this analysis that both sampling rates and Cost Per Unit to be continuous. This simplifies cases where the sampling rate of the monitoring units can be tuned resulting in lower resource utilization for lower sampling rates (e.g. $f_S(\frac{Cost\ Per\ Unit}{Cost_{Base}}) = \left(\frac{Cost\ Per\ Unit}{Cost_{Base}}\right)^2$). In addition, we assume that the deployment scheme chosen can be modeled with a *Gompertz* function of $a = 1$, $b = -0.2$ and $c = -0.05$ (see an illustration of this function in Figure 4.3). In this case, nullifying the partial derivative of Equation 4.9 would produce:

$$\begin{aligned} \frac{\partial \omega_1}{\partial Cost\ Per\ Unit} &= 0 \longrightarrow W(-0.2\cdot\gamma) = \ln(\gamma) + \frac{1}{W(-0.2\cdot\gamma)} \\ \frac{\partial \omega_2}{\partial Cost\ Per\ Unit} &= 0 \longrightarrow W_{-1}(-0.2\cdot\gamma) = \ln(\gamma) + \frac{1}{W_{-1}(-0.2\cdot\gamma)} \end{aligned} \tag{4.10}$$

subsequently implying:

$$\gamma_{opt} \approx 1.77356$$

(in this example, the optimal value of γ for ω_2 has a non-zero imaginary component).

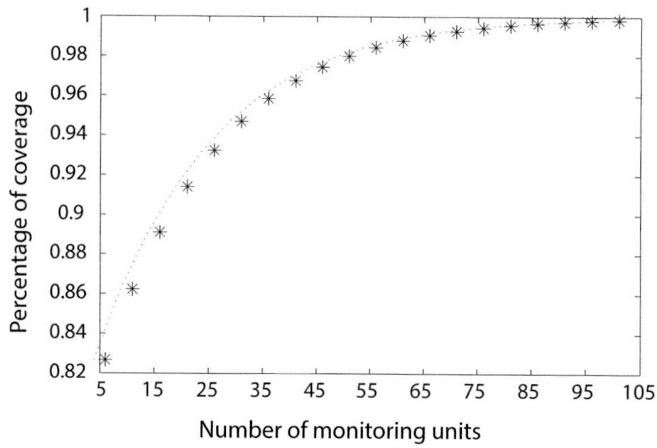

Figure 4.3: An illustration of the *Gompertz* function $e^{-0.2\cdot e^{-0.05\cdot t}}$, used to simulate the increase in monitoring coverage as a function of the increase in number of drones, assumed in the case study of Section 4.6.

Using this optimal value of γ we would now get:

$$\gamma_{opt} = \frac{1}{a\cdot b\cdot c}\cdot\frac{Cost\ per\ Unit}{C_{ATTACK}\cdot Sampling} = \frac{100\cdot Cost_{Base}^2}{C_{ATTACK}\cdot Cost\ per\ Unit} = 1.77356$$

and from this we receive:

$$Cost\ per\ Unit_{opt} = \frac{100\cdot Cost_{Base}^2}{1.77356\cdot C_{ATTACK}} \approx 56.384\cdot\frac{Cost_{Base}^2}{C_{ATTACK}} \tag{4.11}$$

An illustration of Equation 4.11 is shown in Figure 4.4.

From Equation 4.11 we can obtain for each kind of threat the optimal type of units that should be used, in order to maximize the Normalized Benefit of the system. In this example we also see a linear connection between the optimal cost of the drones and the overall estimates of the damages from a threat.

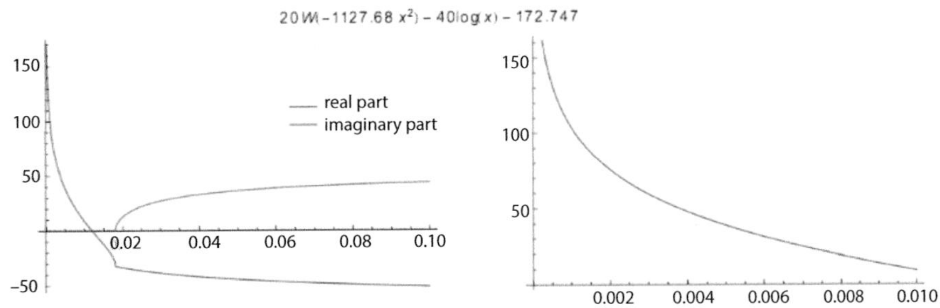

Figure 4.4: An illustration of Equation 4.12, showing the optimal number of drones as a function of the ratio between $Cost_{Base}$, the cost of the best drone model available, and the estimation of the overall damages of a successful attack (the ratio is denoted by x). The left chart contains both the real and imaginary parts of the solution. The right chart shows the segment between 0 and 0.01, where the imaginary part equals zero.

Assigning this back into Equation 4.5, we can get the optimal number of units, for each potential threat (An illustration is shown in Figure 4.4):

$$n_1 = 20 \cdot W\left(-1127.68 \cdot \frac{Cost_{Base}^2}{C_{ATTACK}^2}\right) - 172.747 - 40 \cdot \ln\left(\frac{Cost_{Base}}{C_{ATTACK}}\right)$$

$$n_1 = 20 \cdot W\left(-1, -1127.68 \cdot \frac{Cost_{Base}^2}{C_{ATTACK}^2}\right) - 172.747 - 40 \cdot \ln\left(\frac{Cost_{Base}}{C_{ATTACK}}\right) \quad (4.12)$$

Using Equations 4.11 and 4.12 we can now produce the optimal model of drones to be deployed, and in what numbers, for any given threat[1]: for example, if the best available drone model costs $5,000 per unit then for threats of potential damages that are estimated as $600,000 (which reflect a ratio of $\frac{1}{120}$ between the cost of best unit and the incident's damages) we should use approximately 25 units of the type that cost $1,100 each (and deploy them according to the

[1] Our model assumes continuous selection of drones' types. In reality of course there is a finite number of drones model. Therefore, after producing the optimal value for the cost of a single drone, we would select the two available models closest in cost to this optimal value (namely, the more expensive cheaper one, and the cheapest more expensive one), assign their cost in the model, and select the one for which the merit function is higher.

selected deployment scheme described above). However, for threats estimated at \$10,000,000 in total damages (i.e. a ratio of $\frac{1}{2000}$) an optimal monitoring system should be comprised of approximately 160 units, of the type that costs \$200 each.

Note that the previous example depends on the assumption of a quadratic relation between the cost and quality of the monitoring units, as well as on the assumption regarding the deployment scheme (i.e. the parameters of the *Gompertz Model*). For any change in any of these assumptions, new corresponding solutions to the optimal monitoring problem can be easily generated using the model.

4.7 Case Study II—Real World Transportation Network Monitoring

In the previous sections we have shown an analytic method for optimizing the number and types of monitoring units for any given deployment scheme (defined by the values of the Gomperz function's parameters a, b and c that set the relation between the amount of monitoring resources and the increase in monitoring efficiency).

In this section we discuss several methods for deploying a given number of monitoring units. The goal in all three methods is the same—maximizing the probability that a target randomly moving in one of the network's edges, is detected.

The most trivial method for deploying n monitoring units is of course, the random deployment method. In other words, positioning the units in randomly selected nodes in the network. As we will soon see, this method achieves a surprisingly good performance in some cases. However, this methods can be improved by taking into account certain features of the network, resulting in an improved performance in many cases.

In the coming sections we will see several such improvements, all based on a traffic-oriented improvement of the standard *Betweenness Centrality* method.

We then show that all methods shown in this section can be efficiently approximated using the *Gompertz* function $y = ae^{be^{ct}}$. Hence, any of these methods can be represented by a set of values of a, b and c, as required by the proposed model. In addition, note that the last deployment method we present here achieves remarkable results (in terms of prediction of the real traffic flow, by the network's topology analysis). Therefore, this method can be assumed as a standard deployment scheme for transportation networks, regardless of the scope of this work.

The data used for the evaluation of this analysis was derived from a large number of location-reports, collected from cell phones in Israel, and is described and discussed in length in Section 4.8. Parts of the following analysis can also be found in [91] and [32].

4.7.1 Betweenness Centrality vs. Traffic Flow

Betweenness Centrality (BC) stands for the ability of an individual node to control the communication flow in the networks and is defined as the total fraction of shortest paths between each pair of vertices that pass through a given vertex [35, 62]. In recent years Betweenness was extensively applied for the analysis of various complex networks [38, 104] including among other social networks [96, 112], computer communication networks [58, 118], and protein interaction networks [43]. Holme [69] has shown that Betweenness is highly correlated with congestion in particle hopping systems. Extensions of the original definition of BC are applicable for directed and weighted networks [44, 114] and for multilayer networks where the underlying infrastructure and the origin-destination overlay are explicitly defined [90].

Let $G = (V, E)$ be a directed transportation network where V is the set of junctions and E is the set of directed links as described in Section 4.8. Let $\sigma_{s,t}$ be the number of shortest paths between the origin vertex $s \in V$ and the destination vertex $t \in V$ (in some applications the shortest path constraint can be relieved to allow some deviations from the minimal distance between the two vertices). In the rest of this paper we will refer to the shortest or near shortest paths between two vertices as *routes*. Let $\sigma_{s,t}(v)$ be the number of routes from s to t that pass through the vertex v. The Betweenness Centrality can hence be expressed by the following equation:

$$BC(v) = \sum_{s,t \in V} \frac{\sigma_{s,t}(v)}{\sigma_{s,t}}. \tag{4.13}$$

Note that in this definition we include the end vertices (s and t) in the computation of Betweenness since we assume that vehicles can be inspected also at their origin and at their destination point.

After computing the Betweenness Centrality for the given transportation network, we can easily see that the distribution of Betweenness Centrality follows a power law (Figure 4.5). Long tail distributions such as the power law suggest that there is a non negligible probability for the existence of vertices with very high Betweenness Centrality. This is in contrast to the exponential flow distribution depicted in Figure 4.18. The different nature of these two distributions suggests that BC as defined above will overestimate the actual traffic flow through nodes especially for the most central vertices.

Next we would like to check the correlation between BC and traffic flow. Although the correlation is significant the square error is very low ($R^2 = 0.2021$) as shown in Figure 4.6 (a). Every point in this Figure represents a vertex with the x-axis corresponding to the measured traffic flow and y-axis corresponding to the computed BC.

We now discuss augmented variants of the Betweenness Centrality measure that significantly improve the correlation with the traffic flow.

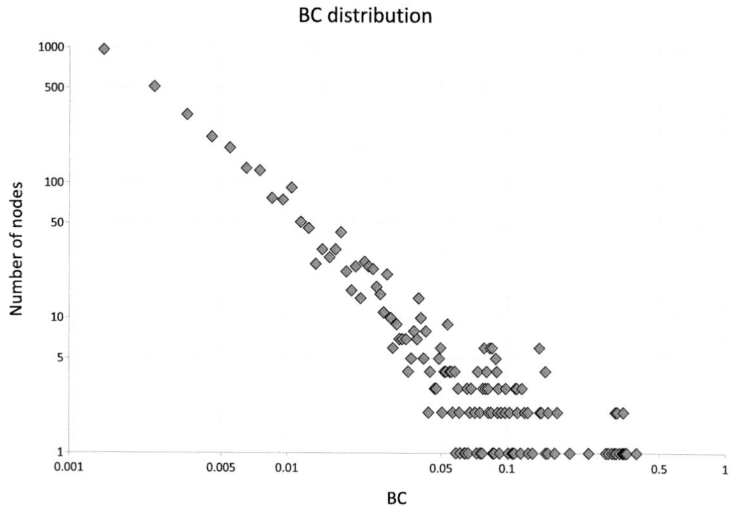

Figure 4.5: Power law distribution of Betweenness Centrality.

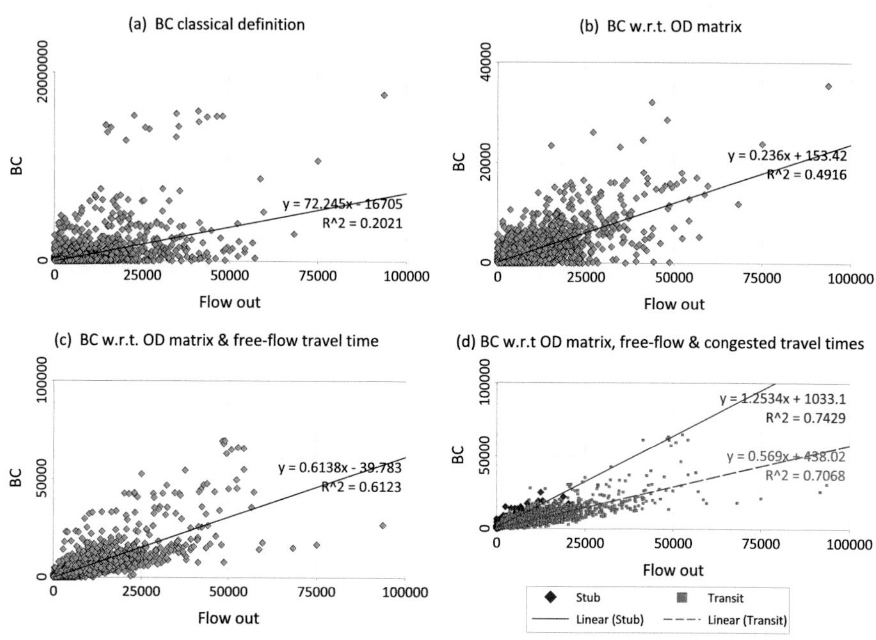

Figure 4.6: Correlation of flow through nodes and Betweeness Centrality.

4.7.2 Origin-Destination Based Betweenness Centrality

The BC definition according to Equation 4.13 assumes equal weights of routes between every pair of vertices in the network. In other words every vertex acts as an origin and a traffic destination. We would like to utilize the measured origin-destination (OD) flow matrix in order to prioritize network regions by their actual use. For this, we shall use the following altered definition for Betweenness, as suggested in [90]:

$$BC(v) = \sum_{s,t \in V} \frac{\sigma_{s,t}(v)}{\sigma_{s,t}} \cdot OD_{s,t} \tag{4.14}$$

where OD is the actual measured origin-destination matrix. This method produces a better correlation ($R^2 = 0.4916$) between the theoretic (BC) and the measured traffic flow (see Figure 4.6 (b)).

4.7.3 *Improving Betweenness Centrality Using Travel Properties*

Shortest Routes based on Time to Travel

In order to further improve our ability to estimate the predicted network flow using the network's topology, we note that both BC calculation methods (Equations 4.13 and 4.14 above) assume that routes are chosen according to shortest path strategy based on hop counting. In this section, we retain the shortest path assumption but use weighted links for calculating the Betweenness score. One option is to use the length of the road segments as their weights for the shortest path calculations. Shortest path algorithms (such as Dijkstra's or Bellman-Ford's) are able to consider only one distance weight on links when computing the shortest path to a destination. We shall therefore assume that the primary heuristic guiding people when they chose a route is the time required to reach their destination, and recompute the BC on the directed transportation, weighting links by their free-flow travel time.

Let $BC^{ft}(v)$ denote the Betweenness of a node v computed w.r.t. the free-flow travel time. Figure 4.6 (c) shows significant improvements in the correlation between the measured traffic flow and the theoretical BC^{ft} values computed w.r.t the OD matrix and free-flow travel time link weights ($R^2 = 0.6123$). We can see that there are few nodes whose flow was significantly underestimated by the BC measure. Notice that there are also several nodes whose flow was actually overestimated. This can be explained by the fact that people do not travel strictly via shortest paths, but may have various deviations. In particular the deviations form shortest paths are affected by the time and day of the week.

***Peak-Hours Aware* Betweenness Centrality**

It is a reasonable assumption that during peak hours travelers will choose to avoid congested roads and choose their routes based on the congested travel time rather than the free-flow travel times. Let $BC^{ct}(v)$ denote the Betweenness of a node v computed w.r.t. the congested time. Computing Betweenness using only the congested travel time weights results in $R^2 = 0.7096$. Although peak hours are a relatively small fraction of the day, most vehicles travel during these hours. This is the reason for higher correlation of BC^{ct} with the measured traffic flow.

We shall now combine both the Betweenness Centrality computed w.r.t. the free-flow travel time and the congested time by taking a weighted average, namely:

$$BC(v) = \alpha \cdot BC^{ft}(v) + (1-\alpha) \cdot BC^{ct}(v)$$

where α denotes the relative fraction of vehicles traveling during the free-flow periods. The resulting centrality index can achieve higher correlation with the measured average traffic flow. The maximal correlation of $R^2 = 0.7285$ is obtained for $\alpha = 0.25$ as shown in Figure 4.7.

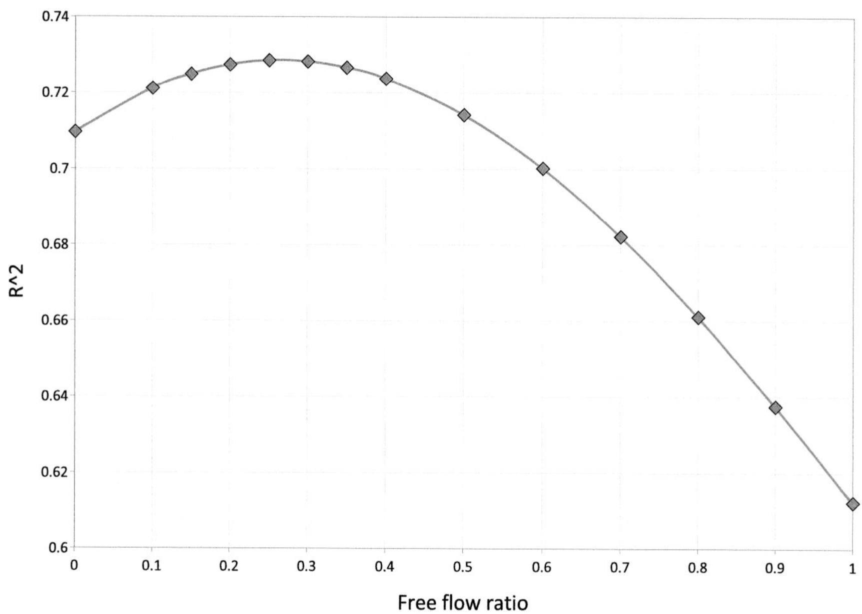

Figure 4.7: Squared error (R^2) as the function of the free flow traffic fraction (α).

Separating *Stub Nodes* from *Transit Nodes*

Nodes in the dataset were divided into two groups : *stub nodes* and *transit nodes*. A Stub node (i.e., centroid) is a node that is an origin or a destination of the traffic (as seen in the Origin-Destination matrix). These nodes account for approximately 10% of the network's nodes. All other nodes (namely, nodes that generate insignificant or no outgoing or incoming routes) are called Transit nodes, as they only forward traffic and do not generate or consume it.

Figure 4.6 (d) presents the correlation that is received when the two groups of nodes are being processed separately. Specifically, the results show a $R^2 = 0.7068$ for the Transit nodes and a $R^2 = 0.7429$ for the Stub nodes.

Mobility Oriented Betweenness Centrality

As each type of road has a different functional class, we shall further improve our flow prediction by examining the Betweenness values achieved when calculating it for each group separately. The results of the correlation that are achieved using this method are presented in Figure 4.8. We can clearly see that for the more important roads (namely, those with lower type number, representing a more infrastructural role in the road transportation network) this technique yields R^2 values that are consistently above 0.74, reaching 0.83 for of types 2 and 9 (note that roads of type 90 are fictive roads with infinite capacity that were artificially added in order to connect distinct regions in the network).

It should be noted that each node may have incoming roads of different types. Each plot corresponds to a set of nodes whose max incoming road type is as specified. In addition, the BC calculations were not made for each set of nodes separately—BC was computed for the complete network, while the correlations were computed separately for each type.

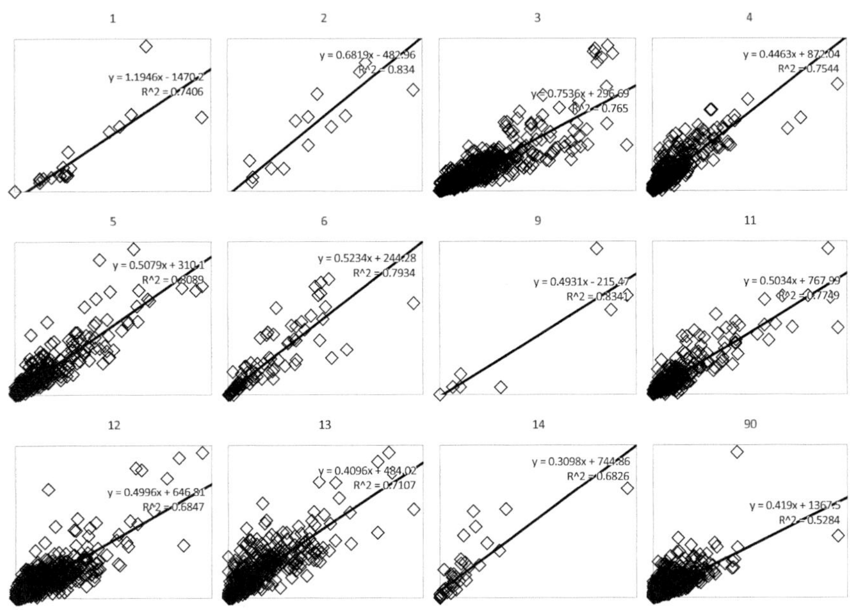

Figure 4.8: Correlation of flow through nodes and Betweenness computed separately for different *types* of links.

4.7.4 Optimizing the Locations of Surveillance and Monitoring Stations

In this section we use the Group variant of the shortest path Betweenness Centrality (GBC) [57] as an estimate for the utility of collaborative monitoring. In other words, we are interested in verifying that, given some mobile agent that we are interested in intercepting, we position the monitoring stations in a way that maximizes the chance the agent would be captured, considering the traffic patterns of the transportation network. In this case, however, significant computational complexity issues arise, rendering the generation of an optimal solution impractical in real time by conventional tools that are based mostly on behavioral based modeling. Using Group Between Centrality we propose a way to generate efficient approximations of the optimal solution to this optimization problem.

GBC of a given group ($M \subseteq V$) of vertices accounts for all routes that pass through *at least one* member of the group. Let $\sigma_{s,t}$ and $\sigma_{s,t}(M)$ be the number of routes from s to t and the number of routes from s to t passing through at least one vertex in M respectively:

$$GBC(M) = \sum_{s,t \in V} \frac{\sigma_{s,t}}{\sigma_{s,t}(M)} \cdot OD_{s,t} \tag{4.15}$$

GBC can be efficiently computed using the algorithm presented in [88].

Assuming the routes are weighted by the origin destination flow in transportation networks, GBC will account for the net number of vehicles that are expected to pass by the monitors during an hour. This net number is different from the total number of vehicle passing by the monitors since the same vehicle can pass by several monitors during a single trip. For example, searching for a suspected escaping terrorist car, one would like to avoid stopping the same vehicle twice and increase the number of distinct vehicles that were inspected. It is therefore important to maximize the GBC value of the set of inspection stations given the number of stations deployed.

Several combinatorial optimization techniques can be used to find a group of nodes of given size that has the largest GBC. In the following discussion we refer to a greedy approximation algorithm for the monitors location optimization problem (Greedy) [52], a classical *Depth First Branch and Bound* (DFBnB) heuristic search algorithm [77], and recently proposed *Potential Search* [103].

The *Greedy* approximation algorithm chooses at every stage the node that has the maximal contribution to the GBC of the already chosen group. The approximation factor of the *Greedy* algorithm as reported in [52] is $e - \frac{1}{e}$.

Both the heuristic search algorithms *DFBnB* and the *Potential Search* provably find the group having the maximal GBC. The *Greedy* algorithm and *DFBnB* were previously compared in [89] in the context of monitoring optimization in computer communication networks. Given the fact that finding a group of a given size having the maximal GBC is a hard problem, the greedy algorithm is good enough for any practical purpose (the hardness of the problem can be proven by a straightforward reduction from the Minimal Vertex Cover problem so that the problem of maximizing GBC is NP-Complete). Figure 4.9 presents the results of selecting one to 39 inspection locations using the greedy algorithm.

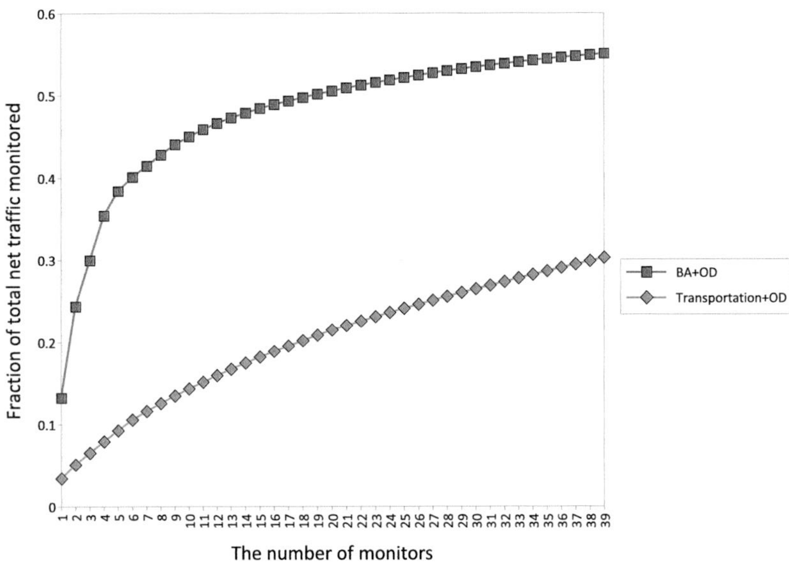

Figure 4.9: The total net traffic flow that passes by monitors as a function of the number of monitors. As expected the marginal value of additional monitors gradually decreases as more of them are added reaching potential traffic coverage of 30% when 39 monitoring stations are deployed.

In certain cases (such as in various homeland security applications) deployment of monitoring systems is often done under tight timing conditions, as a result of new intelligence information. Therefore, any optimization method should provide close-to-real-time capabilities. In this context, it is interesting to note that both the *DFBnB* and the *Potential* algorithms are anytime search algorithms [119]. Their execution can be stopped at any point of time, yielding the best solution found so far. Therefore, in the following experiments we limit the search time to one hour, simulating a quasi-real-time optimization constraint. Still, as can be seen in Figure 4.10 the running time of the *Greedy* algorithm is by far lower than one hour, for the entire Israeli transportation system.

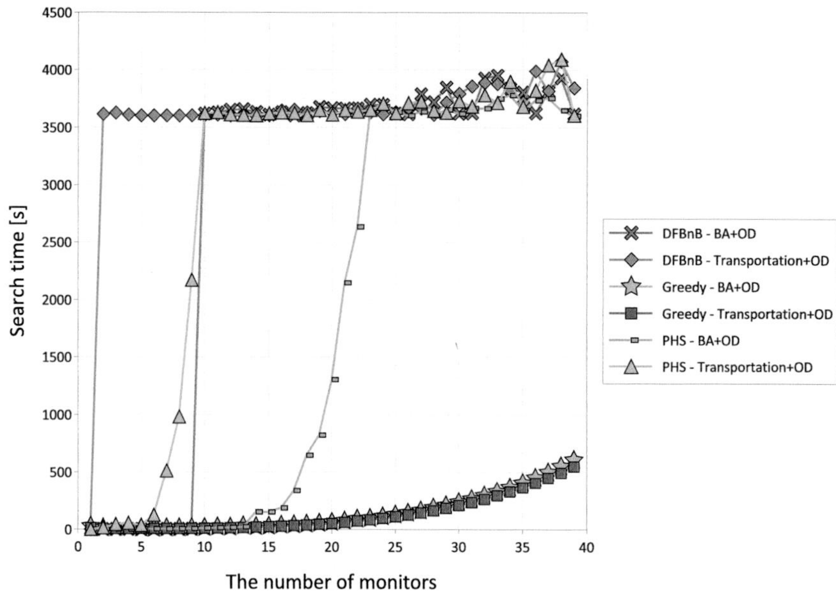

Figure 4.10: The time (in seconds) that the search algorithms were executed in as a function of the number of monitors. Note that due to the highly complex nature of the search algorithms, they present a phase transition around a tipping point (specific to each algorithm) in the number of monitors to be optimized.

When *DFBnB* and *Potential Search* algorithms cannot complete the search process within the given time bounds they produce a close to optimal solution and an estimate of its optimality (i.e. certificate). The certificate is computed by dividing the best solution found so far by the upper bound on the optimal solution. The upper bound is computed using admissible heuristic functions and is maintained by the search algorithms for efficiently pruning the search space. Figure 4.11 shows that *Potential Search* produces higher certificates for its solutions within the one hour time bound for all sizes of the monitors deployed.

4.7.5 Modeling Various Deployment Schemes as a Gompertz Function

Figure 4.12 demonstrates the performance of our monitoring method, by showing the percentage of traffic monitored as a function of the number of monitors, for several deployment schemes: (a) Group Betweenness, (b) Betweenness, and (c) Random deployment. The benefits of the proposed method can clearly be seen from this chart.

Remarkably, using a GBC based deployment strategy it is possible to cover the vast majority of the traffic in the analyzed network (detecting any threat in

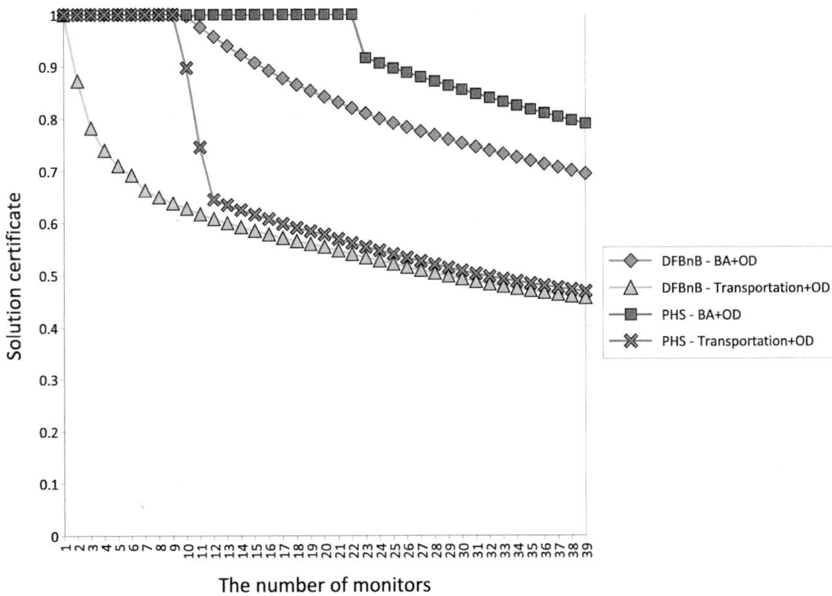

Figure 4.11: The minimal quality of the solution (fraction of the upper bound) as a function of the number of monitors.

a probability of approximately 0.7) using only 70 monitors. This result is most probably due to the relatively small number of origins and destinations in the analyzed network. 680 origins/destinations account for a little bit more than 10% of the network nodes.

BC based strategy produces relatively high quality deployments for a small number of monitors (less than five). However, when 10 or more monitors need to be located random deployment is on average as effective as choosing the most central intersections. Moreover, for a large numbers of monitors (more than 70-80) random deployment, although the simplest strategy, achieves coverage results that are very similar to choosing the most central intersections. This result may seem surprising but in fact it is absolutely reasonable. Central intersections tend to lay on the arterial roads and usually are quite close to each other. This results in reduced marginal utility of each additional junction joining the deployment.

We demonstrate Equation 4.3 using the results presented in Figure 4.12. For this, we need to calculate the regression of the simulated measurements presented in Figure 4.12 to the *Gompertz* function. This is presented in Figure 4.13. The regression yields the following results:

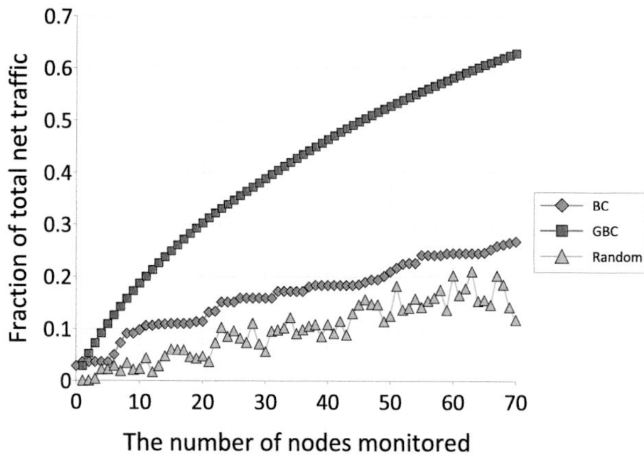

Figure 4.12: The figure presents the results of deployment optimization performed on the Israeli transportation network with average travel times computed using a state of the art traffic assignment model. The Flows and utility of the deployment were estimated using *Betweenness Centrality* and *Group Betweenness Centrality* models, and compared also to the random deployment model. However, the *BC* algorithm had chosen the locations for monitoring units according to the most central intersection based on their BC values, the *GBC* deployment was a greedy algorithm that tried to maximize the net-number of vehicles passing by the monitors. The benefits of the *GBC* strategy is clearly shown, as well as the ability to extrapolate this correlation between number of monitoring units and monitored traffic percentage, in order to find the minimal number of monitoring units required in order to guarantee certain levels of coverage.

$$\begin{aligned} \textbf{Random deployment}: M(n) &= 0.24e^{-2.73e^{-0.03n}} \\ \textbf{BC deployment}: M(n) &= 0.51e^{-2.14e^{-0.02n}} \\ \textbf{GBC deployment}: M(n) &= 0.89e^{-2e^{-0.04n}} \end{aligned} \quad (4.16)$$

At this point, we can assign the values of a,b and c in Equation 4.3 and get the optimal number of monitoring units, for any ratio between the cost of a single monitoring unit and the expected cost of a successful attack. This is presented in Figure 4.14, where the benefit of the GBC deployment scheme can clearly be seen, as it enables the use of more expensive monitoring units (which makes little sense in when the cost of using them is greater than the potential damage of a missed detection—due to low monitoring efficiency).

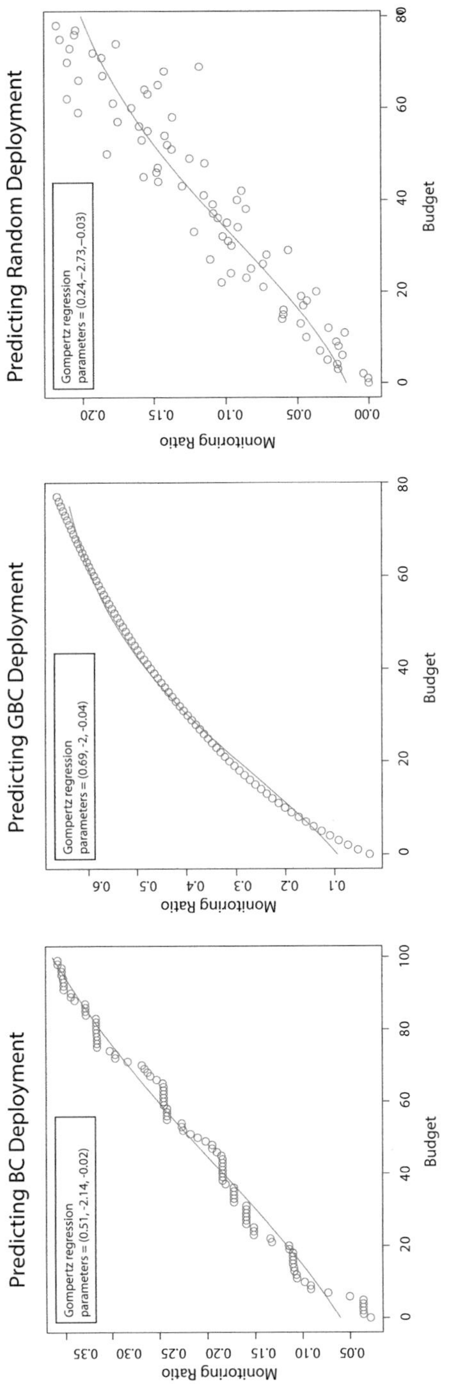

Figure 4.13: *Gompertz* regression of the measurements presented in Figure 4.12—Random deployment, BC deployment and GBC deployment.

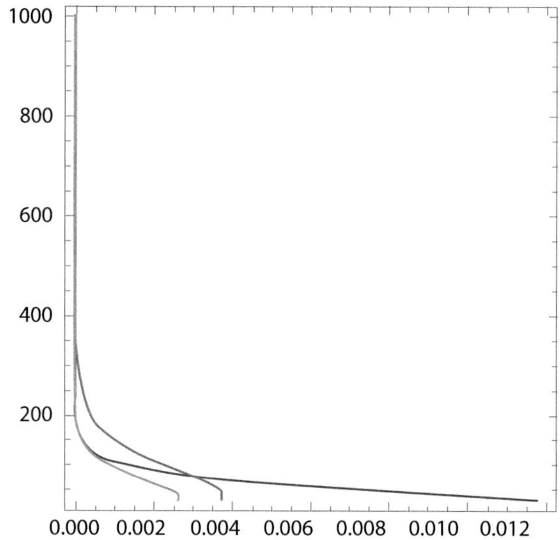

Figure 4.14: The optimal number of monitoring units as a function of the ratio between the cost of a single monitoring unit, and the cost of a successful attack, for the *Gompertz* regressions presented in Figure 4.13 for random deployment (lowest curve), BC deployment (upper curve), and GBC deployment (middle curve, that straches furthest to the right). Note how both the random deployment and the BC deployment schemes can be used only for very low-cost monitoring units, whereas the GBC deployment scheme enables the use of much more expensive monitoring units (due to its increased efficiency in guaranteeing high monitoring coverage).

4.8 Transportation Network Dataset

In this section we evaluate our proposed model using a real-world transportation dataset, containing data regarding the Israeli roads structure, as well as the traffic through them (simulating the potential targets that need to be detected or monitored). An image illustrating the data appears in Figure 4.15. This section describes the dataset and discusses its network properties—from which the optimal deployment of the drones swarm can be derived.

The widespread use of cellular phones in Israel enables the collection of accurate transportation data. Given the small size of the country, all cellular companies provide national wide coverage. As shown in [40], the penetration of cellular phones in the Israeli market is very high, including lower income households, and specifically among individuals in the age group of 10 to 70 years (the main focus of travel behavior studies). Such penetration enables a comprehensive study of travel behavior that is based on the mobility patterns of randomly selected mobile phones in the Israeli transportation system. This data was shown in [40] and [117] to provide high quality coverage of the network, tracking 94% of the trips (defined as at least 2 km in urban areas, and 10 km in rural areas). The resulting data contained a wealth of traffic properties for a network of over 6,000 nodes, and 15,000 directed links. In addition, the network was accompanied with an Origin Destination (OD) matrix, specifying start and end points of trips.

The network was created for the National Israeli Transportation Planning Model. In urban areas the network contains arterial streets that connect the interurban roads. For each link of the network, there is information about the length (km), hierarchical type, free-flow travel time (min), capacity (vehicles per hour), toll (min), hourly flow (vehicles per hour), and congested travel time (min). The hourly flows and congested travel times were obtained from a traffic assignment model that loads the OD matrix on the network links.

4.8.1 Network Structure

Based on the dataset described above we have created a network structure, assigning running indices from 1 to 6716 to the nodes (junctions). We have examined the directed variant of the network where each road segment between two junctions was represented as either one or two directed links between the respective nodes.

To gain a basic understanding of the network, we first extracted and studied several of its structural properties (see Table 4.1). We have partitioned the network into structural equivalence nodal classes and bi-connected components and computed their Betweenness Centrality indices (Betweenness Centrality is an important network feature that measures the portion of shorter paths between all the nodal pairs in the network, that pass through a particular node. See more details in [62, 79, 81]). Structurally equivalent vertices have exactly the same neighbors and this set of these vertices is called a structural equivalence class. As can be seen from Table 4.1 the number of structural equivalence classes is roughly the number of vertices in the network and the largest class size is three. This means that there are no "star-like" structures in the network and alternative paths between any two vertices are either longer than two hops or have other links emanating from the intermediate vertices. On the other hand the number of biconnected components in the network is low compared to the number of nodes, meaning that there are significant regions of the network that can be cut out by merely disconnecting a single node.

4.8.2 Congestions

In this paper we define the impact of congestion as the difference between the time to travel through a congested link and the free-flow travel time. Congestion of a junction can be either inbound or outbound. Inbound congestion is the sum

Table 4.1: Structural properties (Israeli transportation network).

Nodes	6716
Edges (undirected representation)	8374
Edges (directed representation)	15823
Number of structural equivalence classes	6655
Largest equivalence class	3
Number of bi-connected components (BCC)	931
Average BCC size	8.2
Largest BCC	5778

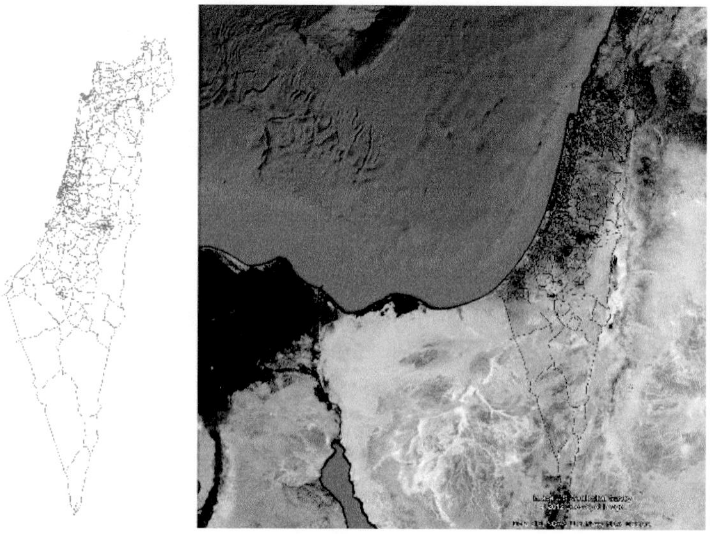

Figure 4.15: A map of the Israeli transportation network that was used for this paper.

of all congestions on inbound links of some junction. Figure 4.16 presents the distribution of congestion on network nodes (junctions). Power law nature of this distribution means that a vast majority of nodes are not congested but there are a few nodes whose congestion can be arbitrarily large. The use of the *Wardrop's User Equilibrium* [63] also implies a low number of yet significant deviations between the routes chosen by travelers during free-flow and during congestions. In Section 4.7.3 we use this fact to merge between two routing strategies.

4.8.3 Flow

The analyzed dataset contains traffic flow through links provided as the number of vehicles per hour. We compute the total inbound flow through a node by

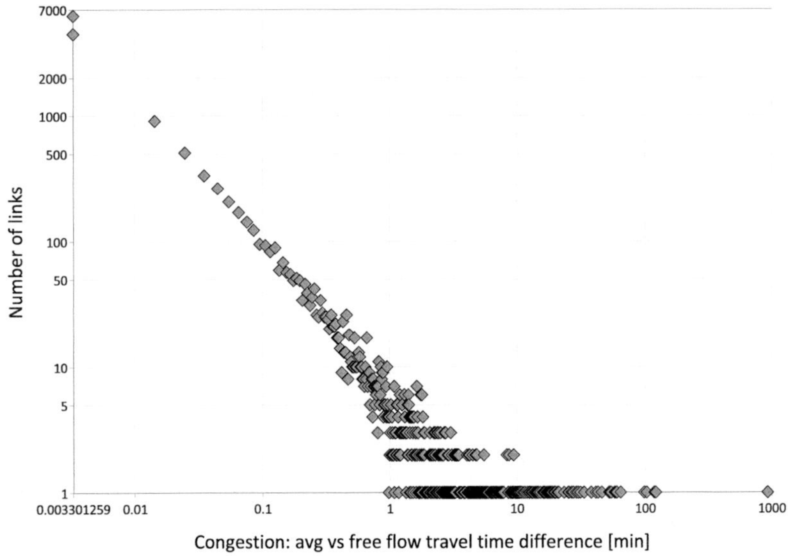

Figure 4.16: Power law distribution of congestion.

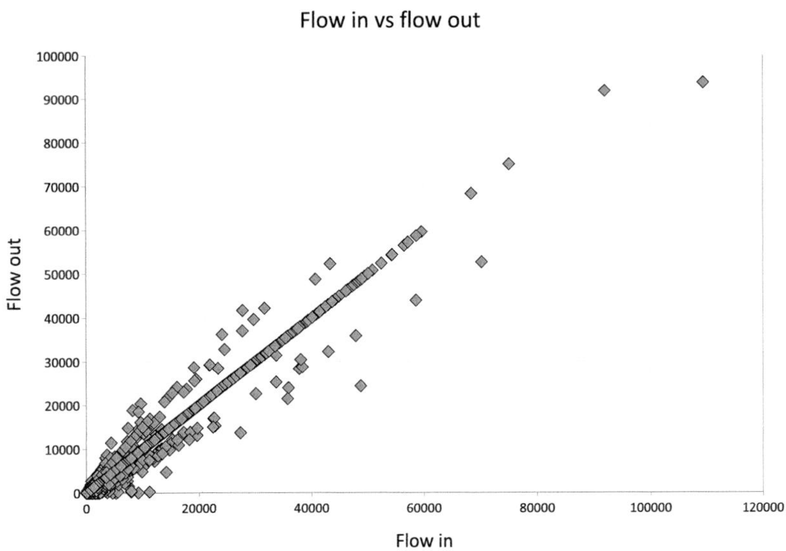

Figure 4.17: Incoming vs. outgoing flow for each node.

summing flows on all of its inbound links, where outbound flow is computed symmetrically. Unless a specific junction is a traffic source or destination we expect the inbound flow to be equal to the outbound flow. Figure 4.17

demonstrates the correlation between inbound and outbound flow. We see that a vast majority of the nodes are located on the main diagonal, however, there are some deviations, caused by the fact that the data represents average measurements that were made in a substantial period of time.

Figure 4.18 presents the distribution of inbound flows on network nodes. This distribution is exponential, meaning that a vast majority of nodes have little flow through them. However, in contrast to network congestion, there are no "unbounded fluctuations", i.e. the flow through the most "busy" junctions is not as high as can be expected from the power law distribution of betweenness and congestions (Figures 4.16 and 4.5). In fact, congestions significantly limit the flow through the busiest junctions, which subsequently is the reason we do not see the long tail in flow distribution.

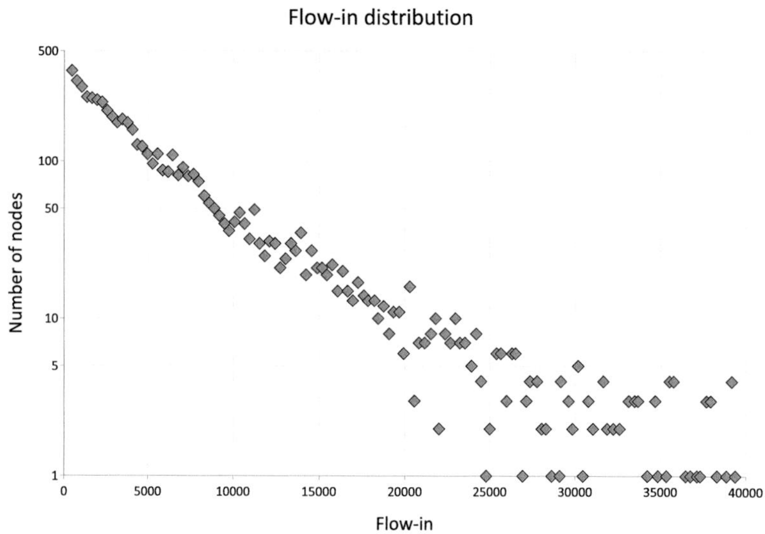

Figure 4.18: Exponential distribution of traffic flow through nodes.

4.9 Conclusions

In this paper we have discussed the problem of optimizing the type, number and locations of surveillance drones trying to detect maneuvering targets that move on top of a pre-defined transportation network. We have presented a model for analytically generating an optimal monitoring strategy, flexible enough to support various models of drones (of different costs and performance specifications), as well as any deployment scheme used. For each threat, the model produces an optimal strategy, based on the estimation of the threat's overall potential damages.

We have validated our model using a variety of deployment schemes, among which was an extremely efficient scheme was developed, which is a transportation oriented variant of the Network Betweenness measure. We have then evaluated the model, as well as this deployment scheme, using a comprehensive dataset that covers the Israeli transportation network, and demonstrated how the optimal locations of a reasonable number of drones can be approximated with high accuracy. The method proposed in this paper can now be used in order to generate highly efficient dynamic monitoring strategies, for a multitude of real-time threats, of different operational parameters.

A more theoretical approach to this problem that studies the complexity of all possible strategies for a collaborative monitoring of a given area at which an unknown number of targets dynamically maneuver can be found in [17]. An additional similar variant to this problem is the search for pollutants' emitting vehicles, where the merit function is derived from environmental considerations [91]. It is interesting to mentioned that in those variants as well, the topological properties of the network along which the "targets" can move significantly influence the ability of drones to track them, as was pointed out in [9, 10].

In future works we plan to conduct a thorough study of the way this method can be used for resource allocation scenarios involving several types threats, using heterogenous deployment strategies that would utilize drones of different characteristics simultaneously. In addition, we intend to conduct a further analysis of the resilience of the method, under uncertainty (regarding the topology, potential threats, and noises in the deployment scheme).

Finally, it is interesting to note that the problem of finding an optimal (and optionally dynamic) monitoring strategy is related to other kinds of monitoring problems, such as "reverse-monitoring"—maximizing the exposure of a certain signal, or campaign to a population [33], or monitoring for evading land targets by a flock of drones. In this problem, however, the fact that the paths of the drones are unconstrained (as they are airborne) makes the calculation of a near-optimal monitoring strategy fairly easy [13, 14].

References

[1] E.U. Acar, Y. Zhang, H. Choset, M. Schervish, A.G. Costa, et al. Path planning for robotic demining and development of a test platform. In: *International Conference on Field and Service Robotics*, pages 161–168, 2001.

[2] Matthew M Aid. All glory is fleeting: Sigint and the fight against international terrorism. *Intelligence and National Security*, 18(4): 72–120, 2003.

[3] S. Alpern and S. Gal. *The Theory of Search Games and Rendezvous*. Kluwer Academic Publishers, 2003.

[4] T. Altshuler, Y. Altshuler, R. Katoshevski and Y. Shiftan. Modeling and prediction of ride-sharing utilization dynamics. *Journal of Advanced Transportation*, 2019, 2019.

[5] Y. Altshuler, N. Aharony, M. Fire, Y. Elovici and A Pentland. Incremental learning with accuracy prediction of social and individual properties from mobile-phone data. *CoRR*, 2011.

[6] Y. Altshuler, N. Aharony, A. Pentland, Y. Elovici and M. Cebrian. Stealing reality: When criminals become data scientists (or vice versa). *Intelligent Systems, IEEE*, 26(6): 22–30, Nov.-Dec. 2011.

[7] Y. Altshuler, A.M. Bruckstein and I.A. Wagner. Swarm robotics for a dynamic cleaning problem. In: *IEEE Swarm Intelligence Symposium*, pages 209–216, 2005.

[8] Y. Altshuler, M. Fire, N. Aharony, Y. Elovici and A Pentland. How many makes a crowd? on the correlation between groups' size and the accuracy of modeling. In: *International Conference on Social Computing, Behavioral-Cultural Modeling and Prediction*, pages 43–52. Springer, 2012.

[9] Y. Altshuler, I.A. Wagner and A.M. Bruckstein. Shape factor‘s effect on a dynamic cleaners swarm. In: *Third International Conference on*

Informatics in Control, Automation and Robotics (ICINCO), the Second International Workshop on Multi-Agent Robotic Systems (MARS), pages 13–21, 2006.

[10] Y. Altshuler, I.A. Wagner and A.M. Bruckstein. On swarm optimality in dynamic and symmetric environments. volume 7, page 11, 2008.

[11] Y. Altshuler, I.A. Wagner and A.M. Bruckstein. Collaborative exploration in grid domains. In: *Sixth International Conference on Informatics in Control, Automation and Robotics (ICINCO)*, 2009.

[12] Y. Altshuler, I.A. Wagner, V. Yanovski and A.M. Bruckstein. Multi-agent cooperative cleaning of expanding domains. *International Journal of Robotics Research*, 30: 1037–1071, 2010.

[13] Y. Altshuler, V. Yanovski, I.A. Wagner and A.M. Bruckstein. The cooperative hunters—efficient cooperative search for smart targets using uav swarms. In: *Second International Conference on Informatics in Control, Automation and Robotics (ICINCO), the First International Workshop on Multi-Agent Robotic Systems (MARS)*, pages 165–170, 2005.

[14] Y. Altshuler, V. Yanovsky, A.M. Bruckstein and I.A. Wagner. Efficient cooperative search of smart targets using uav swarms. *ROBOTICA*, 26: 551–557, 2008.

[15] Y. Altshuler, V. Yanovsky, I. Wagner and A. Bruckstein. Swarm intelligence—searchers, cleaners and hunters. *Swarm Intelligent Systems*, pages 93–132, 2006.

[16] Y. Altshuler and A.M. Bruckstein. The complexity of grid coverage by swarm robotics. In: *ANTS 2010*, pages 536–543. LNCS, 2010.

[17] Y. Altshuler and A.M. Bruckstein. Static and expanding grid coverage with ant robots: Complexity results. *Theoretical Computer Science*, 412(35): 4661–4674, 2011.

[18] Y. Altshuler, S. Dolev, Y. Elovici and N. Aharony. Ttled random walks for collaborative monitoring. In: *NetSciCom 2010 (Second International Workshop on Network Science for Communication Networks)*, San Diego, CA, USA, 3 2010.

[19] Y. Altshuler, M. Fire, N. Aharony, Z. Volkovich, Y. Elovici, et al. Trade-offs in social and behavioral modeling in mobile networks. In: *Social Computing, Behavioral-Cultural Modeling and Prediction*, pages 412–423. Springer, 2013.

[20] Y. Altshuler, M. Fire, E. Shmueli, Y. Elovici, A. Bruckstein, et al. Detecting anomalous behaviors using structural properties of social networks. In: *Social Computing, Behavioral-Cultural Modeling and Prediction*, pages 433–440. Springer, 2013.

[21] Y. Altshuler, M. Fire, E. Shmueli, Y. Elovici, A. Bruckstein, et al. The social amplifier—reaction of human communities to emergencies. *Journal of Statistical Physics*, 152(3): 399–418, 2013.

[22] Y. Altshuler, A. Pentland, S. Bekhor, Y. Shiftan and A. Bruckstein. Optimal dynamic coverage infrastructure for large-scale fleets of reconnaissance uavs. *arXiv preprint arXiv:1611.05735*, 2016.

[23] Y. Altshuler, A. Pentland and Alfred M Bruckstein. Collaborative patrolling swarms in stochastically expanding environments. In: *Swarms and Network Intelligence in Search*, pages 155–185. Springer, 2018.

[24] Y. Altshuler, A. Pentland and A.M. Bruckstein. The cooperative hunters–efficient and scalable drones swarm for multiple targets detection. In: *Swarms and Network Intelligence in Search*, pages 187–205. Springer, 2018.

[25] Y. Altshuler, A. Pentland and A.M. Bruckstein. Cooperative "swarm cleaning" of stationary domains. In: *Swarms and Network Intelligence in Search*, pages 15–49. Springer, 2018.

[26] Y. Altshuler, A. Pentland and A.M. Bruckstein. Introduction to swarm search. In: *Swarms and Network Intelligence in Search*, pages 1–14. Springer, 2018.

[27] Yaniv Altshuler, Alex Pentland, and Alfred M Bruckstein. Optimal dynamic coverage infrastructure for large-scale fleets of reconnaissance uavs. In: *Swarms and Network Intelligence in Search*, pages 207–238. Springer, 2018.

[28] Y. Altshuler, A. Pentland and A.M. Bruckstein. The search complexity of collaborative swarms in expanding z^2 grid regions. In: *Swarms and Network Intelligence in Search*, pages 129–153. Springer, 2018.

[29] Y. Altshuler, A. Pentland and A.M. Bruckstein. Swarm search of expanding regions in grids: Lower bounds. In: *Swarms and Network Intelligence in Search*, pages 51–89. Springer, 2018.

[30] Y. Altshuler, A. Pentland and A.M. Bruckstein. Swarm search of expanding regions in grids: Upper bounds. In: *Swarms and Network Intelligence in Search*, pages 91–127. Springer, 2018.

[31] Y. Altshuler, A. Pentland and A.M. Bruckstein. *Swarms and Network Intelligence in Search*. Springer, 2018.

[32] Y. Altshuler, R. Puzis, Y. Elovici, S. Bekhor and A.S. Pentland. On the rationality and optimality of transportation networks defense: a network centrality approach. *Securing Transportation Systems*, pages 35–63.

[33] Y. Altshuler, E. Shmueli, G. Zyskind, O. Lederman, N. Oliver, et al. Campaign optimization through behavioral modeling and mobile network analysis. *Computational Social Systems, IEEE Transactions on*, 1(2): 121–134, 2014.

[34] Y. Altshuler, E. Shmueli, G. Zyskind, O. Lederman, N. Oliver, et al. Campaign optimization through mobility network analysis. *Geo-intelligence and visualization through big data trends*, pages 33–74, 2015.

[35] J.M. Anthonisse. The rush in a directed graph. Technical Report BN 9/71, Stichting Mathematisch Centrum, Amsterdam, 1971.

[36] R.C. Arkin. Integrating behavioral, perceptual, and world knowledge in reactive navigation. *Robotics and Autonomous Systems*, 6: 105–122, 1990.

[37] Desmond Ball, et al. *Burma's Military Secrets: Signals Intelligence (SIGINT) from 1941 to Cyber Warfare*. White Lotus Press, 1998.

[38] M. Barthélemy. Betweenness centrality in large complex networks. *The European Physical Journal B – Condensed Matter*, 38(2):163–168, March 2004.

[39] R. Bejar, B. Krishnamachari, C. Gomes and B. Selman. Distributed constraint satisfaction in a wireless sensor tracking system. In: *Proceedings of the IJCAI-01 Workshop on Distributed Constraint Reasoning*, 2001.

[40] S. Bekhor, Y. Cohen and C. Solomon. Evaluating long-distance travel patterns in israel by tracking cellular phone positions. *Journal of Advanced Transportation*, pages n/a–n/a, 2011.

[41] D.J. Berndt, A.R. Hevner and J. Studnicki. Bioterrorism surveillance with real-time data warehousing. In: *Proceedings of the 1st NSF/NIJ conference on Intelligence and security informatics*, ISI'03, pages 322–335, Berlin, Heidelberg, 2003. Springer-Verlag.

[42] G. Confessore, P. Reverberi and L. Bianco. A network based model for traffic sensor location with implications on o/d matrix estimates. *Transportation Science*, 35(1): 50–60, 2001.

[43] P. Bork, L.J. Jensen, C. von Mering, A.K. Ramani, I. Lee, et al. Protein interaction networks from yeast to human. *Curr. Opin. Struct. Biol.*, 14(3): 292–299, 2004.

[44] U. Brandes. On variants of shortest-path betweenness centrality and their generic computation. *Social Networks*, 30(2): 136–145, 2008.

[45] Z. Butler, A. Rizzi and R. Hollis. Distributed coverage of rectilinear environments. In: *Proceedings of the Workshop on the Algorithmic Foundations of Robotics*, 2001.

[46] Daniel Byman. Why drones work. *Foreign Affairs*, 92(4): 32–43, 2013.

[47] H. Chen, Fei-Yue Wang and D. Zeng. Intelligence and security informatics for homeland security: information, communication, and transportation. *Intelligent Transportation Systems, IEEE Transactions on*, 5(4): 329–341, dec. 2004.

[48] Robert Chesney. Military-intelligence convergence and the law of the title 10/title 50 debate. *Journal of National Security Law and Policy*, 5: 539, 2012.

[49] D. Chevallier and S. Payandeh. On kinematic geometry of multi-agent manipulating system based on the contact force information. In: *The*

Sixth International Conference on Intelligent Autonomous Systems (IAS-6), pages 188–195, 2000.

[50] Lt Kendra LB Cook. The silent force multiplier: the history and role of uavs in warfare. In: *Aerospace Conference, 2007 IEEE*, pages 1–7. IEEE, 2007.

[51] L. Damianos, J. Ponte, S. Wohlever, F. Reeder, D. Day, et al.. Mitap for bio-security: A case study. *AI Magazine*, 23(4): 13–29, 2002.

[52] S. Dolev, Y. Elovici, R. Puzis and P. Zilberman. Incremental deployment of network monitors based on group betweenness centrality. *Inf. Proc. Letters*, 109: 1172—-1176, 2009.

[53] A. d'Onofrio. A general framework for modeling tumor-immune system competition and immunotherapy: Mathematical analysis and biomedical inferences. *Physica D*, 208: 220–235, 2005.

[54] G. Dudek, M. Jenkin, E. Milios and D. Wilkes. Robotic exploration as graph construction. *IEEE Transactions on Robotics and Automation*, 7: 859–865, 1991.

[55] M.G. Bell, S. Grosso and A. Ehlert. The optimisation of traffic count locations in road networks. *Transportation Research Part B*, 40(6): 460–479, 2006.

[56] G.M. Erickson, P.J. Currie, B.D. Inouye and A.A. Winn. Tyrannosaur life tables: An example of nonavian dinosaur population biology. *Science*, 313(5784): 213–217, 2006.

[57] M. G. Everett and S. P. Borgatti. The centrality of groups and classes. *Mathematical Sociology*, 23(3): 181–201, 1999.

[58] M. Faloutsos, P. Faloutsos and C. Faloutsos. On power-law relationships of the internet topology. *SIGCOMM Comput. Comm. Rev.*, 29(4): 251–262, 1999.

[59] H.S. Mahmassani, S.M. Eisenman and X. Fei. Sensor coverage and location for real-time traffic prediction in large-scale networks. *Transportation Research Record*, 2039(1): 1–15, 2007.

[60] A. Felner, Y. Shoshani, Y. Altshuler and A.M. Bruckstein. Multi-agent physical a* with large pheromones. *Journal of Autonomous Agents and Multi-Agent Systems*, 12(1): 3–34, 2006.

[61] Ann Finkbeiner. Military technology: Death by remote control. *Nature*, 534(7609): 618, 2016.

[62] L.C. Freeman. A set of measures of centrality based on betweenness. *Sociometry*, 40(1): 35–41, 1977.

[63] WARDROP J. G. Some theoretical aspects of road traffic research. *Proceedings of the Institution of Civil Engineers*, 1: 325–378, 1952.

[64] B.P. Gerkey and M.J. Mataric. Sold! market methods for multi-robot control. *IEEE Transactions on Robotics and Automation, Special Issue on Multi-robot Systems*, 2002.

[65] Benjamin Gompertz. On the nature of the function expressive of the law of human mortality, and on a new mode of determining the value of life contingencies. *Philosophical Transactions of the Royal Society of London*, 115: 513–583, 1825.

[66] T. Haynes and S. Sen. *Evolving Behavioral Strategies in Predators and Prey*, volume 1042 of *Lecture Notes in Computer Science*, chapter Adaptation and Learning in Multi-Agent Systems, pages 113—126. Springer, Berlin, 1986.

[67] Ian Henderson. Civilian intelligence agencies and the use of armed drones. In: *Yearbook of International Humanitarian Law-2010*, pages 133–173. Springer, 2011.

[68] S. Hettiarachchi and W. Spears. Moving swarm formations through obstacle fields. In: *International Conference on Artificial Intelligence*, 2005.

[69] P. Holme. Congestion and centrality in traffic flow on complex networks. *Advances in Complex Systems*, 6(2): 163–176, 2003.

[70] S. Peeta, C.-H. Chu, S.-R. Hu. Identification of vehicle sensor locations for link-based network traffic applications. *Transportation Research Part B*, 43(8–9): 873–894, 2009.

[71] David Eugene Johnson. *Military capabilities for hybrid war: Insights from the Israel Defense Forces in Lebanon and Gaza*. Rand Corporation, 2014.

[72] P.S. Kanaroglou, M. Jerrett, J. Morrison, B. Beckerman, M.A. Arain, et al. Establishing an air pollution monitoring network for intra-urban population exposure assessment: A location-allocation approach. *Atmospheric Environment*, 39(13): 2399–2409, 2005. 12th International Symposium, Transport and Air Pollution 12th International Symposium, Transport and Air Pollution.

[73] W. Kerr and D. Spears. Robotic simulation of gases for a surveillance task. In: *Intelligent Robots and Systems (IROS 2005)*, pages 2905–2910, 2005.

[74] S. Kirkpatrick and J.J. Schneider. How smart does an agent need to be? *International Journal of Modern Physics*, C 16: 139–155, 2005.

[75] B. Koopman. *Search and Screening: General Principles with Historical Applications*. Pergamon Press, 1980.

[76] B.O. Koopman. The theory of search II, target detection. *Operations Research*, 4(5): 503–531, 1956.

[77] R. E. Korf and W. Zhang. Performance of linear-space search algorithms. *Artificial Intelligence*, 79(2): 241–292, 1995.

[78] Lo H. Lam, W. Accuracy of o-d estimates from traffic counts. *Traffic Engineering and Control*, 31: 358–367, 1990.

[79] J. Lerner. *Network analysis: methodological foundations*, chapter Role Assignments. Springer LNCS 3418, 2005.

[80] Y. Ouyang and X. Li. Reliable sensor deployment for network traffic surveillance. *Transportation Research Part B*, 45: 218–231, 2011.

[81] F. Lorrain and H.C. White. Structural equivalence of individuals in social networks. *The Journal of Mathematical Sociology*, 1(1): 49–80, 1971.

[82] M.J. Mataric. Designing emergent behaviors: From local interactions to collective intelligence. In: J.Meyer, H.Roitblat and S.Wilson, editors, *Proceedings of the Second International Conference on Simulation of Adaptive Behavior*, pages 432–441. MIT Press, 1992.

[83] P.M. Morse and G.E. Kimball. *Methods of operations research*. MIT Press and New York: Wiley, 1951.

[84] A.T. Murray, K. Kim, J.W. Davis, R. Machiraju and R. Parent. Coverage optimization to support security monitoring. *Computers, Environment and Urban Systems*, 31(2): 133–147, 2007.

[85] K. Passino, M. Polycarpou, D. Jacques, M. Pachter, Y. Liu, et al. *Cooperative Control for Autonomous Air Vehicles*, chapter Cooperative Control and Optimization. Kluwer Academic, Boston, 2002.

[86] J.T.K. Ping, A.E. Ling, T.J. Quan and C.Y. Dat. Generic unmanned aerial vehicle (uav) for civilian application-a feasibility assessment and market survey on civilian application for aerial imaging. In: *Sustainable Utilization and Development in Engineering and Technology (STUDENT), 2012 IEEE Conference on*, pages 289–294. IEEE, 2012.

[87] R. Puzis, Y. Altshuler, Y. Elovici, S. Bekhor, Y. Shiftan and A.S. Pentland. Augmented betweenness centrality for environmentally-aware traffic monitoring in transportation networks.

[88] R. Puzis, Y. Elovici and S. Dolev. Fast algorithm for successive computation of group betweenness centrality. *Phys. Rev. E*, 76(5): 056709, 2007.

[89] R. Puzis, Y. Elovici and S. Dolev. Finding the most prominent group in complex networks. *AI Comm.*, 20: 287–296, 2007.

[90] R. Puzis, M. D. Klippel, Y. Elovici and S. Dolev. Optimization of nids placement for protection of intercommunicating critical infrastructures. In: *EuroISI*, 2007.

[91] R. Puzis, Y. Altshuler, Y. Elovici, S. Bekhor, Y. Shiftan, et al. Augmented betweenness centrality for environmentally-aware traffic monitoring in transportation networks. *Journal of Intelligent Transportation Systems*, 17(0): 91–105, 2013.

[92] G. Rabideau, T. Estlin, T. Chien and A. Barrett. A comparison of coordinated planning methods for cooperating rovers. In: *Proceedings of the American Institute of Aeronautics and Astronautics (AIAA) Space Technology Conference*, 1999.

[93] I. Rekleitis, V. Lee-Shuey, A. Peng Newz and H. Choset. Limited communication, multi-robot team based coverage. In: *IEEE International Conference on Robotics and Automation*, April 2004.

[94] Petri Rouvinen. Diffusion of digital mobile telephony: Are developing countries different? *Telecommunications Policy*, 30(1): 46–63, 2006.

[95] Niklas Schörnig. Unmanned warfare: Towards a neo-interventionist era? In: *The Armed Forces: Towards a Post-Interventionist Era?*, pages 221–235. Springer, 2013.

[96] J. Scott. *Social Network Analysis: A Handbook.* Sage Publications, London, 2000.

[97] E. Shmueli, I. Mazeh, L. Radaelli, A.S. Pentland and Y. Altshuler. Ride sharing: A network perspective. In: *Social Computing, Behavioral-Cultural Modeling, and Prediction*, pages 434–439. Springer, 2015.

[98] B. Shucker and J.K. Bennett. Target tracking with distributed robotic macrosensors. In: *Military Communications Conference (MILCOM 2005)*, volume 4, pages 2617–2623, 2005.

[99] D.A. Smale, G.A. Kendrick, E.S. Harvey, T.J. Langlois, R.K. Hovey, et al. Regional-scale benthic monitoring for ecosystem-based fisheries management (EBFM) using an autonomous underwater vehicle (AUV). *ICES Journal of Marine Science*, 69(6): 1108–1118, 05 2012.

[100] S. Somin, G. Gordon and Y. Altshuler. Network analysis of erc20 tokens trading on ethereum blockchain. In: *Unifying Themes in Complex Systems IX: Proceedings of the Ninth International Conference on Complex Systems*, page 439. Springer, 2018.

[101] S. Somin, G. Gordon and Y. Altshuler. Social signals in the ethereum trading network. *arXiv preprint arXiv: 1805.12097*, 2018.

[102] L. Steels. Cooperation between distributed agents through self-organization. In: Y.DeMazeau and J.P.Muller, editors, *Decentralized A.I - Proc. first European Workshop on Modeling Autonomous Agents in Multi-Agents world*, pages 175–196. Elsevier, 1990.

[103] R. Stern, R. Puzis and A. Felner. Potential search: a bounded-cost search algorithm. In: *AAAI 21st International Conference on Automated Planning and Scheduling (ICAPS)*, 2011.

[104] S.H. Strogatz. Exploring complex networks. *Nature*, 410: 268–276, March 2001.

[105] S.M. Thayer, M.B. Dias, B.L. Digney, A. Stentz, B. Nabbe, et al. Distributed robotic mapping of extreme environments. In: *Proceedings of SPIE*, volume 4195 of *Mobile Robots XV and Telemanipulator and Telepresence Technologies VII*, 2000.

[106] A. Thorndike. Summary of antisubmarine warfare operations in world war ii. Summary report, NDRC Summary Report, 1946.

[107] Gábor Vásárhelyi, Csaba Virágh, Gergő Somorjai, Tamás Nepusz, Agoston E Eiben and Tamás Vicsek. Optimized flocking of autonomous drones in confined environments. *Science Robotics*, 3(20): eaat3536, 2018.

[108] P. Vincent and I. Rubin. A framework and analysis for cooperative search using uav swarms. In: *ACM Simposium on applied computing*, 2004.

[109] Paul Voosen. Drones reveal earthquake hazards hidden in the abyss, 2019.

[110] I.A. Wagner, Y. Altshuler, V. Yanovski and A.M. Bruckstein. Cooperative cleaners: A study in ant robotics. *The International Journal of Robotics Research (IJRR)*, 27(1): 127–151, 2008.

[111] I.A. Wagner and A.M. Bruckstein. From ants to a(ge)nts: A special issue on ant—robotics. *Annals of Mathematics and Artificial Intelligence, Special Issue on Ant Robotics*, 31(1–4): 1–6, 2001.

[112] S. Wasserman and K. Faust. *Social network analysis: Methods and applications*. Cambridge, England: Cambridge University Press., 1994.

[113] M.P. Wellman and P.R. Wurman. Market-aware agents for a multiagent world. *Robotics and Autonomous Systems*, 24: 115–125, 1998.

[114] D.R. White and S. P. Borgatti. Betweenness centrality measures for directed graphs. *Social Networks*, 16: 335–346, 1994.

[115] C. Yang, L. Gan and H. Yang. Models and algorithms for the screen line-based traffic-counting location problems. *Computers and Operations Research*, 33(3): 836–858, 2006.

[116] J. Zhou and H.Yang. Optimal traffic counting locations for origin-destination matrix estimation. *Transportation Research Part B*, 32(2): 109–126, 1998.

[117] Charles Solomon-Leonid Kheifits Yehuda J. Gur, Shlomo Bekhor. Intercity person trip tables for nationwide transportation planning in israel obtained from massive cell phone data. *Transportation Research Record: Journal of the Transportation Research Board*, 2121: 145–151, 2009.

[118] S.H. Yook, H. Jeong and A.-L. Barabasi. Modeling the internet's large-scale topology. *Proceedings of the National Academy of Science*, 99(21): 13382–13386, Oct. 2002.

[119] Shlomo Zilberstein. Using anytime algorithms in intelligent systems. *AI Magazine*, 17(3): 73–83, 1996.

[120] M. Zwang, S. Somin and A. 'Sandy' Pentland and Yaniv Altshuler. Detecting bot activity in the ethereum blockchain network, 2018.

Chapter 5

Socioeconomic Patterns of Twitter Activity

Jacob Levy Abitbol[1] and Alfredo J. Morales[2]
[1]Lyon East Medical School
[2]Massachusetts Institute of Technology, Media Lab

5.1 Introduction

Income and wealth are relevant contributors of social behaviors in terms of culture, beliefs, health and education among others [1, 13, 20, 28, 67, 79]. Accessing socioeconomic information is crucial for designing public policies [38, 39]. It represents a challenge for governments and corporations to gather socioeconomic information given the costly efforts associated to the deployment of large-scale surveys and the national census. The recent access to datasets collected from social media and other electronic platforms enables the direct observation of individuals and their behavior [43]. The application of new algorithms to these data sources has consolidated the inference of socioeconomic status into a mature and innovative field of research.

Throughout history, governments have systematically quantified natural and social systems as a foundation for annual planning, tax policy formulation, territorial organization, and maintenance of order. The adoption of standardized metrics and measurements is pivotal for effective planning. However, this approach often simplifies the description of social and natural systems to a level that is comprehensible to policymakers, as noted for example in [10]. This simplification can sometimes omit critical details necessary for the healthy functioning of systems, such as intricate relationships and background elements crucial for systemic stability [72]. The advent of new data sources offers a more nuanced observation of the complexities and diversities inherent in social behaviors and relationships. This advancement presents a unique opportunity to develop plans and policies that are more aligned with the intricate realities of these phenomena, potentially enhancing their effectiveness and relevance.

The data obtained from social media has enabled unprecedented analysis of the complexity of societies [43]. Recent studies show patterns of social behaviors across multiple scales of observation, ranging from individual preferences up to the structure and dynamics of self-organized groups and collectives [4, 5, 11, 29, 31]. Example applications of these analyses include movements of the stock market based on collective sentiments and moods [9, 50, 65, 66], the prediction of electoral results [59, 75], detection of emergencies, riots and vicil unrest [8], the political polarization of societies [57, 74], and the relationship between health and shopping preferences [6, 53, 84]. Moreover, access to geolocated data enables the modeling and prediction of human mobility patterns and online communication [3, 22, 30, 70].

Socioeconomic status often manifests through distinct behavioral patterns characteristic of different income groups. It is a well-established fact that income levels significantly influence residential choices, contributing to urban segregation based on wealth, race, ethnicity, and other factors. Leveraging social media data allows for a more comprehensive understanding of social segregation by capturing not only spatial divisions but also the nuances of interactions and behaviors. Prior research has demonstrated that neighborhood segregation in urban areas significantly influences the locales people frequent [16, 83], their online interactions [19], and even the topics they discuss [58]. These studies highlight how the compartmentalization of urban spaces effectively fragments social networks, thereby influencing the flow of information and the propagation of behaviors among individuals. Given that learning often occurs through imitation, this segregated structure of social networks fosters distinct social behaviors, encompassing even sentiments and emotions [15]. This phenomenon underlines the deep interconnection between socioeconomic factors and the intricate web of social behaviors and interactions.

In this chapter we analyze Twitter activity and observe patterns of behavior that are characteristic of different socioeconomic groups. We apply methods from the suite of machine learning and information theory, including predictors and dimensional reduction, and expose the patterns the algorithms use to make their inferences. The chapter is organized as follows: Section 5.2 contains a summary of related studies in the field of income prediction. Section 5.3 includes a description of the data and the methods we use to collect and analyze it. In Section 5.4 we present results from predicting income based on mobility and hashtag behavior. In Section 5.5 we show the structure of the conversational space by means of dimensional reduction. In Section 5.6 we show signature patterns of socio-economic groups according to the diversity of their interactions. Finally, we summarize our results in Section 5.7 and conclude in Section 5.8.

This chapter is based on a work previously published in [2].

5.2 Related Work

Methods for inferring demographic information from observations of social behaviors have been increasingly developed since the recent introduction of social media data, which combined with more traditional sources such as census records, enable the creation of a richer picture of society [7, 42]. Until recently, researchers could only access data from surveys or questionnaires, which by definition are limited in size, scope and frequency, given the difficulties in their deployment and collection. By observing the behaviors of individuals and mapping that information to metadata from the Census or surveys, researchers may obtain patterns of behavior which are characteristic of certain demographic groups and allow the inference of traits from unlabeled individuals.

Numerous diverse features have been used in order to predict demographic traits of individuals from the data generated by the usage of multiple types of electronic communication. These advances enable a further characterization of the population and prediction of individual attributes such as age [17], occupation [32, 41, 68, 69], political affiliation [82], personality traits [71], and income [52, 69]. The properties of Twitter activity and network of followers have been used to estimate gender and ethnicity [18], unemployment [51] and language [24].

An important predictor of income is human mobility patterns. Previous research has shown that the diversity of human mobility is an indicator of economic development across multiple regions [23]. Aggregated data produced by using mobile phones [26, 78] and geo-located social media outlets [36] have been crucial in advancing the analysis of human mobility patterns, which are predictable given the regularity of commuting [62] and visitation destinations.

Another basis for income prediction is language usage and online content. The relationship between income and language dates back to the early development of socio-linguistics. Occupations determine both the socio-economic status of individuals and their use of language [12]. Twitter users have been accurately mapped to professional occupations based on the content of their tweets [41, 68, 69]. In this type of analysis, the data must be manually annotated prior to further predictions and such a process requires extensive field knowledge which is not easily accessible. More recent works

[58], advance these limitations by automatically finding collective topics of conversation which are mapped to socio-economics by means of aggregate behavior.

Extending the scope of demographic inference, recent studies have begun exploring the relationship between blockchain activity and socio-economic factors, particularly within the Ethereum network. Ethereum, known for its versatility in supporting a wide array of tokens and decentralized applications, offers a unique dataset in the form of transaction histories and token movements. These blockchain-based activities provide novel insights into user behaviors and economic interactions that are not captured by traditional social media or mobile phone data [60].

Research into Ethereum's network has shown that the patterns of token usage and transaction networks can reflect economic behavior and possibly infer socio-economic status. For instance, the frequency and diversity of token transactions, the network structure of token exchanges, and the distribution of holdings across different tokens can all serve as indicators of economic activity and financial status [76]. Moreover, the decentralized and pseudonymous nature of blockchain transactions presents an opportunity to study economic behaviors in a more unobstructed and naturalistic setting, compared to traditional social media platforms where user behavior might be influenced by self-presentation biases [73].

The integration of Ethereum blockchain data into the study of socio-economic status offers a complementary perspective to traditional data sources. It enables a more comprehensive analysis of economic behaviors, transcending the limitations of geographical and cultural boundaries inherent in social media and mobile phone usage patterns [61, 77]. This blockchain-based approach not only enriches our understanding of the digital economy but also aligns with the evolving nature of financial transactions and asset management in the digital age. As such, it provides a valuable addition to the methodologies for inferring socio-economic status and contributes to a more nuanced and multi-faceted understanding of economic behaviors in the contemporary world.

5.3 Data and Methods

The data has been collected using the Application Programming Interface (API) provided by Twitter for streaming content [81]. The stream API provides around 3% of the overall Twitter activity in real time [44] and over 90% of publicly available tweets with geo-location [64]. The geo-location feature provides precise coordinates of individuals as they post messages.

Twitter is a social media platform whose users post texts with links to figures and videos either from smartphones and other personal computers. Twitter activity has been analyzed to understand the geography of human sentiments [25], content share networks [31] and dynamics of social influence [56]. It has also been used to advance the understanding of global patterns of human mobility [46], activity [45] and languages [55].

Previous studies have analyzed biases in geo-located Twitter users [21, 35, 54]. The Twitter population trends younger, wealthier and urban. However, the under-aged population is under-represented and the wealth of individuals seems irrelevant in most American cities [58]. Despite these observations, more recent studies show that opinions collected on social media around relevant topics do not differ from traditional surveys [37].

We characterize users by the set of locations where they tweet from and assume their home location according to the neighborhood most frequently visited at night during weekdays. These methods are consistent with other studies using mobile phone [63] and Twitter data [62]. Individuals are labeled according to the income of their home location neighborhood based on Census data. The income labels are used to build predictors and characterize the collective activity of the different socio-economic groups. The results are shown here.

5.4 Mobility and Hashtag Space

Users are characterized by both visited locations and hashtags posted. For this purpose we create two independent feature spaces: one representing human mobility and another representing hashtag activity. In both cases, samples (or rows) represent individuals. In the mobility feature space, the features (or columns) indicate neighborhoods and the non-zero values show the locations individuals visit and tweet from. In the hashtag feature space, columns represent hashtags used by at least 5 people and the non-zero elements indicate whether individuals posted those hashtags. Each feature space is transformed using TF-IDF [48]. This transformation is often used to classify documents and highlights local information as opposed to globally used terms.

We apply two learning algorithms to each feature space. The first algorithm is a MLP regression that predicts income as a numeric value. The other one is based on predicting the income quantile using a MLP classifier. We apply them to both mobility and hashtag space. For this purpose, we divide the sample into a training set with 75% of the individuals and a test set with the remaining 25%. We create multiple samples in order to analyze the performance of the predictors behavior as a random variable. Bootstrapping the performance enables a more robust understanding of the prediction quality.

Figure 5.1 shows the results of the prediction both numerically (left panels) and categorically (right panels). The top panels show the results of the human mobility feature space. The bottom panels show results of hashtag usage feature space. The results of the regressor are shown in the left panels as scatter plots showing the predicted values (y-axis) against the real ones (x-axis). The scatter plots show the overlapped results of the multiple samples we create to bootstrap the algorithm behavior. For each sample we calculate the Pearson correlation. We present the distribution of these correlations in Figure 5.2 (top left). The correlations are high with an average of $r = 0.8$ for the mobility feature space and $r = 0.55$ for the hashtag one. In both cases a considerable part of the variance is explained by the algorithm.

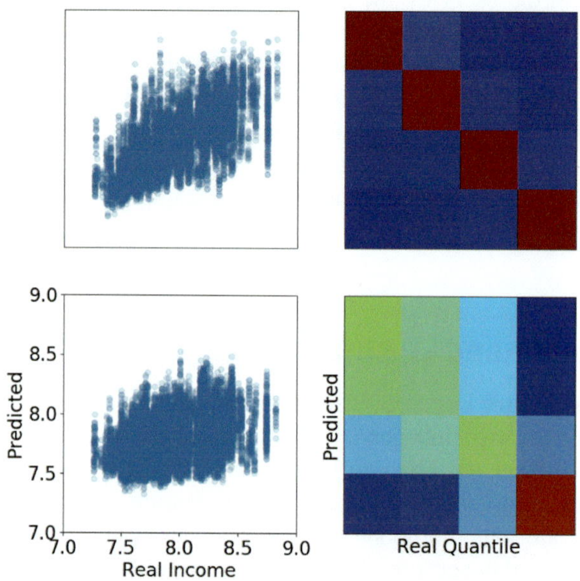

Figure 5.1: Income prediction based on mobility patterns (top) and hashtag usage (bottom). Left panels show scatter plots of actual (x-axis) and predicted (y-axis) income using regression. Right panels show the prediction of income quantiles using a classifier. Each quantile represents 25% of the population sorted by income (from left to right, and top to bottom). The matrix elements quantify the number of guesses for each quantile pair (confusion matrix).

The right panels in Figure 5.1 show the categorical prediction of socio-economic quantiles rather than the numeric values. The matrices show the number of times that a real quantile is predicted to be another quantile. The diagonal indicates the number of True-Positives and values off the diagonal indicate either the errors and miss-classifications. The matrices show a strong diagonal structure indicating a very good prediction. In the case of mobility (top matrix) the diagonal is almost perfect. In the case of hashtags (bottom matrix) the results are more diffused. However, the wrong predictions are close to their

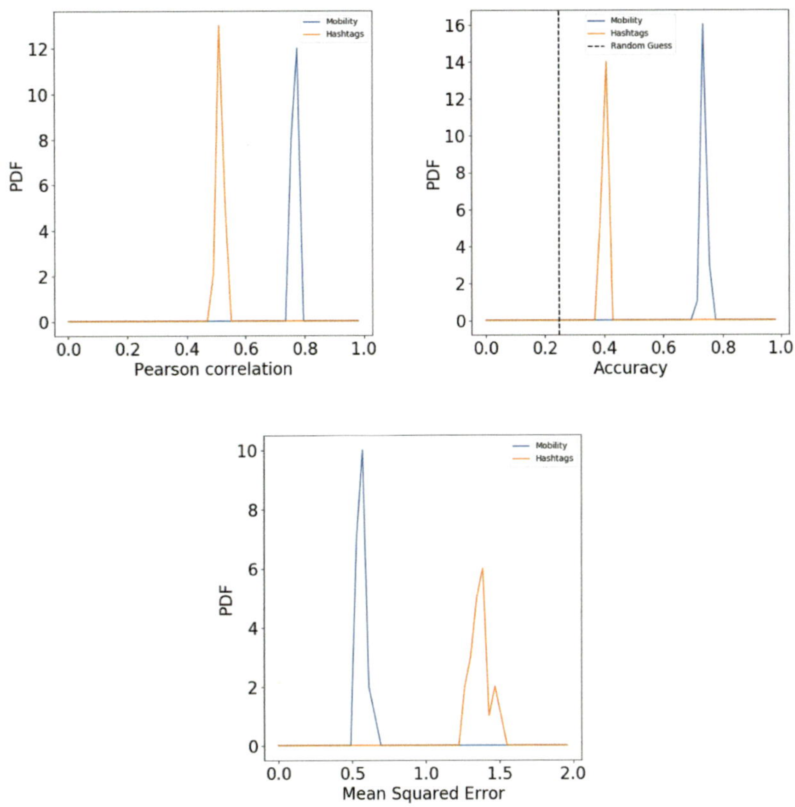

Figure 5.2: Top left: Pearson correlation between predicted and real income values using Regression. Top right: Accuracy of the classifier used to predict income quantiles. Dashed line shows random guess. Bottom: Mean square error of the income quantile prediction using classifiers. Units are income quantiles. In all panels, distributions represent bootstrapping results. Colors indicate mobility or hashtag usage feature space.

original values and not homogeneously distributed among quantiles. This indicates that although there are miss-classifications they are still close to their original values.

The error structure presented in the classification matrices can be interpreted as measures of behavioral distance between the individuals at different quantiles (right panels in Figure 5.1). The more miss-classifications among individuals of different income quantiles, the closer their behavior. Previous studies also use similar prediction accuracies as proxies of cultural distances [13]. In this case, the upper right and bottom left corners of the matrix are colored by dark blue, showing the least amount of confusion. This means that the top and bottom socioeconomic quantiles have the most differentiated behavior and therefore are easier to classify.

The bootstrap of the prediction accuracy is shown in Figure 5.2 (top right). Both hashtag (orange) and mobility (blue) are significantly higher than the error guess (dashed line). Therefore a considerable part of the hashtag usage variance and mobility patterns are explained by income. Another way to measure the prediction quality is through the Mean Square Error (MSE). Since quantile labels are also numerical, we can estimate the error of the prediction using the average euclidean distance between the real and predicted value. The MSE in Figure 5.2 (bottom) shows that the while mobility errors are the lowest, the hashtag ones are still low–with errors below the 1.5 quantile difference. Studies based on semantic features and topics of conversation report similar predictive performances [49].

The relationship of mobility and communication has been observed using mobile phones [80] and social media data [19]. People tend to communicate with places they have already visited. Moreover, patches in the territory that host certain populations are consistent with their geographic communication at multiple scales, from national levels down the suburban granularity [30]. These analyses have shown that income fragments the human mobility patterns in cities due to neighborhood segregation and therefore also affects the way people interact with each other both offline and online [58]. Furthermore, previous studies have already hinted toward the existing correlation between the socioeconomic status of people and the diversity of locations they visit. Indeed as previously pointed out [27] high SES users tend to have more diverse mobility patterns than low SES ones, which in turn leads to lower predictability of their whereabouts. These results may relate to previous work [47, 85], which explain this trend by means of the positive payoff between commuting further for better jobs, while keeping better housing conditions. This in turn explains why mobility might be used as an indicative predictor of an individual's socioeconomic status.

5.5 Topic Analysis

Conversational patterns differ by people's income. In order to analyze the preferences of conversational patterns by income, we applied topic analysis to tweets and obtained their structure. For this purpose, we characterize users by a probability distribution over a set of pre-defined topics rather than a frequency distribution over all the words of a given dictionary. Topics represent a latent space between users and words, such that certain words create topics and users sample words from the topics they address. Topic analysis creates a new space of reduced dimensions that represent new features. These new features can input the classic algorithms used to predict a user's income based on their tweets.

The topic analysis begins by training a *word2vec* model with the skip-gram architecture and negative sampling on a given collection of tweets. The skip-gram architecture predicts the context of words given their location in sentences. It learns the relationships between words under the assumption that their proximity in the text is not independent of their meaning. The negative sampling method is used for reducing the number of parameters to be inferred in the network. These methods are commonly used for natural language processing. During this process, words are mapped onto a 50-dimensional vector. The words that co-occur in the same tweet will be embedded in vectors that are in proximity to each other. The co-occurring words becomes the basis for deriving conversational topics.

In Figure 5.3 (top panel) we show a 2-dimensional representation of the embedding space using t-SNE for visualization. Words are represented by dots and their proximity is not encoded by the euclidean distance but rather by the cosine similarity value existing between their vectors. By running a spectral clustering algorithm on the word-to-word similarity matrix and setting negative similarity values to null we derived a prefixed number d (here $d = 100$) of clusters of words or topics grouping similar words. These topics were then manually labelled based on which words they contained. In the visualization, some topics have been colored with distinct colors and labeled after manual inspection.

We obtain a distribution of topical interest for users by computing the frequency of use of a given topic over a user's tweets. The individual vectors show the normalized usage frequency of words from each topic. These vectors coarsely represent users' syntax and interests and can be used to cluster individuals based on areas of interest. More importantly, we can use the vectors in order to analyze the income distribution of those who talk about certain topics. In Figure 5.3 (bottom panel) we show the income distribution of the individuals who mention (or not) the topics we obtained from clustering the embedding space. The population that talks about politics, technology, literature and travel has a higher income than the population which chooses not to talk about these issues. Analogously, the population that talks about slang, insults and urban interjection has a significantly lower income than the population that does not use these types of words.

In a previous work [58] we analyzed the collective topics of conversation coming from neighborhoods of different incomes in multiple cities around the world. We noticed two important things that are related with these results. First, consistently with Figure 5.3, wealthier neighborhoods tend to discuss life-style topics such as travels or leisure, while poorer neighborhoods seem characterized for more less sophisticated topics such as sports or TV shows. Second, we noticed that the frequency of visitation between neighborhoods was consistent with the similarity of their conversation topics. Therefore, neighborhoods that are segregated from one another, such as the case of cities that are segregated by

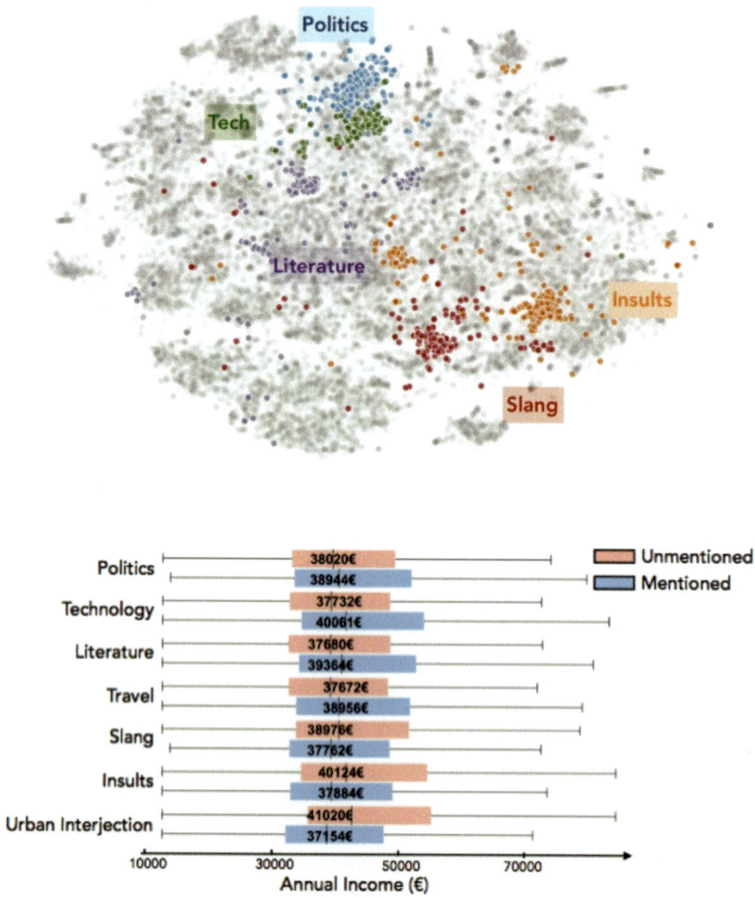

Figure 5.3: Topic model analysis of tweets per individual. Top panel: 2D visualization (t-SNE) of the embedding space obtained from applying *word2vec* on skip-grams and negative sampling. Colors correspond to topics obtained via clustering. Bottom panel shows income distributions of those who talk about the topics identified.

income, tend to also be separated in the space of online conversations. Akin to this strand of research, user features collected from online platforms have also been used in the inference of individual demographic traits. The tradition of relating SES of individuals to their language dates back to the early stages of sociolinguistics where it was first shown that social status reflected through a person's occupation is a determinant factor in the way language is used [12]. A similar line of work was explored by Lampos and authors in a series of seminal works [41, 68, 69]. Through it, the authors applied Gaussian Processes to predict user income, occupation and socioeconomic class based on demographic, psycho-linguistic features and a standardized job classification taxonomy, which mapped Twitter users to their professional occupations. The high predictive performance has proven this concept with $r = 0.633$ for income prediction, and a precision of 55% for 9-ways SOC classification, and 82% for binary SES classification further solidifying the use of semantical features as a mean of predicting individual socioeconomic status.

5.6 Diversity

In our study, we also examine the diversity of collective activities within neighborhoods. To achieve this, we construct interaction vectors for each neighborhood, encapsulating the aggregated behavior of its residents. These vectors are divided into two categories: mobility and online communication. The mobility vectors compile the frequency of visits from individuals in neighborhood i to neighborhood j. For online communication, we analyze the data derived from the mention vectors' mechanism on social media platforms, creating those that track the number of times residents of neighborhood i mention individuals from neighborhood j.

To quantify the diversity of mobility and communication patterns for each neighborhood, we calculate the entropy of these collective behavior vectors. This process involves normalizing the vectors to transform them into probability density functions, ensuring that the sum of each vector equates to one. Subsequently, we compute the entropy of these normalized distributions. This entropy value is then normalized by the entropy of a hypothetical uniform distribution, representing the maximum possible entropy. Consequently, neighborhoods with entropy values approaching 1 exhibit highly diverse patterns in visitation and online interactions. In contrast, neighborhoods with entropy values nearing zero demonstrate more homogeneous behavioral patterns. This method provides a nuanced understanding of the variation in social dynamics across different urban areas.

In Figure 5.4 we present a scatter plot where dots represent neighborhoods colored by income (from red to blue). The x-axis shows the entropy of the mobility vectors and the y-axis shows the entropy of the mentions vectors. There is a direct relationship between the entropy of both vectors. Diverse urban exploration is consistent with diverse online communications (r=0.57). Moreover, a clear separation of behavior by income is manifested. The diversity for both types of behaviors is consistent with the neighborhood income. Wealthier neighborhoods are consistently more diverse than poorer neighborhoods both in terms of mobility (r=0.46) and mentions (r=0.35).

The diversity of social exploration is closely related with the diversity of the information people are exposed to. Those who receive information from multiple sources are more likely to find better opportunities than those who

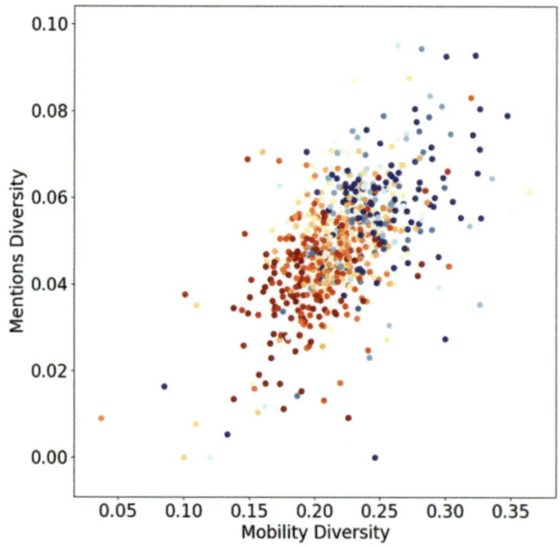

Figure 5.4: Diversity of collective behaviors in terms of urban mobility and online communication patterns. Dots represent neighborhoods colored by income (from red to blue). The x-axis represents the entropy of mobility vectors aggregated by neighborhoods. The y-axis represents the entropy of Twitter mention vectors aggregated by neighborhoods.

receive information from fewer sources. Therefore, while diverse neighborhoods are also richer, they might be richer precisely because they are diverse. Previous work shows that the diversity of urban exploration is consistent with income and age [33, 40]. Our results show that it is also consistent with the diversity of online interactions.

While physical exploration requires resources, online exploration in principle should be considerably less costly. However, the patterns from both offline and online worlds are remarkably similar. This result is further explored in a previous study [19], where multiple cities are compared and consistent results are obtained by means of analyzing multiple data types, including shopping and credit cards. Despite having new methods to interact with one another, people continue to mainly interact with those from their offline lives and behave in similar manners.

In another work [30], we show that people's mobility and communication patterns online create geographical patches that are preset at multiple scales of observation, ranging from sub-urban areas, up to large national regions. The multiscale nature of these regions arises from the structure of weak and strong ties [29]. While strong ties are local and remain in a radius of 5 km approximately, the larger scale ones are diverse and connect distant areas. Previous analyses of social networks show that those long range connections, which are responsible for the spread of information across the whole system, are also unequally distributed among the different income groups.

5.7 Discussion

The task of inferring an individual's socioeconomic status is pivotal in crafting tools that aid policymakers in addressing critical social issues like income inequality, segregation, and poverty. Traditional methods like nation-wide censuses, though comprehensive, are costly and infrequent, limiting their utility in responding to rapidly changing social landscapes. In contrast, social media analysis offers a more cost-effective, scalable, and timely alternative for gathering this crucial information. In this study, we have outlined a methodology for predicting individual socioeconomic status based on linguistic patterns and mobility data derived from social media. Crucially, our findings illuminate the profound impact of societal phenomena such as polarization and segregation on social media usage and interactions. This insight provides a valuable perspective in understanding these complex social processes.

However, to gain a more holistic understanding of these phenomena, research should not be confined solely to social media data. Emerging methodologies that utilize satellite imagery and mobile phone data are proving to be instrumental in uncovering previously hidden aspects of these societal dynamics. For instance, satellite imagery can reveal geographic and infrastructural disparities, while mobile phone data can offer real-time insights into human mobility and communication patterns, both of which are integral to understanding socioeconomic conditions [14, 34].

The integration of these diverse data sources presents an opportunity to develop a more nuanced and comprehensive understanding of socioeconomic phenomena. Nonetheless, the information gleaned from these innovative approaches must be approached with caution. A critical analysis of the inherent biases in these models is essential to ensure their accuracy and reliability. Such scrutiny is particularly important before these methods are applied in contexts where they are most needed, such as in resource-limited or rapidly changing environments.

Further research in this area should aim to refine these methods, address potential biases, and explore the integration of various data sources for a more robust analysis. By doing so, we can enhance our understanding of socioeconomic dynamics and better equip policymakers with the tools and insights needed to effectively address some of the most pressing social challenges of our time.

5.8 Conclusion

In summary, our research demonstrates that Twitter activity can serve as an effective proxy for determining an individual's income. We discovered that patterns of mobility are more predictive of income than hashtags, which aligns with our income estimations based on geographic data. Nonetheless, hashtags still contribute significantly to our understanding, exhibiting a substantial degree of explained variance and maintaining a low quadratic error. This predictive capability arises from the distinct behavioral patterns exhibited by different social classes.

Our findings underscore that the behavior characteristic of various socio-economic strata is not only observable in offline environments but are also reflected in online activities. Individuals from wealthier and poorer backgrounds exhibit distinct behaviors; they frequent different places and engage in conversations about different topics. This segregation in social groups and polarized behaviors in both physical and virtual spaces is a key enabler in predicting their income levels. Traditional and contemporary data sources, coupled with advanced analytical methods, have shed light on this phenomenon. Recent studies suggest a correlation between behavioral differentiation, physical segregation, and the reinforcing dynamics of collective learning. These non-linear processes manifest as unstructured patterns in feature spaces, which algorithms can exploit to make inferences.

The potential of inferring socioeconomic status (SES) from widely available digital traces, such as Twitter activity, is substantial. It paves the way for broader studies exploring population-level correlations of SES with language, space, time, or social network dynamics. The employment of these methods is crucial as it yields fresh insights into how SES influences the societal fabric. This understanding is invaluable, extending its implications to diverse fields such as recommendation systems, public policy, and economic aid distribution. As we continue to refine these methodologies and explore their applications, we can expect to unlock even deeper understandings of the complex interplay between socioeconomic status and various aspects of human behavior and society.

References

[1] J.L. Abitbol, M. Karsai, J.-P. Magué, J.-P. Chevrot, E. Fleury, et al. Socioeconomic dependencies of linguistic patterns in twitter: A multivariate analysis. In: *Proceedings of the 2018 World Wide Web Conference*, WWW '18, pages 1125–1134, Republic and Canton of Geneva, Switzerland, 2018. International World Wide Web Conferences Steering Committee.

[2] J.L. Abitbol and A.J. Morales. Socioeconomic patterns of twitter user activity. *Entropy*, 23(6): 780, 2021.

[3] T. Altshuler, Y. Altshuler, R. Katoshevski and Y. Shiftan. Modeling and prediction of ride-sharing utilization dynamics. *Journal of Advanced Transportation*, 2019, 2019.

[4] Y. Altshuler, N. Aharony, A. Pentland, Y. Elovici, M. Cebrian, et al. Stealing reality: When criminals become data scientists (or vice versa). *Intelligent Systems, IEEE*, 26(6): 22–30, Nov.–Dec. 2011.

[5] Y. Altshuler, M. Fire, N. Aharony, Y. Elovici, A. Pentland, et al. How many makes a crowd? on the correlation between groups' size and the accuracy of modeling. In: *International Conference on Social Computing, Behavioral-Cultural Modeling and Prediction*, pages 43–52. Springer, 2012.

[6] Y. Altshuler, W. Pan and A. Pentland. Trends prediction using social diffusion models. In: *International Conference on Social Computing, Behavioral-Cultural Modeling and Prediction*, pages 97–104. Springer, 2012.

[7] Y. Altshuler. Recent developments in the theory and applicability of swarm search. *Entropy*, 25(5): 710, 2023.

[8] Y. Altshuler, M. Fire, E. Shmueli, Y. Elovici, A. Bruckstein, et al. The social amplifier—reaction of human communities to emergencies. *Journal of Statistical Physics*, 152(3): 399–418, 2013.

[9] Y. Altshuler, A.S. Pentland and G. Gordon. Social behavior bias and knowledge management optimization. In: *Social Computing, Behavioral-Cultural Modeling, and Prediction*, pages 258–263. Springer, 2015.

[10] W.R. Ashby. *An Introduction to Cybernetics*. Chapman & Hall Ltd, 1961.

[11] E. Bakshy, S. Messing and L.A. Adamic. Exposure to ideologically diverse news and opinion on Facebook. *Science*, 348(6239): 1130–1132, 2015.

[12] B. Bernstein. Language and social class. *The British Journal of Sociology*, 11(3): 271–276, 1960.

[13] M. Bertrand and E. Kamenica. Coming apart? Cultural distances in the United States over time. Working Paper 24771, National Bureau of Economic Research, 2018.

[14] J. Blumenstock, G. Cadamuro and R. On. Predicting poverty and wealth from mobile phone metadata. *Science*, 350(6264): 1073–1076, 2015.

[15] J. Bollen, H. Mao and A. Pepe. Modeling public mood and emotion: Twitter sentiment and socio-economic phenomena. In: *Fifth International AAAI Conference on Weblogs and Social Media*, 2011.

[16] N. Bora, Y.-H. Chang and R. Maheswaran. Mobility patterns and user dynamics in racially segregated geographies of US cities. In: *Social Computing, Behavioral-Cultural Modeling and Prediction*, pages 11–18. Springer International Publishing, 2014.

[17] B.P. Chamberlain, C. Humby and M.P. Deisenroth. Detecting the age of twitter users. *CoRR*, abs/1601.04621, 2016.

[18] A. Culotta, N. Kumar and J. Cutler. Predicting the demographics of twitter users from website traffic data. *AAAI Conference on Artificial Intelligence*, 2015.

[19] X. Dong, A.J. Morales, E. Jahani, E. Moro, B. Lepri, et al. Segregated interactions in urban and online space. *EPJ Data Science*, 9(1): 20, 2020.

[20] J.S. Duesenberry. *Income, Saving, and the Theory of Consumer Behavior.* Taylor & Francis, Ltd., 1949.

[21] M. Duggan and J. Brenner. The demographics of social media users, Pew Research. Technical report, Pew Research, Washington, DC, 2013.

[22] N. Eagle, A. Pentland and D. Lazer. Inferring social network structure using mobile phone data. *Proceedings of the National Academy of Sciences (PNAS)*, 106: 15274–15278, 2009.

[23] N. Eagle, M. Macy and R. Claxton. Network diversity and economic development. *Science*, 328(5981): 1029–1031, 2010.

[24] J. Eisenstein, B. O'Connor, N.A. Smith and E.P. Xing. Diffusion of lexical change in social media. *PLOS ONE*, 9(11): 1–13, 11 2014.

[25] W.M. Golder and M.W. Macy. Diurnal and seasonal mood vary with work, sleep, and daylength across diverse cultures. *Science*, 333(6051): 1878–1881, 2011.

[26] M.C. González, .A. Hidalgo and A.-L. Barabási. Understanding individual human mobility patterns. *Nature*, 453(7196): 779–782, 2008.

[27] M.C. Gonzalez, C.A. Hidalgo and A.-L. Barabasi. Understanding individual human mobility patterns. *Nature*, 453(7196): 779–782, June 2008.

[28] M. Hasanuzzaman, S. Kamila, M. Kaur, S. Saha and A. Ekbal. Temporal orientation of tweets for predicting income of users. In: *ACL*, 2017.

[29] L. Hedayatifar, A. J. Morales and Y. Bar-Yam. Geographical fragmentation of the global network of twitter communications. *Chaos: An Interdisciplinary Journal of Nonlinear Science*, 30(7): 073133, 2020.

[30] L. Hedayatifar, R.A. Rigg, Y. Bar-Yam and A.J. Morales. US social fragmentation at multiple scales. *Journal of The Royal Society Interface*, 16(159): 20190509, 2019.

[31] A. Herdagdelen, W. Zuo, A. Gard-Murray and Y.r Bar-Yam. An exploration of social identity: The geography and politics of news-sharing communities in Twitter. *Complexity*, 19(2): 10–20, 2013.

[32] T. Hu, H. Xiao, T.T. Nguyen and J. Luo. What the language you tweet says about your occupation. *CoRR*, abs/1701.06233, 2017.

[33] S. Isaacman, R. Becker, R. Caceres, S. Kobourov, M. Martonosi, et al. Ranges of human mobility in Los Angeles and New York. In: *2011 IEEE International Conference on Pervasive Computing and Communications Workshops (PERCOM Workshops)*, pages 88–93, March 2011.

[34] N. Jean, M. Burke, M. Xie, W.M. Davis, D.B. Lobell, et al. Combining satellite imagery and machine learning to predict poverty. *Science*, 353(6301): 790–794, 2016.

[35] Y. Jiang, Z. Li and X. ye. Understanding demographic and socioeconomic bias of geotagged Twitter users at the county level. *Cartography and Geographic Information Science*, 46(3): 228–242, 2019.

[36] R. Jurdak, K. Zhao, J. Liu, M. AbouJaoude, M. Cameron, et al. Understanding human mobility from twitter. *PLOS ONE*, 10(7): 1–16, 2015.

[37] K. Kalimeri, M.G. Beiro, A. Bonanomi, A. Rosina, C. Cattuto, et al. Evaluation of biases in self-reported demographic and psychometric information: Traditional versus Facebook-based surveys. *arXiv preprint arXiv: 1901.07876*, 2019.

[38] A. Kibekbaev and E. Duman. Benchmarking regression algorithms for income prediction modeling. *Information Systems*, 61:40–52, 2016.

[39] L. Koskinen, T. Nurminen and J. Salonen. *Modelling and predicting individual salaries: a study of Finlands unique dataset*. Eläketurvakeskus, 2005.

[40] M.-P. Kwan and T. Schwanen. Geographies of mobility. *Annals of the American Association of Geographers*, 106(2): 243–256, 2016.

[41] V. Lampos, N. Aletras, J.K. Geyti, B. Zou, I.J. Cox, et al. Inferring the socioeconomic status of social media users based on behaviour and language. In: *Lecture Notes in Computer Science*, pages 689–695. Springer International Publishing, 2016.

[42] D. Lazer, A. Pentland, L. Adamic, S. Aral, A.-L. Barabasi, et al. Social science: Computational social science. *Science*, 323(5915): 721–723, 2009.

[43] D. Lazer, A. Pentland, L. Adamic, S. Aral, A.-L. Barabasi, et al. Computational social science. *Science*, 323(5915): 721—-723, 2009.

[44] K. Leetaru, S. Wang, G. Cao, A Padmanabhan, E. Shook, et al. Mapping the global Twitter heartbeat: The geography of Twitter. *First Monday*, 18(5), 2013. doi: 10.5210/fm.v18i5.4366.

[45] M. Lenormand, B. Gonçalves, A. Tugores and J.J. Ramasco. Human diffusion and city influence. *Journal of The Royal Society Interface*, 12(109), 2015.

[46] M. Lenormand, A. Tugores, P. Colet and J.J. Ramasco. Tweets on the road. *PLoS ONE*, 9(8): e105407, 08 2014. doi: 10.1371/journal.pone.0105407.

[47] Y. Léo, E. Fleury, C. Sarraute and M. Karsai. Socioeconomic Correlations and Stratification in Social Communication Networks . In: *NetSci'16*, Seoul, South Korea, May 2016. Peer-reviewed abstract submission.

[48] J. Leskovec, A. Rajaraman and J.D. Ullman. *Mining of Massive Datasets*. Cambridge University Press, New York, NY, USA, 2nd edition, 2014.

[49] J.L. Abitbol, M. Karsai and E. Fleury. Location, occupation, and semantics based socioeconomic status inference on twitter. *2018 IEEE International Conference on Data Mining Workshops (ICDMW)*, Nov 2018.

[50] Y.-Y. Liu, J.C. Nacher, T. Ochiai, M. Martino, Y. Altshuler, et al. Prospect theory for online financial trading. *PloS one*, 9(10): e109458, 2014.

[51] A. Llorente, M. Garcia-Herranz, M. Cebrian and E. Moro. Social media fingerprints of unemployment. *PLOS ONE*, 10(5): 1–13, 05 2015.

[52] S. Luo, F. Morone, C. Sarraute, M. Travizano, H.A. Makse, et al. Inferring personal economic status from social network location. *Nature Communications*, 8: 15227, 2017.

[53] A. Madan, S.T. Moturu, D. Lazer and A.S. Pentland. Social sensing: obesity, unhealthy eating and exercise in face-to-face networks. In: *Wireless Health 2010*, WH '10, pages 104–110, New York, NY, USA, 2010. ACM.

[54] A. Mislove, S. Lehmann, Y.-Y. Ahn, J.-P. Onnela, N. Rosenquist, et al. Understanding the demographics of Twitter users. In: *Proceedings of the 5th International AAAI Conference on Weblogs and Social Media*, pages 554—-557, Palo Alto, CA, 2011. The AAAI Press.

[55] D. Mocanu, A. Baronchelli, N. Perra, B. Gonçalves, Q. Zhang, et al. The Twitter of Babel: Mapping world languages through microblogging platforms. *PLoS ONE*, 8(4): e61981, 04 2013. doi: 10.1371/journal.pone.0061981.

[56] A. Morales, J. Borondo, J.C. Losada and R.M. Benito. Efficiency of human activity on information spreading on Twitter. *Social Networks*, 39: 1–11, 2014.

[57] A.J. Morales, J. Borondo, J.C. Losada and R.M. Benito. Measuring political polarization: Twitter shows the two sides of Venezuela. *Chaos: An Interdisciplinary Journal of Nonlinear Science*, 25(3): 033114, 2015.

[58] A.J. Morales, X. Dong, Y. Bar-Yam and A.S. Pentland. Segregation and polarization in urban areas. *Royal Society Open Science*, 6(10): 190573, 2019.

[59] A.J. Morales, J.C. Losada and R.M. Benito. Users structure and behavior on an online social network during a political protest. *Physica A: Statistical Mechanics and its Applications*, 391(21): 5244—5253, 2012.

[60] A.J. Morales, S. Somin, Y. Altshuler and A.S. Pentland. Patterns of user behavior and token adoption on ERC20. 2023.

[61] A.J. Morales, S. Somin, Y. Altshuler and A.S. Pentland. User behavior and token adoption on erc20. *arXiv preprint arXiv:2005.12218*, 2020.

[62] A.J. Morales, V. Vavilala, R.M. Benito and Y. Bar-Yam. Global patterns of synchronization in human communications. *Journal of the Royal Society Interface*, 14(128): 20161048, 2017.

[63] A.J. Morales, W. Creixell, J. Borondo, J.C. Losada, R. Benito, et al. Characterizing ethnic interactions from human communication patterns in ivory coast. *Networks and Heterogeneous Media*, 10(1), 2015.

[64] F. Morstatter, J. Pfeffer, H. Liu and K.M. Carley. Is the sample good enough? Comparing data from Twitter's streaming API with Twitter's firehose. In: *Proceedings of The 7th International AAAI Conference on Weblogs and Social Media*, Palo Alto, CA, 2013. The AAAI Press.

[65] V.S. Pagolu, K.N. Reddy, G. Panda and B.Majhi. Sentiment analysis of twitter data for predicting stock market movements. In: *2016 International Conference on Signal Processing, Communication, Power and Embedded System (SCOPES)*, pages 1345–1350. IEEE, 2016.

[66] W. Pan, Y. Altshuler and A. Pentland. Decoding social influence and the wisdom of the crowd in financial trading network. In: *Privacy, Security, Risk and Trust (PASSAT), 2012 International Conference on and 2012 International Confernece on Social Computing (SocialCom)*, pages 203–209. IEEE, 2012.

[67] V.G. Perry and M.D. Morris. Who is in control? the role of self-perception, knowledge, and income in explaining consumer financial behavior. *Journal of Consumer Affairs*, 39(2): 299–313, 2005.

[68] D. Preot, V. Lampos and N. Aletras. An analysis of the user occupational class through Twitter content. In: *Proc. of the 53rd Annual Meeting of the Association for Computational Linguistics*, pages 1754–1764, 2015.

[69] D. Preoţiuc-Pietro, S. Volkova, V. Lampos, Y. Bachrach, N. Aletras, et al. Studying user income through language, behaviour and affect in social media. *PLOS ONE*, 10(9): 1–17, 09 2015.

[70] R. Puzis, Y. Altshuler, Y. Elovici, S. Bekhor, Y. Shiftan, et al. Augmented betweenness centrality for environmentally-aware traffic monitoring in transportation networks.

[71] H.A. Schwartz, J.C. Eichstaedt, M.L. Kern, et al. Personality, gender, and age in the language of social media: The open-vocabulary approach. *PLOS ONE*, 8: 1–16, 09 2013.

[72] J.C. Scott. *Seeing like a state: How Certain Schemes to Improve the Human Condition have Failed.* Yale University Press, 1998.

[73] S. Somin, Y. Altshuler, G. Gordon, A. Pentland, E. Shmueli, et al. network dynamics of a financial ecosystem. *Scientific reports*, 10(1): 1–10, 2020.

[74] S. Somin, Y. Altshuler, E. Shmueli. Remaining popular: power-law regularities in network dynamics. *EPJ Data Science*, 11(1): 61, 2022.

[75] S. Somin, Y. Altshuler, A.S Pentland and E. Shmueli. Beyond preferential attachment: falling of stars and survival of superstars. *Royal Society Open Science*, 9(8): 220899, 2022.

[76] S. Somin, G. Gordon and Y. Altshuler. Network analysis of erc20 tokens trading on ethereum blockchain. In: *Unifying Themes in Complex Systems IX: Proceedings of the Ninth International Conference on Complex Systems*, page 439. Springer, 2018.

[77] S. Somin, G. Gordon and Y. Altshuler. Social signals in the ethereum trading network. *arXiv preprint arXiv: 1805.12097*, 2018.

[78] C. Song, Z. Qu, N. Blumm and A.-L. Barabási. Limits of predictability in human mobility. *Science*, 327(5968): 1018–1021, 2010.

[79] M. Spengler, R.I. Damian and B.W. Roberts. How you behave in school predicts life success above and beyond family background, broad traits, and cognitive ability. *Journal of Personality and Social Psychology*, 4(114), 2018.

[80] J.L. Toole, C. Herrera-Yaqüe, C.M. Schneider and M.C. González. Coupling human mobility and social ties. *Journal of The Royal Society Interface*, 12(105):20141128, 2015.

[81] Twitter. Twitter Streaming Application Programming Interface, 2015. [Online; posted 2015].

[82] S. Volkova, G. Coppersmith and B.Van. Durme. Inferring User Political Preferences from Streaming Communications. In: *Proceedings of the Annual Meeting of the Association for Computational Linguistics (ACL)*, 2014.

[83] Q. Wang, N.E. Phillips, M.L. Small and R.J. Sampson. Urban mobility and neighborhood isolation in America's 50 largest cities. *Proceedings of the National Academy of Sciences*, 10.1073/pnas.1802537115, 2018.

[84] M.J. Widener and W. Li. Using geolocated twitter data to monitor the prevalence of healthy and unhealthy food references across the us. *Applied Geography*, 54: 189–197, 2014.

[85] Y. Xu, A. Belyi, I. Bojic and C. Ratti. Human mobility and socioeconomic status: Analysis of singapore and boston. *Comput. Environ. Urban Syst.*, 72: 51–67, 2018.

Chapter 6

From Microbes to Methane: AI-Based Predictive Modeling of Feed Additive Efficacy in Dairy Cows

Yaniv Altshuler[1,3], Tzruya Calvão Chebach[2,3] and Shalom Cohen[3]
[1]Massachusetts Institute of Technology (yanival@mit.edu)
[2]Tel Aviv University(tzruyac@mail.tau.ac.il)
[3]Metha Artificial Intelligence(dr.shalom@metha.ai)

6.1 Introduction

Ruminants, thanks to the intricate symbiotic relationship with their resident microbiota, have the unique ability to breakdown complex polysaccharides like cellulose and hemicellulose, which constitute the primary components of their plant-based diet. This process is facilitated by the host animal's provision of a stable environment, facilitating continuous mixing, deconstruction, and fermentation of ingested plant material. This, in turn, results in the production of short-chain fatty acids which serve as a digestible energy source for the host animal [62, 63].

The assembly and development of the rumen microbiota is a multifactorial process, influenced by several host and environmental factors. These include the host's age, diet [82], genetic makeup [80], and herd origin [25], all of which play a pivotal role in defining the microbiota's compositional layout [45]. Moreover, the stochastic colonization events of the rumen during early life stages can leave lasting imprints on the ruminant microbiome's structure [82].

While this symbiotic relationship allows ruminants to thrive on fibrous diets, it also has an environmental cost. The ruminant's digestive process is a significant contributor to the emission of methane, a potent greenhouse gas, which accounts for about 14% of total greenhouse emissions and has a global warming potential 28 times higher than carbon dioxide (CO_2). Notably, livestock are estimated to contribute to nearly 30% of all anthropogenic methane emissions [62].

Efforts to mitigate the environmental impact of dairy farming have given rise to several innovative strategies. One such approach involves the utilization of microbial biomarkers to identify cows with high methane emission rates, thereby enabling targeted management strategies aimed at reducing methane emissions and fostering environmental sustainability [73]. A more refined strategy involves characterizing microbial gene abundances as proxies for methane emissions, focusing specifically on metabolic pathways expected to exhibit variation between low and high methane emitters [94].

The study described in this chapter (previously published as [9]) adopts an unbiased metagenomic approach to create a model that determines the most suitable feed additive customized for individual herds, allowing for precision application based on individual microbiome profiles. This methodology acknowledges the significant variation and multitude of contributing factors that lead to the diverse responses observed among ruminants. We capitalize on the rich biological information stored in the rumen microbiome, transforming the microbiome of a select few cows in each herd into a living 'sensor', and allowing us to predict the most effective feed additive tailored to that specific herd.

We devised a two-stage trial design targeting the prediction of the efficacy of methane-reducing additives using cows' microbiome data. The initial stage engages an unsupervised machine learning process, trained on a diverse dataset that includes microbiome samples from a wide spectrum of cows across various farms. The following stage makes use of a smaller subset of cows, whose methane emissions have been documented periodically, to implement supervised learning. This stage is aimed at constructing a predictive model that associates microbiome profiles with the effectiveness of feed additives.

By directly tapping into the microbiome, we bypass conventional variables such as weather and diet, which, though traditionally deemed critical, pose a challenge in establishing a clear, direct association with feed additive efficacy. Our approach presents an objective and comprehensive solution designed to effectively mitigate methane emissions in livestock farming.

The innovation of this approach lies not only in the use of the microbiome as a predictive tool, but also in our capacity to make sense of its complex raw data. The rumen microbiome, rich in diversity and complexity, presents a significant analytical challenge that until now has hindered its utility in such applications. To tackle this, we employed a data-driven approach powered by state-of-the-art artificial intelligence technology, creating a pioneering model that acknowledges the extensive variation and plethora of factors contributing to diverse responses observed among ruminants. By leveraging the power of the microbiome and artificial intelligence, our approach offers a promising avenue for future environmentally-conscious livestock management.

Our approach is noteworthy for its scalability, applicability beyond mere methane reduction, and potential for continuous improvement as more data accumulates. This manuscript elucidates our detailed trial design and deliberates upon its prospective influence on environmental sustainability, farm economy, and the expansive realm of precision agriculture.

6.2 Materials and Methods

6.2.1 Microbiome Acquisition and Ruminal Fluid Sampling

Rumen cannulation (RC) and stomach tubing (ST) stand as the two most prevalent techniques for the study of ruminal fermentation and microbial community composition in both large and small ruminants [64, 67]. While some researchers posited that samples acquired using ST may be less indicative of certain rumen parameters like pH, VFA concentrations, or bacterial communities compared to RC [23], ST remains a valuable technique when exploring molar proportions of VFA, protozoa count, dH2, methane, and ammonia concentrations [26, 31, 84, 95].

For our study, we employed ST, specifically the passive ruminal fluid collection technique, for its distinct advantages. Cannulation, although precise, is both expensive and invasive. It typically necessitates a smaller cohort of animals due to its inherent complexity. In contrast, ST allowed us to collect ruminal fluid from a large cohort of intact animals. This not only induces less stress in the animals but also facilitates sampling in commercial dairies rather than solely experimental facilities.

Opting for stomach tubing as our primary technique for ruminal fluid extraction also aligns well with anticipated expansions to studies involving other ruminants. In research focused on small ruminants, a methodological comparison between surgical rumen cannulation and stomach tubing found the latter to be a more viable, safer, and pragmatic choice for sampling rumen contents in sheep and goats for the study of ruminal fermentation [74]. Notably, stomach tubing facilitates the collection of a diverse bacterial community and effectively mirrors most results garnered from cannula-based sampling.

The utilization of stomach tubing for sample collection is particularly conducive to the comprehensive, data-driven approach we adopted in our study. Recent research [35] illustrated that, as anticipated, stomach tubing garners more diverse and varied microbial samples than those derived from cannulated animals. This heightened diversity is invaluable for our analytics. A richer microbial profile provides a broader spectrum of data, enabling us to capture a more nuanced and holistic understanding of the complex interactions and processes occurring within the rumen. Crucially, despite this increase in diversity, stomach tubing samples remain highly representative of the fermentation processes and the methanogenic microbiota present in the rumen. This ensures our analyses are grounded in a more complete representation of the rumen environment, maximizing the reliability and precision of our results.

In our ST sampling method, we utilized a 300-cm long polyvinyl chloride orogastric tube (2.9 cm O.D. and 2.5 cm I.D.) with 4 holes perforated at its distal 30 cm. During each sampling event, the cow's head was restrained, and the tube, when attached to a 50 cm speculum, was gently inserted through the esophagus into the rumen. Once in place, the tube remained lower than the cow's head, allowing ruminal fluid to passively accumulate. We discarded the initial 10 ml of fluid to mitigate potential saliva contamination, after which we collected 50 ml of ruminal fluid in a sterile conical tube for further analyses. To prevent cross-contamination between samples, both the tube and the speculum were meticulously rinsed and bleached after each use.

6.2.2 Sample Processing and Sequencing

The rumen fluid samples collected were instantaneously frozen on-site with dry-ice and dispatched the very same day to a storage facility, where they were securely stored at a temperature of –80°C.

When ready for shipment, the frozen rumen fluid samples were brought to room temperature by defrosting them in a ZYMO DNA/RNA Shield buffer (ZR-R1100-250). This process safeguards the samples from degradation during transport. They were then transported at room temperature to a specialized DNA service facility.

The DNA was extracted from the rumen fluid samples using the ZymoBIOMICS 96 MagBead DNA Kit (Zymo Research, Cat. no. D4308), in strict adherence to the provided manufacturer's protocol.

Subsequently, library preparation for sequencing was executed strictly adhering to the guidelines provided by the Illumina DNA Prep (Illumina, Cat. no. 20060059). The assembled libraries were then subjected to paired-end sequencing with 2 × 150 base reads employing an Illumina NovaSeq 6000 instrument. For this purpose, the NovaSeq 6000 S4 Reagent Kit v1.5 (300 cycles) (Illumina, Cat. no. 20028312) was used.

Notably, all processes, inclusive of DNA extraction, library preparation, and sequencing, were conducted at ZYMO Research based in Freiburg, Germany.

Upon receiving the results of the sequencing process, the integrity of the raw FASTQ sequences was assessed by employing the FASTQC software for quality control, and subsequently, the data was fine-tuned by BBDUK with a set of customized parameters for optimal refinement.

Specifically, Illumina adapters were meticulously eliminated from the 3' end of the sequence, and all reads that contained fewer than 100 bp were systematically discarded. This data filtering and refinement process resulted in an average yield of approximately 15.7 million reads per sample (see more detailed in Figure 6.1). The average Phred Score for all samples was higher than 35, the adapter content was less than 0.5% with duplication rate of 25% (see Figure 6.2). The complete sequencing quality control report is available online and will be provided by the authors upon request.

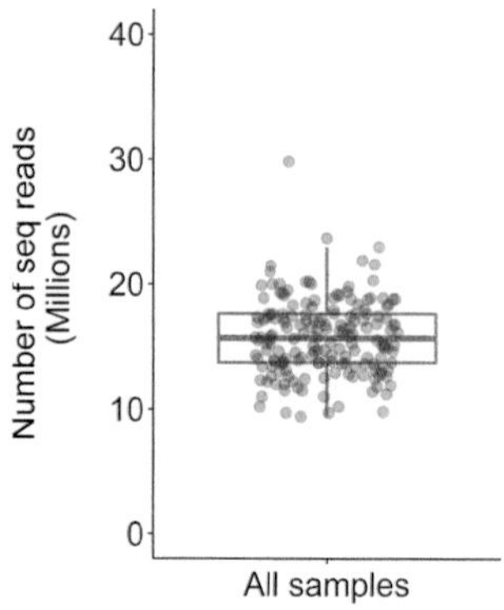

Figure 6.1: Evaluation of the depths of sequenced samples: an in-depth examination of the sampling depth across various sequencing processes.

6.2.3 Animals

In this study, we examined 2,190 Israeli Holstein cows. Currently, Israel has approximately 102,200 dairy cows, the vast majority of which are of the Israeli Holstein breed. Roughly 70% of these cows are found in Kibbutz herds, which are large units within cooperatively owned and managed farms. The remaining 30% are part of Moshav herds, which are smaller, family-owned farms. According to the Israeli Herd Book annual report of 2022 [97] the Israeli cow produced an average of 12,442 kg of milk (production/cow/305 days), of which 3.32% is protein and 3.89% is fat. The annual milk yield per cow recorded here

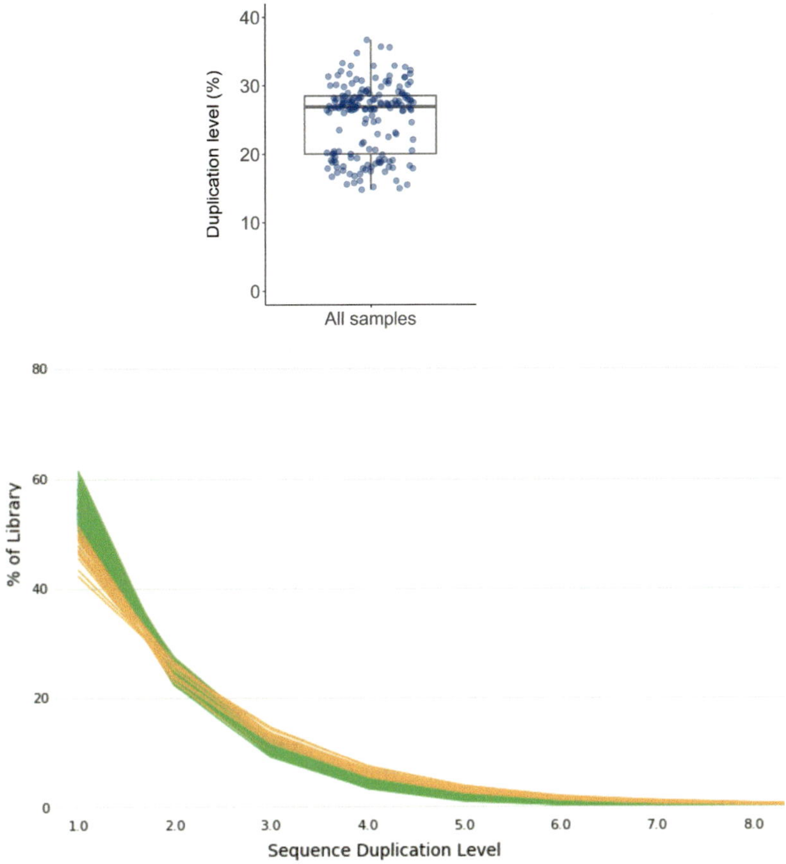

Figure 6.2: Analysis of sequence duplication levels, illustrating the extent of duplication across various sequencing processes. Notably, the majority of samples exhibit a low count of duplicates, signifying the high quality of the sequencing process.

is among the highest globally. For comparison's sake, the per-cow milk production in the USA in 2022 stood at 10,839 kg, as referenced in [4].

The participating study sites housed between 400 and 900 cows each. No significant correlation was found between the farm size and the average efficacy of the feed additives. The microbiome sample from 10 cows used for prediction constituted at least 1.1% of the total herd size. Given the absence of a significant correlation between farm size and the prediction engine's accuracy, we postulate that our method should be pertinent for sample sizes of at least 0.5%. For example, this suggests that a sample from 15 cows could representatively predict the expected efficacy for a herd of up to 3,000 cows.

6.2.4 Global Applicability: Overcoming Geographical and Nutritional Variabilities

While the methodology outlined in this study was initially implemented in commercial dairies across Israel, it possesses a broad applicability that extends well beyond this geographical boundary. This universality stems from several key factors that were integral to our research design and execution.

First, our trials were deliberately conducted across a diverse range of geographical settings within Israel, encompassing areas from arid deserts to cooler mountainous regions and more temperate plains. This variety in environmental conditions mirrors the broad spectrum of climatic zones found in many countries where extensive dairy farming is practiced. Hence, the efficacy of our methodology in these varied Israeli locales strongly suggests its potential adaptability and effectiveness in similar geographical contexts worldwide.

Secondly, our research accounted for various nutritional regimes implemented in these farms. The diet of dairy cows in Israel, much like in numerous other countries, includes by-products from the food industry, such as pulp, gluten meal, and citrus fruits. This aspect of our study is particularly significant as it demonstrates the adaptability of our methodology to different feeding practices and regimes, which are a common feature in global dairy farming.

Furthermore, while our primary focus lies in the realm of the microbiome and its mathematical interactions, it is important to consider the potential variability of microbiomes across different countries. However, we posit that regional differences in the rumen microbiome are unlikely to significantly impact the applicability of our method. This is because our approach hinges on a mathematical framework that is largely data-agnostic. The core of our methodology is the analysis of the power-law and network dynamics within the microbiome, a process that is not inherently limited by geographical variations in microbiomic composition.

In essence, the requirement of our methodology is straightforward: to sequence the microbiome and measure methane emissions. From these datasets, regardless of their geographic origin, our model is capable of making accurate predictions. The mathematical nature of this approach underlines its potential global applicability, as the fundamental principles of power-law dynamics and network analysis are universally applicable.

6.2.5 Methane Detection and Quantification

A multitude of methane sensing techniques exists in the market today, reflecting the diverse array of applications they serve—from safety monitoring in mining and natural gas industries, air quality surveillance in urban areas, to greenhouse gas emission tracking for climate research. Originally, these devices were not specifically conceived for agricultural settings, let alone for monitoring ruminant animals like cows. However, recognizing the crucial role of livestock in methane emissions, our study undertook a comprehensive evaluation of the available sensing technologies. After thorough scrutiny, which considered aspects such as accuracy, durability, suitability for large scale deployment, and adaptability to the unique conditions of a farm environment, we were able to select those sensors most apt for our specific needs in measuring bovine methane emissions.

6.2.5.1 Existing Technologies for Methane Detection and Measurement

The following presents an overview of the principal technological methodologies employed today in commercially available methane sensors and their potential suitability for ruminant enteric methane emissions.

Infrared Sensors

These sensors measure methane concentration by detecting the specific wavelengths absorbed by methane. They tend to be reliable and require low maintenance. Some commercial examples include the ExplorIR-M 5% CO_2 Sensor and the SGX Sensortech's IR Methane Sensor [14]. While these sensors are quite accurate, their placement and exposure to environmental conditions could impact the readings in a free-range cattle environment.

Semiconductor Sensors

These sensors measure methane by detecting the change in resistance of a semiconductor material exposed to different methane concentrations. These are usually less expensive than infrared sensors, but they tend to have a shorter lifespan and require more maintenance. Figaro's TGS2611 [83] is an example of a semiconductor methane sensor. Their low-cost could be advantageous for wide-scale deployment across large cattle farms.

Catalytic Sensors

These sensors measure the methane concentration by detecting the heat produced when methane reacts with a catalyst. However, these sensors might not be ideal for methane measurement from cows due to their susceptibility to poisoning and their requirement for oxygen to function. An example of such a commercial sensor is the Honeywell XCD Methane Gas Detector. This fixed gas detector is designed to provide comprehensive monitoring of combustible gas levels in various environments and is known for its reliability and accuracy.

Electrochemical Sensors

These sensors measure methane by detecting the current generated when methane is oxidized. While they are sensitive and compact, they tend to have a shorter lifespan than other sensor types. The ALTAIR Pro Single-Gas Detector [78] is an example of an electrochemical sensor, designed for worker safety in mind, with a primary goal to monitor potentially harmful gases in confined spaces.

Photoacoustic Spectroscopy Sensors

Instruments like the INNOVA 1412i Photoacoustic Gas Monitor [72] use the principle of photoacoustic spectroscopy to measure methane emissions. These are highly accurate but can be more expensive and might be more suited to laboratory settings or small scale, intensive research studies.

Laser-based Sensors

Sensors like the LI-COR's LI-7700 Open Path CH_4 Analyzer [59] use laser technology to measure methane concentrations in the open air. These are highly accurate and can cover a large area, making them suitable for large farms, but they are often also significantly more expensive.

Sulfur Hexafluoride (SF_6) Tracer

This technique is commonly used for measuring methane emissions in ruminants. It involves the animal inhaling a small quantity of SF_6, and the concentration of SF_6 and methane in the exhaled air is measured, allowing for the implicit calculation of the animal's methane production rate [61]. This technique is widely used in research settings due to its accuracy, but it requires specific equipment and technical expertise, making it less suitable for widespread commercial use. An example is the SF6 Sulfur Hexafluoride Gas Analyzer by Nova Analytical Systems, specifically designed for such applications.

6.2.5.2 Assessment of Methane Emissions in Prior Ruminant Research

In recent years, the quest to reduce emissions from enteric methane fermentation has garnered increasing attention. This has sparked significant efforts towards devising techniques that not only accurately represent in-field situations but also minimize the disturbance to the animals [22, 76].

While conventional industrial methane sensors have been repurposed for measuring ruminant methane emissions (see infra-red-based sensing in [14], laser-based studies in [24], and a combination of photoacoustic spectroscopy and infrared sensors in [71]), there has also been progress in developing methods specifically tailored for bovine applications. A comprehensive comparison of these techniques is elaborated in [30]. Below, we provide a succinct summary of the predominant methodologies, highlighting their respective advantages and drawbacks:

Respiration Chambers

The breathing or calorimetric chamber has been the traditional benchmark for measuring CH_4 emissions from ruminants in various settings [13]. This method's chief aim is to quantify the energy generated through an animal's regular metabolic processes. Such chambers play a pivotal role in exploring strategies to curtail CH_4 emissions. They function by monitoring the concentrations of gases in the animal's exhaled air within a regulated environment. However, the use of the calorimetric chamber is generally confined to the analysis of a single animal due to construction costs and the need for specialized operational skills [89].

Head Chamber

This method employs an airtight box, encircling the ruminant's head, with a curtain or sleeve around the neck to restrict air exchange between the internal and external atmospheres of the chamber [13]. The box should be adequately sized to allow unhindered head movements and access to feed and water. Compared to the calorimetric chamber, the prime benefit of this approach lies in its cost-effectiveness (in comparison to the respiration chamber). Similar to the calorimetric chamber, measurements must be performed individually on trained animals.

Face Chamber

The face mask, akin to the calorimetric and head chambers, presents another approach for measuring CH_4 from ruminants [48]. This method involves fitting a mask onto the animal's head to gather air exhaled through the airways. The animal requires a brief acclimation period to the equipment, typically spanning seven days, with six-minute sessions each day [68]. During this time, the animal is not allowed to eat or drink, and the analyses are conducted similarly to those in an open calorimetric chamber.

Polyethylene Tunnel

This method utilizes a structure reminiscent of an agricultural greenhouse, erected on a pasture with dual layers of inflatable polyethylene walls and a large entrance. It serves as a simpler alternative to the calorimetric chamber in terms of operation and data collection. Inside this tunnel, air is consistently drawn in, allowing for continuous collection of air samples from an exhaust port for gas analysis or gas chromatography [49]. This method is typically employed to assess CH_4 emissions in areas of fresh forage, allowing animals to behave naturally and controlling selected forage within the confined tunnel space. This technique's benefits include the animals' unrestricted movement within the tunnel and the relatively low acquisition and installation costs. However, it is impractical to control the tunnel's temperature during periods of high ambient temperature. Most studies using this method have focused on sheep due to pasture space constraints [13]. Additionally, this technique is unsuitable for experiments evaluating various treatments.

Sulfur Hexafluoride (SF_6) Tracer

The sulfur hexafluoride (SF6) method involves a small permeation capsule, essentially a metal tube with a porous plate at one end, filled with SF_6. Initially, the capsule is placed in a thermostatic water bath for a month before it is inserted into the animal's rumen. The animal is fitted with a halter that has a capillary tube connected to a PVC yoke. Over a specified duration, this apparatus collects exhaled gases. After a vacuum is applied, the sample is sent to the lab for gas chromatography analysis. A valve in the PVC yoke ensures the collection of exhaled air at a steady rate. This collection system is calibrated to stop once the sample fills approximately half of the system's storage capacity, typically within 24 hours. This method allows the animals to move freely and engage in normal grazing activities, removing the necessity for confining them in cages or barometric chambers. Nevertheless, the animals need training to acclimate to the equipment, and the PVC tubing requires daily replacement [61].

Automatic Feeder Technique (GreenFeed)

The GreenFeed technique operates by recognizing an electronic tag on the animal as it begins to feed. The system then measures the gases emitted every second during the feeding process, allowing for the monitoring of individual emission rates over time. Given that approximately 90% of gases produced by ruminants are released through eructation via the mouth and nostrils, this system generates a highly reliable dataset for research on GHG reduction strategies [37]. Upon insertion of its head into the feeder, the animal is identified via an electronic tag using radio frequency technology (RFID). A fan then activates to draw in the air exhaled through the animal's nostrils and mouth. Sensors within the equipment measure gas concentrations, the volume of emitted gas, and other environmental parameters. Despite its advantages, GreenFeed presents considerable challenges that may limit its use. Its high cost can be prohibitive, especially in larger studies, making implementation unfeasible in many research centers. Additionally, the time required to acclimate animals, particularly Zebu and other native commercial breeds, to the equipment should be considered when planning studies utilizing this technique [41].

6.2.5.3 *Selected Sensor*

Though several dedicated sensing mechanisms have been developed specifically for monitoring methane emissions from cows, such as respiration chambers, polyethylene tunnels, head chambers, face masks or automatic feeders, these tools share common disadvantages. Primarily, their use results in an alteration of the cows' natural behavior, making the captured data less representative of the animals' day-to-day emission patterns. The disruption of normal behavior is due to the intrusive nature of these devices, which typically require direct contact with the animals or confinement within a restricted space.

These methods also lack scalability. In larger farms with hundreds or thousands of cows, the application of these techniques becomes a logistical challenge, limiting their utility in extensive real-world scenarios. The expenses associated with these techniques further dampen their practicality—the high costs involved in the construction, maintenance, and operation of these devices often make them economically unfeasible for most farms. Furthermore, their use typically demands trained cows which accustom to the devices, imposing an additional layer of complexity to the measurement process.

Given these limitations, we elected to leverage an industrial methane sensor for our study. Industrial sensors are known for their high degree of accuracy and sensitivity, essential features for reliable data collection. More importantly, their non-intrusive nature allows the cows to behave naturally, ensuring that the data gathered is reflective of standard methane emissions under typical conditions. This non-intrusiveness also means the cows require no special training or conditioning to tolerate the device. Being designed for industrial applications, these sensors are robust, cost-effective, and scalable, enabling their usage across larger herds without a significant uptick in operational complexity. Thus, by using an adapted industrial sensor, we hope to bypass many of the hurdles associated with dedicated cow methane sensors and collect reliable and representative data on methane emissions from ruminants.

SEM5000 by Geotech: Detailed Specifications and Primary Advantages

The ATEX Gas Analyser Geotech SEM5000 is a robust, hand-held device specifically designed to measure methane concentrations. Built for use in challenging environments, this device has some key capabilities and advantages that make it well-suited for methane measurement in ruminants like cows. The SEM5000 utilizes laser-based technology for detecting and quantifying methane levels with a range of 0 ppm to 100% volume (laser based sensors were demonstrated to be of ideal use for methane measurements in ruminants [75]). It boasts of a rapid response time, delivering results in seconds. The device also has an in-built GPS, which allows for geo-tagging of measurements and the creation of gas concentration maps. The key advantages of this sensor in methane measurement for cows are as follows:

- **Non-invasive and very low stress method:** Unlike some techniques that require animal confinement or behaviour modification, the SEM5000 allows for measurements to be taken in a non-invasive manner, reducing stress on the animals and ensuring data gathered represents their natural behaviour.
- **Accuracy and precision:** The laser technology used in the SEM5000 delivers high-accuracy and precision readings, reducing the potential for errors and increasing the reliability of the data.
- **Portability and robustness:** Given its hand-held design and robust construction, the SEM5000 can be used in a variety of field conditions, making it a practical tool for monitoring methane emissions in grazing environments.
- **Scalability:** The SEM5000 allows for high-throughput data collection and can be used to measure methane emissions from a large number of animals over a short period, making it a more scalable solution than other techniques.
- **Cost:** Notably, the SEM5000 Methane Detector is a cost-effective choice, priced at nearly one-tenth the cost of the Greenfeed system, offering efficient and affordable monitoring of methane emissions.

Figure 6.3: The Geotech SEM5000 Methane Detector, a hand-held device utilizing laser-based technology for highly accurate and rapid methane concentration measurements. Its compact design enhances portability and usability under various field conditions, and its fast response time coupled with high data throughput make it an ideal tool for extensive in-field ruminant measurements.

6.2.5.4 Comparative Performance Evaluation of the SEM5000 Sensor

In order to ascertain the reliability and robustness of the SEM5000 sensor for ruminant methane measurement, we devised a comprehensive study. Our objective was to show that, although more affordable and scalable than other commonly used technologies, the SEM5000 can deliver equivalent levels of accuracy. For comparative benchmarking, we selected the Li-Cor LI-7810

Table 6.1: Main technical specifications of the Geotech SEM5000 methane detector.

Specification	Value
Range	0 to 10,000 ppm
Resolution	0.1 ppm
Accuracy	0.7 ppm
Technology	laser based
Response time	< 2.5 sec.
Rate	2 sec. per reading
Battery life	10 hours
Flow	1 litter per minute
Weight	1.3 kg

laser-based [99] system and the GreenFeed system [37]—two established technologies in the field. Li-Cor LI-7810, while robust and accurate [47, 51, 90], is considerably more expensive, and GreenFeed, though regarded as the gold standard [36, 41, 42], presents limitations in scalability, usability and cost.

Comparative Analysis Methodology

Our comparison methodology involved conducting two distinct sets of tests. In the first set, we concurrently measured the same cows using the SEM5000 and GreenFeed devices. This allowed for a direct comparison between these two methods on the same animal subjects. We compared the mean and median methane emissions from each cow as measured by both sensors. Given the heightened concern over high methane emissions in the context of mitigation, we also analyzed the average of the top 25% of readings. To further validate the consistency between the two sensors, we employed the Mann-Whitney U Test [65], a non-parametric method designed to determine if two sets of readings originate from the same source.

In the second set of tests, we used the SEM5000 and Li-Cor 7810 devices in parallel over a period of four weeks, measuring 48 different cows. This extended period of observation, as well as the large number of specimens being measured, provided a comprehensive set of data to compare the performance of the SEM5000 sensor to the well-established Li-Cor 7810 laser-based system. For each cow, we determined the regression between measurements from the two sensors. Subsequently, we used the Bland–Altman method [15], a technique tailored for evaluating the concordance of two sensors measuring identical data. As a final step, we computed the Root Mean Square Error (RMSE) to quantify the differences between the two sensors.

The order of measurements was randomized in both sets of tests to minimize potential bias.

Results

Our preliminary findings provide compelling evidence supporting the suitability of the SEM5000 sensor for ruminant methane measurement. The comparative data suggest that the SEM5000 sensor achieves high levels of accuracy, comparable to that of the pricier Li-Cor and GreenFeed systems. A detailed analysis of these results is presented in the forthcoming figures, demonstrating the commendable performance of the SEM5000 sensor.

Table 6.2 presents a comparison between measurements taken using the SEM5000 and the Greenfeed sensors for four cows. The results demonstrate that while the SEM5000 sensor exhibits a marginally higher data variance, the key properties, such as average and median emissions levels align closely. The observed differences not only meet the stringent criteria set out by Verra's VM41 protocol [79] and the CDM Meth Panel Guidance on Addressing Uncertainty in the Estimation of Emissions Reductions for CDM Project Activities but also fall below the 15% threshold for sensors' compliance defined in the IPCC 2006 Guidelines, Volume 2, Chapter 2, Tables 2.2 to 2.6 [27]. This assertion of compliance is further validated by the Mann-Whitney U Test results [65], which suggest that (for each cow) both data streams likely derive from the same source. This analysis establishes the credibility of the SEM5000 sensor for measuring enteric methane emissions, affirming its readings to be as dependable as those from the Greenfeed system. The results from the SEM500 and the GreenFeed system, when measuring the same cow, are depicted in Figure 6.4. Both measurements showcase comparable emissions levels and temporal dynamics.

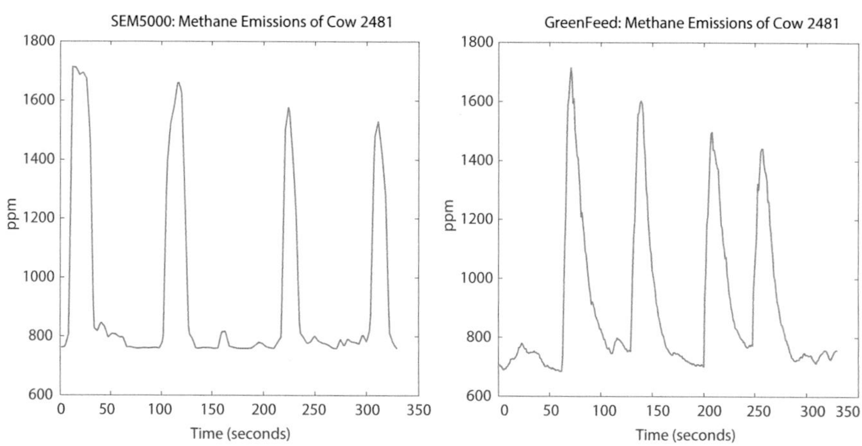

Figure 6.4: Methane levels (in parts per million) for the same cow, as measured at different times by the SEM5000 sensor and the GreenFeed system. The readings exhibit similar dynamics and values.

Table 6.2: Comparison between measurements taken using the Greenfeed system and the SEM5000 sensor, for four different cows (values are shown as Greenfeed/SEM5000 for each cow). The close alignment in these metrics between the two sensors underscores their comparable performance. The Mann-Whitney U Test [65], a non-parametric statistical test, is employed to determine whether two independent samples were drawn from a population with the same distribution. In this context, the test's results suggest a high probability that the measurements from both sensors are from the same distribution. This conclusion further suggests that data captured using the SEM5000 can, with a high degree of confidence, be used as a proxy for results from the Greenfeed system, paving the way for broader and more flexible deployment of these sensors in methane measurement campaigns.

Metric	Cow 2071	Cow 2299	Cow 2481	Cow 2849	Avg. % Change
Mean CH_4	159/167	123/129	909/934	127/133	4.8%
Median CH_4	84/87	70/73	773/782	66/68	3.6%
Mean of top 25% CH_4 readings	410/418	295/326	1318/1397	335/328	5.6%
STD of CH_4 readings	196/235	123/168	261/326	161/221	29.6%
Mann-Whitney U Test p-value	0.065245	0.022432	0.000019	0.009071	N/A

Figures 6.5 and 6.6 further bolster the credibility of the SEM5000 sensor for enteric methane measurements. In these figures, methane emissions from 48 distinct cows, measured over a span of 4 weeks by both sensors, are showcased. A discernible strong correlation between the measurements from the two sensors is evident. Coupled with the robust correlation, the Bland-Altman analysis further corroborates these observations. The Bland-Altman method is primarily utilized to assess the agreement between two different measurement techniques, determining the consistency and discrepancy in results. Its affirmation in this context emphasizes the reliability and similarity of readings between the two sensors.

6.2.5.5 Periodic Methane Measurement: Duration and Consistency

One of the core strengths of our research methodology revolves around the repeated measurements of each cow throughout an extended 12-week period post-treatment. This strategy is deeply rooted in the need to validate the persistent efficacy—or potential lack thereof—of the additives. Over time, factors such as changes in feed quality, external environmental conditions, or the cow's inherent physiology might impact methane emission levels. By measuring emissions repeatedly over several weeks, we ensure that the observed effects (or non-effects) of the additive remain consistent.

In addition, to enhance data accuracy and volatility reduction, methane

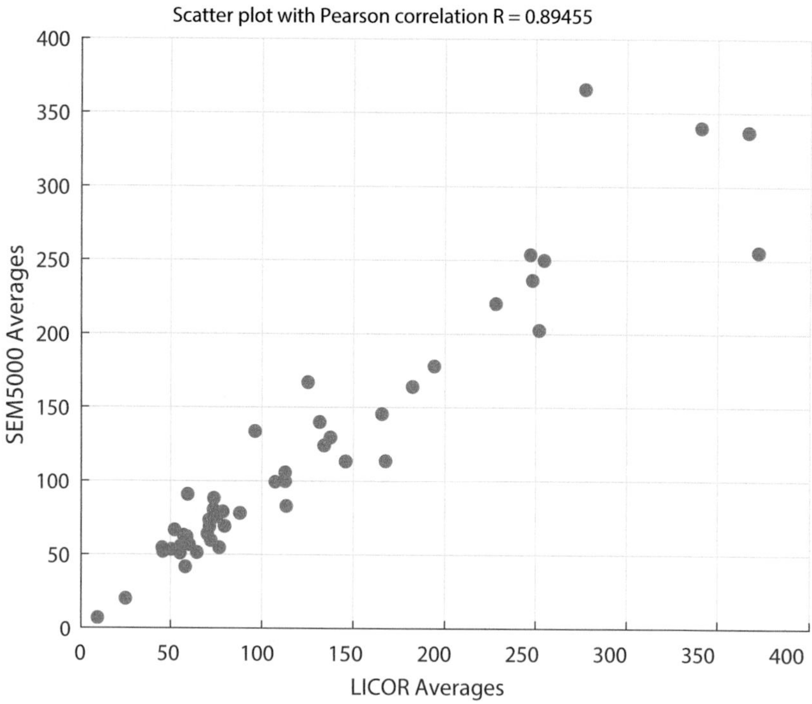

Figure 6.5: Scatter plot comparing readings from the SEM5000 sensor and the LICOR 7810 laser-based sensor. Each point represents the averaged methane emissions from a unique cow, as measured by both sensors. The Pearson regression line is also depicted, with an accompanying R^2 value, indicating the strength and direction of the linear relationship between the two sets of measurements. This high correlation implies that the two sensors produce consistent and comparable results, further validating the reliability and accuracy of the SEM5000 in measuring enteric methane emissions. The average discrepancy (noise) between measurements from the two sensors was found to be 14%, with a median discrepancy of 10%. Additionally, the Root Mean Square Error (RMSE) between their measurements stood at 28.

emissions were consistently measured at the same hour of the day for each test session, thereby minimizing the impact of diurnal variations.

The importance of extended measurement periods is not just an internal research principle but is also underscored by the standards set by external bodies. Specifically, the VM41 protocol [79] for enteric methane measurement and reduction, established by the Verra agency, mandates that projects measure emissions for at least 8 weeks to be compliant with its guidelines [75]. This timeframe is recommended to ensure a thorough assessment of the additive's performance. In acknowledgment of the significance of these guidelines, and

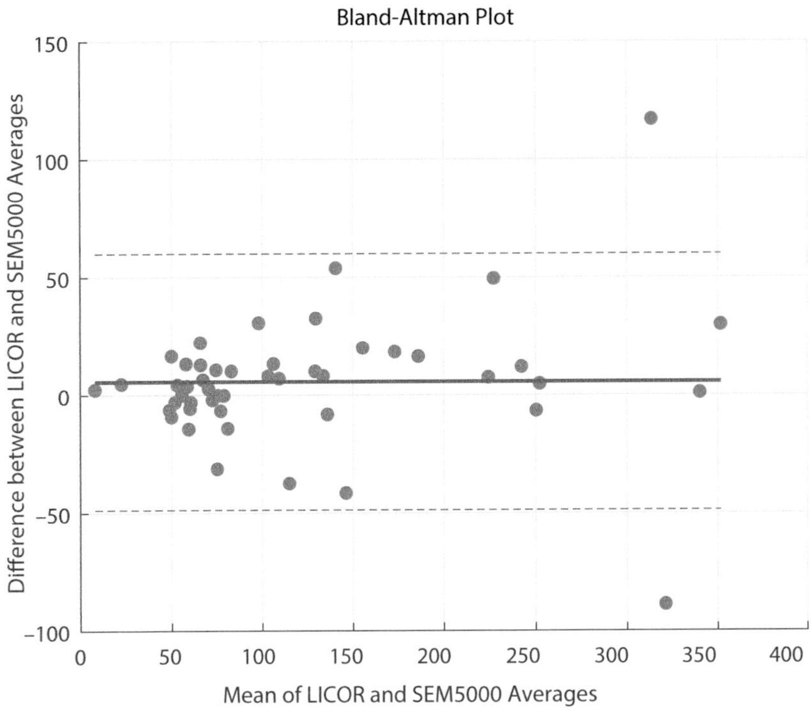

Figure 6.6: Bland-Altman analysis comparing the SEM5000 and the LICOR 7810 laser-based sensor. The Bland-Altman test[15] is used to assess the agreement between two different instruments measuring the same parameter. In the plot, the difference between the two sensors' readings is plotted against their average. The central line represents the mean difference, while the outer lines depict the limits of agreement, which are calculated as the mean difference $\pm$ 1.96 times the standard deviation of the differences. Apart from two outliers, all data points lie within the confidence limits, indicating that the measurements from the two sensors are largely in agreement and can be used interchangeably for most practical purposes [32].

with an aim to enhance the statistical robustness of our results, we opted for an extended 12-week measurement duration for our study.

At each time point, the efficacy of the feed additive was calculated by comparing the current available measurements of the cows and measured on that day to their emission levels recorded before the commencement of the trial. This implies that the exact composition of cows measured may differ between consecutive farm visits. However, this does not compromise the accuracy of the efficacy calculation at each point. The reason being, the cows in both the treatment and control groups are always compared to their individual baseline emissions level, which was established prior to the initiation of the trial. This procedure ensures that the efficacy determination is based on individualized comparisons, thereby maintaining the overall reliability of the results.

We also took steps to manage the potential variability in our measurements due to cow evasiveness. The percentage of cows that managed to evade measurement was consistently kept below 20%, reducing the overall impact of this phenomenon on our data set. Consequently, we believe that the influence of this variability on our overall findings is minimal. This aspect of the study underscores the complexities of field research in livestock environments and the need for adaptable research methodologies.

6.2.5.6 Methane Reading Methodology

When measuring methane emissions from cows, the variability in the data due to diverse observation durations and transient spikes presents significant challenges. Our methodology needed to effectively address these discrepancies to yield reliable and consistent metrics.

Ambient Noise Filtering

Methane measurements can capture ambient readings, especially before and after the actual approach to the cow. To filter out these irrelevant readings and hone in on the cow's emissions, we considered only values above 5 parts per million (ppm). This threshold ensures we primarily focus on the cow's emissions, excluding most ambient interference.

Data Consolidation and Noise Reduction

To condense the varied readings from each visit into a single representative number and simultaneously mitigate the noise (like sudden spikes due to burping), we employed the median value. As a measure of central tendency, the median is inherently robust against outliers, offering a more stable representation of the cow's typical methane emission.

Formally, given a set of methane readings for a specific cow c taken on a sequence of time stamps $\{R_{c,t_1}, R_{c,t_2}, \ldots, R_{c,t_N}\}$ on a specific day d, the consolidated value for that day and cow is computed as:

$$\hat{R}_{c,d} = \text{median}\left(\{R_{c,t_i} | R_{c,t_i} > 5 \text{ ppm}\}\right)$$

By implementing this methodology, we ensure a single, consistent methane reading per cow for each visit, establishing a dependable foundation for subsequent comparative analysis across various visits and cows.

6.2.6 Feed Additives

The landscape of feed additives, designed to mitigate methane emissions, is rich and varied, with each product leveraging a unique biological strategy. These formulations are designed to interact with the bovine digestive process in various ways to reduce the production of methane, a major byproduct [5]. The efficacy of these additives is largely determined by the specific biological pathway they target, underlining the need for personalized application based on each farm's specific conditions and requirements [12]. From methane inhibitors and direct-fed microbials to natural plant extracts and chemical compounds, the range of solutions showcases the vast scope of scientific novelty directed towards curbing this environmental concern. In this study we have tested the following widely used and commercially available additives.

6.2.6.1 Agolin

Ruminant (Agolin) is a commercially available blend of essential oils (coriander seed oil, eugenol, geranyl acetate, and geraniol) which has been demonstrated to reduce greenhouse gas emissions in dairy cows and improve energy corrected milk and feed efficiency [18] at a daily dose of 0.8 to 1 gram per animal. Agolin increases milk production in cows producing moderate milk yield (30 kg/d), however, this response depends on duration of feeding (5 to 8 weeks minimum). Some observed a consistent and convincing 2–3% increase in milk yields or ECM [19, 29]. Agolin is shown to inhibit ruminal methane production or intensity by 8% on average while no apparent change in dry matter intake (DMI) nor on milk composition was described. The exact mode of action is yet to be elucidated [29].

6.2.6.2 Relyon

Manufactured by Phibro Animal Health, this tannins flavonoid and essential oils-based additive was shown to mitigate ruminal methene emission in cows by 13% on average, while no change in milk yield or its composition was observed [40, 70]. While more rigorous scientific studies are desirable to substantiate Relyon's promising role in also enhancing feed conversion and stimulating appetite in ruminants, the preliminary results presented to date are encouraging.

6.2.6.3 Kexxtone (Elanco)

Kexxtone is a Monensin containing intraruminal bolus for administration 3–4 weeks pre-calving to help the peri-parturient dairy cow/heifer maintaining an appropriate energy balance and thereby preventing many peri-parturient metabolic based diseases [10]. The Kexxtone bolus releases Monensin for a period of 95 days in the rumen [28].

Ionophores such as Monensin improve methane mitigation by enhancing digestive efficiency to favor propionate production over acetate, which reduces H_2 for methanogens. This methanogenesis inhibition becomes more pronounced in diets with higher fat content [34, 92]. Meta-analyses of

Monensin conclude an effect on methanogenesis inhibition of up to 10% reduction on average in dairy [29, 58].

6.2.6.4 Allimax

Allimax bolus (Garlic, Allicin) has been developed for the purpose of alternative anti-microbial activity in dairy. The natural extract Allicin, which is the main active ingredient of the sulfur-containing organic compounds in garlic, has anti-inflammatory, anticancer, antioxidant, and antibacterial properties [57]. However, the specific mechanism underlying its effect on mastitis in dairy cows needs to be further studied [6, 17]. The supplementation of Allicin has been observed to elevate the levels of propionate and butyrate during partial incubation periods, suggesting its potential role in curtailing methane emissions [91]. Even though compelling in vitro evidence demonstrates the ability of Allicin to mitigate methane emissions by up to 38% [29, 40, 93], in vivo studies confirming these findings remain scarce to date [6, 50].

6.2.7 From Microbial Data to Additive Efficacy Prediction

This section offers a concise overview of the methodology used in this study. Detailed discussions and formal mathematical delineations concerning data processing, training, validation, and the employed algorithms can be found in Sections 6.3 and 6.5.

Input

This study was carried out in partnership with 34 trial sites. We collected 15 microbiome samples from each site, which were subsequently subjected to deep shotgun metagenomic sequencing. At every site, one or more feed additives were administered to distinct groups, each consisting of 20 cows. Additionally, a control group of 20 cows was established at each site, receiving no treatment. Throughout a period of 3 months, starting from the onset of the trial, we periodically recorded methane emissions from each cow. This consistent monitoring allowed us to determine the normalized efficacy of each additive across the various sites.

Unsupervised Detection of Microbial DNA Patterns

In our research, we directly utilized the raw sequencing data—strings of 100-150 nucleotides—without delving into specific identification of microbes or strains. Our distinctive DNA analytics algorithm, with a network-oriented approach, processed the data from all gathered microbiome samples (refer to Section 6.5). This analysis yielded a plethora of DNA patterns. Each pattern has been analytically validated to be improbable to emerge spontaneously in random microbial genetic samplings, making them statistically likely to correlate with a phenotypic trait, regardless of its relevance to the study's goals.

The initial phase of our method involves an unsupervised data analysis, serving as an effective dimensionality reduction technique. This process can efficiently handle billions of raw sequences, each 100–150 bases in length. It distills these into a computationally tractable number of clusters containing "statistically significant" substrings. Importantly, this approach sidesteps any bias toward predetermined feature spaces, data preprocessing, or the inherent semantics of the problem.

This phase can be updated as new data becomes available, leading to the identification of new DNA patterns that contribute to the system's predictive capabilities for the same or new properties.

Filtering the Microbial DNA Patterns Using Semantic Labels

For each DNA pattern (actually a collection of 100–150 long DNA bases), we can filter only those whose frequency (i.e., the number of times they occur in a sample) correlates strongly with the property we aim to predict (i.e., an additive's efficacy, defined by its methane reduction capacity, normalized for the control group on the same farm). This phase is executed once for each group of labels (i.e., once per additive).

Output

The process yields a collection of DNA sequence groups, which are statistically validated to correlate with our target property. These segments, termed "microbiome markers", subsequently serve as a reference against which samples from new farms are contrasted. This comparison produces an anticipated efficacy score ranging from 0 (no efficacy) to 1 (maximal efficacy) for the given additive.

6.3 Field Study Design

6.3.1 Definitions

Below are the definitions of groups and annotations used in the description of the study:

- F: The set of all farms participating in the study, $F = \{f_1, f_2, \dots, f_N\}$.
- F_A: The subset of farms selected for testing a specific additive A.
- $C_{u,i}$: The set of "Learning Microbiome Cows" (LMCs) for each farm $f_i \in F$, selected for unsupervised learning of the microbiome.
- P_X: This represents the collection of "microbial genetic patterns" derived from each microbiome sample X. Each sample gives rise to a distinct

network G_X comprised of a constant set of $M = |V|$ nodes and unique edges, E_X. Patterns are extracted both from individual network analysis and from the superposition of networks, enabling exploration of a broad spectrum of combinatorial possibilities based on various criteria.

- P_S: Represents the patterns derived from a combination of networks, where S is the set of samples considered for superposition.
- $C_{t,i}$ and $C_{v,i}$: The partition of the LMCs into a "Microbiome Train Group" and a "Microbiome Validation Group" for each farm $f_i \in F_A$.
- $C_{m,i}$: The 40 cows selected for methane measurement in each farm $f_i \in F_A$.
- $C_{mc,i}$, $C_{mt,i}$, $C_{mv,i}$: The division of $C_{m,i}$ into three groups—"Control Methane Cows", "Train Methane Cows" and "Test Methane Cows" (or "Validation Methane Cows").
- $\hat{R}_{c,d}$: the level of methane emission for a cow c for a day d (calculated as the median of methane readings above a certain "ambiance threshold").
- $M_{\text{pre}}(c)$ and $M_{\text{post}}(c)$: The pre-additive and post-additive methane levels for a cow c. Notably, since the cows were measured multiple times over the 12-week period following the introduction of the additive, we obtain multiple $\hat{R}_{c,d}$ values for each cow (each corresponding to a respective day d, whereas D_{pre} and D_{post} represent the days of pre-additive treatment and post-additive treatment respectively), of which we take the mean:

$$M_{\text{post}}(c) = \text{mean}\left(\{\hat{R}_{c,d} | d \in D_{\text{post}}\}\right)$$

$$M_{\text{pre}}(c) = \text{mean}\left(\{\hat{R}_{c,d} | d \in D_{\text{pre}}\}\right)$$

  This repeated sampling further strengthens the statistical significance of the measurements.
- $T_{pre,i}$, $T_{post,i}$, $C_{pre,i}$, $C_{post,i}$: The mean pre-additive and post-additive methane levels for the treatment and control cows in a farm f_i.
- η_{A,f_i}: The methane efficacy for farm f_i and additive A, calculated as:

$$\eta_{A,f_i} = \frac{T_{post,i}/T_{pre,i}}{C_{post,i}/C_{pre,i}} = \frac{T_{post,i} \cdot C_{pre,i}}{T_{pre,i} \cdot C_{post,i}}$$

Following are some key notes regarding the groups and their properties:

- The Learning Microbiome Cows ($C_{u,i}$) and the cows selected for methane measurement ($C_{m,i}$) in each farm are disjoint, i.e. $C_{u,i} \cap C_{m,i} = \emptyset$ for each farm $f_i \in F_A$.
- The "Microbiome Train Group" ($C_{t,i}$) and "Microbiome Validation Group" ($C_{v,i}$) are also disjoint for each farm $f_i \in F_A$, i.e. $C_{t,i} \cap C_{v,i} = \emptyset$.

- Similarly, the "Control Methane Cows", "Train Methane Cows" and "Test Methane Cows" groups are pairwise disjoint for each farm $f_i \in F_A$, i.e. $C_{mc,i} \cap C_{mt,i} = \emptyset$, $C_{mc,i} \cap C_{mv,i} = \emptyset$, and $C_{mt,i} \cap C_{mv,i} = \emptyset$.
- The efficacy of an additive A in a farm f_i is a time series measurement, represented as a set $\eta_{A,f_i} = \{e_1, e_2, \ldots, e_n\}$ where each e_k is calculated from the ratio of mean pre-additive and post-additive methane levels for the treatment and control cows.
- The division of cows into "control methane cows" (CMC), "train methane cows" (TMC), and "test methane cows" (TeMC) has been optimized to minimize bias. This has been achieved by exhaustively examining all possible allocations of cows into the three groups, and selecting the assignment that minimizes the maximum discrepancy among the distributions of age (AGE_i), days in lactation (DIL_i), and average milk yield (AMY_i) across the groups. Let Gr_{CMC}, Gr_{TMC}, and Gr_{TeMC} represent the groups of cows. The condition for the optimal assignment can be formally expressed as: For all i, and for any two distinct groups Gr_x and Gr_y from $\{Gr_{CMC}, Gr_{TMC}, Gr_{TeMC}\}$, we choose the assignment that minimizes the following quantity:

$$\max \left\{ \begin{array}{l} |P(AGE_i|Gr_x) - P(AGE_i|Gr_y)|, \\ |P(DIL_i|Gr_x) - P(DIL_i|Gr_y)|, \\ |P(AMY_i|Gr_x) - P(AMY_i|Gr_y)| \end{array} \right\}$$

  This guarantees that the selected division of cows into groups ensures the least possible bias across all characteristics, given the distributions of age, days in lactation, and average milk yield among the cows.

6.3.2 Training the Model

This stage is executed per each feed additive A. Note that whereas not all farms are included in this stage (since feed additive A may have been tested by only a subset of the available farms), all of the data patterns extracted in the unsupervised phase is used to train its efficacy prediction model. We denote the subset of farms used for this stage as $F_A \subseteq F$.

For each farm $f_i \in F_A$, we further partition the LMCs into a "Microbiome Train Group" $C_{t,i} \subseteq C_{u,i}$ and a "Microbiome Validation Group" $C_{v,i} = C_{u,i} \setminus C_{t,i}$.

Also for each farm f_i, we introduce a set of 40 cows $C_{m,i}$ for methane measurement. We divide this set into three groups: "Control Methane Cows" $C_{mc,i}$, "Train Methane Cows" $C_{mt,i}$ and "Test Methane Cows" $C_{mv,i}$.

Let $M_{pre}(c)$ and $M_{post}(c)$ denote the pre-additive and post-additive methane levels for a cow c, respectively. $M_{pre}(c)$ and $M_{post}(c)$ are calculated as the median of methane levels over 30 to 120 seconds, excluding values smaller than 5 parts per million.

The methane efficacy η_{A,f_i} for farm f_i and additive A is calculated as follows:

- $T_{pre,i} = \frac{1}{|C_{mt,i}|}\sum_{c\in C_{mt,i}} M_{pre}(c)$, the mean pre-additive methane for the treatment cows in the farm,
- $T_{post,i} = \frac{1}{|C_{mt,i}|}\sum_{c\in C_{mt,i}} M_{post}(c)$, the mean post-additive methane for the treatment cows in the farm,
- $C_{pre,i} = \frac{1}{|C_{mc,i}|}\sum_{c\in C_{mc,i}} M_{pre}(c)$, the mean pre-additive methane for the control cows in the farm,
- $C_{post,i} = \frac{1}{|C_{mc,i}|}\sum_{c\in C_{mc,i}} M_{post}(c)$, the mean post-additive methane for the control cows in the farm.

Having the overall efficacy calculated as:

$$\eta_{A,f_i} = \frac{T_{post,i} \cdot C_{pre,i}}{T_{pre,i} \cdot C_{post,i}}$$

This efficacy, along with the microbiome samples from $C_{t,i}$, is used for the supervised learning process, serving as the label for all microbiome cows at farm f_i. Specifically, the efficacy derived from the training cows $C_{mt,i}$, is utilized as labels for the features generated by the training microbiome cows $C_{t,i}$. Likewise, the efficacy determined for the validation cows, $C_{mv,i}$, is employed as labels for the features from the validation microbiome cows $C_{v,i}$.

6.3.3 Flowchart

The following figures offer a visual breakdown of our study's flowchart. Figure 6.7 outlines the foundational design tailored for one feed additive. This design is reiterated for various additives, with both the control group and the microbiome group being reused. Figure 6.8 showcases the unsupervised learning phase. Meanwhile, the supervised learning segment, followed by validation, is illustrated in Figure 6.9.

6.3.4 Microbiome Markers Used in This Study

As highlighted in Section 6.2.7 and expounded upon in Section 6.5, our proposed AI-driven analytical methodology interprets sequenced microbial data in tandem with corresponding attribute labels, crafting a predictive model applicable for subsequent microbiome samples. Versatile in its design, this approach can formulate a "microbiome marker" for any given attribute presented as a label. In the context of this study it is associated with cows exhibiting high efficacy towards a specific feed additive. However, in future works this could equally pertain to other biological attributes such as heightened survival rates against certain diseases and so on. This biomarker comprises two

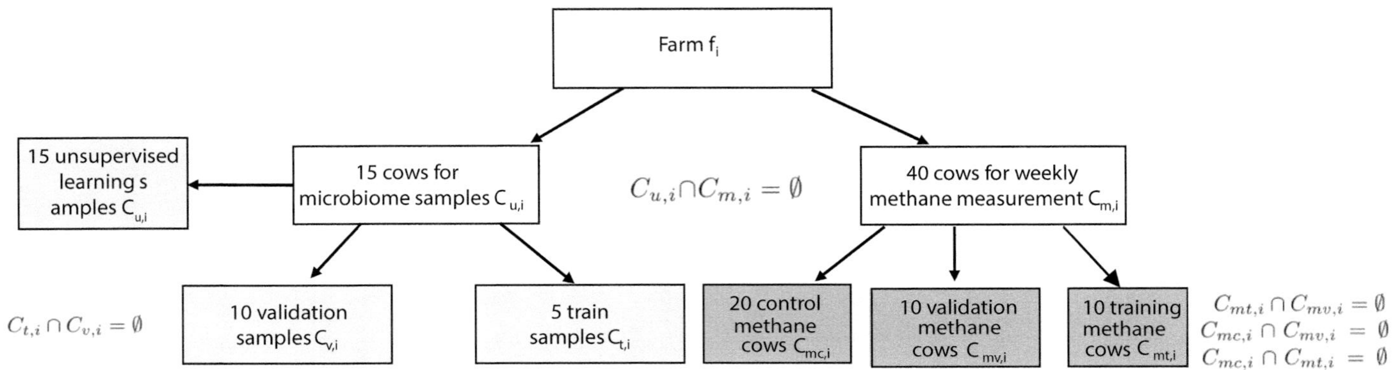

Figure 6.7: Schematic representation of the field study design tailored for a single feed additive. The initial unsupervised phase identifies genetic patterns from all 15 microbiome samples of a farm and is detailed further in Figure 6.8. Crucially, this phase is conducted once and is applicable across all additives. Of the 15 microbiome samples, 5 are designated for model training while the remaining 10 are earmarked for validation. The microbiome samples form the feature set of the model, with labels being generated based on the average performance of a distinct group of methane cows. One of the pivotal strengths of our design is the assurance that the genetic markers identified by the model are not just emblematic of the microbiome cow group but resonate with the broader farm context. This robustness is reinforced by two layers of separation involving the methane cows: the initial distinction from the microbiome cows and subsequently ensuring the cows used for model training differ from those in the validation, each group being wholly independent. This separation takes place both among the microbiome cows as well as the methane measurement cows.

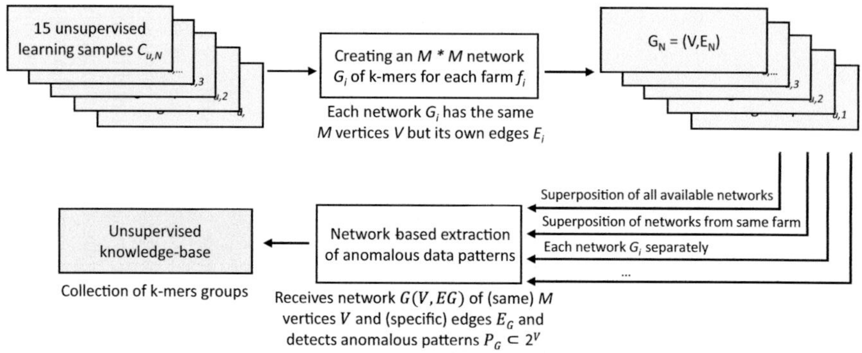

Figure 6.8: Illustration detailing the unsupervised learning phase. This foundational stage is executed just once, harnessing all accessible microbiome samples. The essence of this phase lies in constructing networks from the genetic material, necessitating varied combinations of these samples. Initially, individual samples stand as solitary data sources to form their respective networks. Subsequently, a comprehensive network is created by pooling samples from an entire farm, resulting in a denser structure that potentially offers a more holistic representation of farm-level features. Multiple other networks can emerge, shaped by diverse criteria like geography, weather conditions, or even an aggregation of all available data. Intriguingly, the formation of these intricate networks is computationally straightforward. By overlaying selected foundational networks, a superposition network is birthed – akin to executing a simple boolean OR operation on their edges. The establishment of networks specific to each sample paves the way for a nimble and robust integration of fresh data or new samples. It is crucial to emphasize the adaptability of this phase: it is indiscriminate to data sources, allowing for the amalgamation of microbiome samples from diverse entities like cows, sheep, soil, or even humans. Furthermore, its generic nature ensures that identical genetic markers are relevant across varied label groups linked to any targeted biological condition or trait.

sets of short DNA sequences, their prevalence in microbiome samples serves as a predictor of the target attribute. The first set, termed the "top list" (or "positive list"), features DNA segments indicative of a high likelihood of association with the desired biological condition. Conversely, the "bottom list" (or "negative list") captures DNA segments that exhibit a low probability of such an association.

The explicit sets of DNA segments identified from the microbiome data and methane measurements used in this study are presented in Tables 6.3, 6.4, 6.5 and 6.6. Future studies and commercial projects can leverage these lists to predict the efficacy of the additives evaluated in this study. Given these lists and microbiome samples from cows $c_1, c_2, \ldots$ in a farm f we formally define the prediction score for the efficacy of additive A in f as follows:

1. For each cow c_i in farm f identify the top-1000 most popular k-mers.

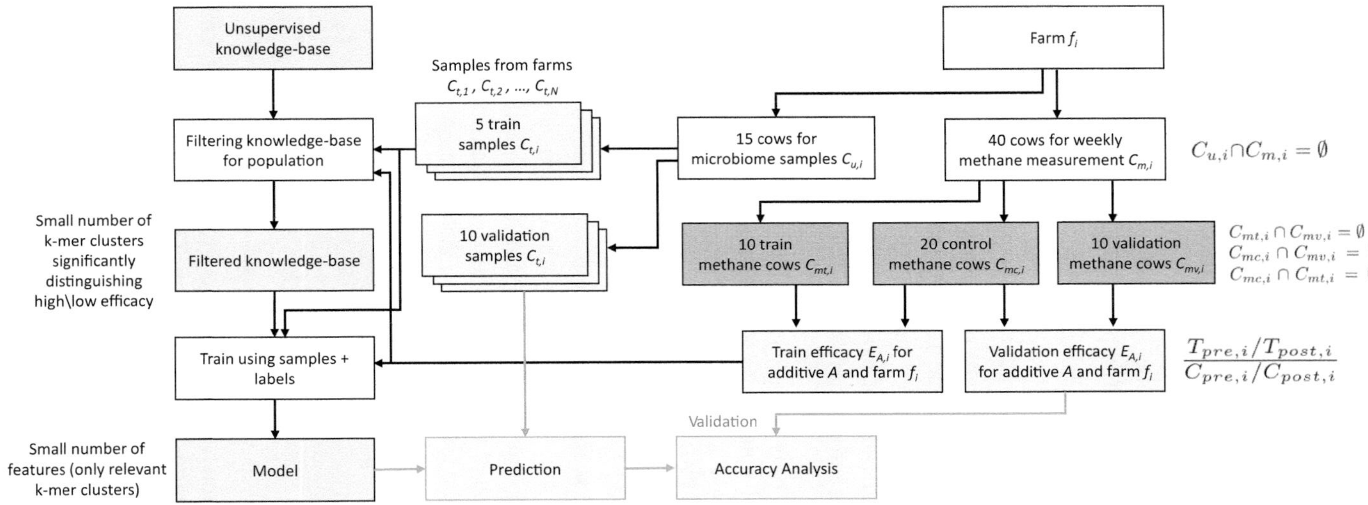

Figure 6.9: An illustration of the supervised learning phase paired with validation, highlighting the detection of a given additive's expected efficacy. It is important to emphasize the distinction between the microbiome cows, which are used as feature sets, and the methane cows, which contribute to label creation for both training and validation. Additionally, note the deployment of distinct groups of cows for methane measurements during training and validation. The control group is consistently reused to guarantee consistent data normalization. A comprehensive definition of additive efficacy can be found in Section 6.3.1.

2. Compute the score for cow c_i as:

$$C_{\text{top}} - C_{\text{bottom}}$$

   where:
   - C_{top} is the ratio of the number of k-mers from the "top list" for additive A present in the top-1000 most popular k-mers for cow c_i to the total length of the "top list" for additive A. This results in values ranging from 0 (no presence in top-1000) to 1 (all k-mers in the top-list are present in top-1000).
   - C_{bottom} is defined similarly using the "bottom list" for additive A.

   Consequently, each cow can now have scores between –1 and 1.
3. For farm f compute the average of its cows' scores. This farm-level score will also range between –1 and 1. To normalize this score, add 1 and then divide by 2, yielding values between 0 (indicative of expected low efficacy) and 1 (indicative of expected high efficacy). Scores around 0.5 suggest insufficient information for prediction.

For the scope of this study it is assumed that both the "top list" and "bottom list" consist of k-mers of length $k = 30$. Nonetheless, the analysis detailed in this study can be promptly adapted to encompass k-mers of various lengths. Additionally, k-mers of different lengths that are found to be associated with the efficacy in question can be seamlessly integrated to boost prediction accuracy.

In this study, we opted for a straightforward method to compute both the cow-level and farm-level scores. This decision was made to bolster robustness and minimize the risk of over-fitting. Clearly, the chosen value of 1000 and the counting method for the k-mers, which disregards their specific rank or absolute popularity, can be substituted with a more refined mechanism. Additionally, this approach could be superseded by advanced machine learning techniques that train models on the presence of identified k-mers, potentially improving predictive accuracy.

6.4 Results

For each feed additive A and for each farm $f_i \in F_A$, the microbiome samples from $C_{v,i}$ were used to predict the additive's efficacy. This predicted efficacy was then contrasted with the actual efficacy determined through the analysis of methane emissions from $C_{mc,i}$ and $C_{mv,i}$.

This design allows for general applicability to different additives and use cases, with potential for synergistic improvement as more data is added to the unsupervised learning stage. Importantly, the division of cows into various groups is done in a way that reduces bias for factors like age of cows, their days in lactation, and average milk yield.

Table 6.3: Top and Bottom k-mers markers for feed additive Agolin. See more details in Section 6.3.4

Top K-mers	**Bottom K-mers**
ACGTGATCAGTGCATGATCAGTCACGTGAT	AAAGGTACGAAAATTTTAGCTAATCACAAC
AGGTGTCGCGCGGCTCAGCTGGCGAGTATC	ACCTTGCAAAGGTACGAAAATTTTAGCTAA
AGTATCAGGCAGATGAGCGGGCAGGTGTCG	ACGCGTGGACGCGTGGACGCGTGGACGCGT
AGTGCATGATAGCCACGTGATCAGTGCATG	ATAATAATAATAATAATAATAATAATAATA
ATAGCCACGTGATCAGTGCATGATCAGTCA	ATGACCTTGCAAAGGTACGAAAATTTTAGC
ATCAGTGCATGATAGCCACGTGATCAGTGC	CGTGGACGCGTGGACGCGTGGACGCGTGGA
ATCATGCACTGATCACGTGACTGATCATGC	CTTATACACATCTCGAGCCCACGAGACCTA
ATCATGCACTGATCACGTGGCTATCATGCA	CTTATACACATCTCGAGCCCACGAGACGCT
ATGATAGCCACGTGATCAGTGCATGATCAG	GACGCATGACGCATGACGCATGACGCATGA
ATGATCAGTCACGTGATCAGTGCATGATCA	GCCAAGCTGTTCTTGGCGTAAGATGCAATG
ATGCACTGATCACGTGGCTATCATGCACTG	GCGTAAGATGCAATGGCTGAGAACTTGACT
ATTGGGGATTGGGGATTGGGGATTGGGGAT	GCTGAGAACTTGACTTTCAAGAGTTCTTTT
CAGCTGGCGAGTATCAGGCAGATGAGCGGG	GCTGTTCTTGGCGTAAGATGCAATGGCTGA
CGCGGCTCAGCTGGCGAGTATCAGGCAGAT	GTTCTTGGCGTAAGATGCAATGGCTGAGAA
CGTGATCAGTGCATGATAGCCACGTGATCA	GTTGAGAGTTGAGAGTTGAGAGTTGAGAGT
CTCATCTGCCTGATACTCGCCAGCTGAGCC	GTTGATGACCTTGCAAAGGTACGAAAATTT
GCATGATCAGCCACGTGATCAGTGCATGAT	TAAGATGCAATGGCTGAGAACTTGACTTTC
GCGAGTATCAGGCAGATGAGCGGGCAGGTG	TAGGCCAAGCTGTTCTTGGCGTAAGATGCA
GCTCAGCTGGCGAGTATCAGGCAGATGAGC	TCATGCGTCATGCGTCATGCGTCATGCGTC
GGCAGGTGTCGCGCGGCTCAGCTGGCGAGT	TCTCTTATACACATCTACGCTGCCGACGAC
GGGATTGGGGATTGGGGATTGGGGATTGGG	TCTTATACACATCTCCAGCCCACGAGACTT
GTGCATGATCAGTCACGTGATCAGTGCATG	TCTTATACACATCTCGAGCCCACGAGACTT
GTGTGTGTGTGTGTGTGTGTGTGTGTGTGT	TCTTATACACATCTTGACGCTGCCGACGAC
TCATGCACTGATCACGTGGCTGATCATGCA	TGCAAAGGTACGAAAATTTTAGCTAATCAC
TGTCGCGCGGCTCAGCTGGCGAGTATCAGG	TGCAATGGCTGAGAACTTGACTTTCAAGAG
	TGTCAAGCGGCAACCGATCGGTTACGCTGA
	TTATCTCATTGCTTTTCACCTCACACATTT
	TTCAAGAGTTCTTTTCTCTTTCTGATTGCC
	TTCACCTCACACATTTCAGTGTCAAGCGGC
	TTCAGTGTCAAGCGGCAACCGATCGGTTAC
	TTGACTTTCAAGAGTTCTTTTCTCTTTCTG
	TTGCTTTTCACCTCACACATTTCAGTGTCA
	TTGGCGTAAGATGCAATGGCTGAGAACTTG

6.4.1 Additive Efficacy Measurements

The following Tables 6.7, 6.8, 6.9, and 6.10 provide a comprehensive summary of the results obtained from measuring methane emissions across various farms and for different feed additives, as measured for the test group of cows $C_{mv,i}$, and normalized by the control group $C_{mc,i}$. Specifically, each table represents one unique additive. The columns in the table are as follows:

- **Farm**: Identifier for each farm where measurements were taken.
- CH_4 **Change Treatment**: This column displays the mean percentage change in methane emissions for cows that received the feed additive. For example, a value of 0 indicates no change compared to the values

Table 6.4: Top and Bottom k-mers markers for feed additive Allimax. See more details in Section 6.3.4

Top K-mers	**Bottom K-mers**
AAACATGGGCAGGCCTATGAAACCCACCGC	AAAATTAGATAAATTTAAAGAAGTTAAAGA
AAAGAGAGGTGAGAAACATGGGCAGGCCTA	AACATTATTAGTATTAAAATTAGATAAATT
AAATTAATGTTTATATATGTTAAATTAATG	AATAATAATAATAATAATAATAATAATAAT
AACGCTGTACAAGAAGCGCCTGAACACCGA	AATCCCCAATCCCCAAAACCCAAAACCCAA
ACGCATGACGCATGACGCATGACGCATGAC	AATGGGGATTGGGGATTGGGGATTGGGGAT
ATATGTTAAATTAATGTTTATATATGTTAA	AATTGGGGATTGGGGATTGGGGATTGGGGA
ATGACGCATGACGCATGACGCATGACGCAT	AGATAAATTTAAAGAAGTTAAAGAAGAACA
ATGCGTCATGCGTCATGCGTCATGCGTCAT	AGTATTAAAATTAGATAAATTTAAAGAAGT
ATGGGCAGGCCTATGAAACCCACCGCAGTC	ATTAAAATTAGATAAATTTAAAGAAGTTAA
CAGGCCTATGAAACCCACCGCAGTCAAGAA	ATTAGATAAATTTAAAGAAGTTAAAGAAGA
CCAGACCCTCAGCGACATCGGAACGACCGC	ATTAGTATTAAAATTAGATAAATTTAAAGA
CGTCATGCGTCATGCGTCATGCGTCATGCG	ATTATTAGTATTAAAATTAGATAAATTTAA
CTCTGCTCTGCTCTGCTCTGCTCTGCTCTG	ATTATTATTATTATTATTATTATTATTATT
CTCTTATACACATCTCGAGCCCACGAGACA	ATTGGGCCCAATCCCCAATCCCCAAACCCC
GAGAAACATGGGCAGGCCTATGAAACCCAC	ATTGGGGATTGGGGATTGGGGATTGGGCCC
GGTGAGAAACATGGGCAGGCCTATGAAACC	CCAATCCCCAAAACCCAAAACCCCAAACCC
GTTAAATTAATGTTTATATATGTTAAATTA	CCAATCCCCAATCCCCAATACCCAAAACCC
TCTTTTTCTTTTTCTTTTTCTTTTTCTTTT	CCAATCCCCAATCCCCAATCCCCAATCCCC
TTTTCTTTTTCTTTTTCTTTTTCTTTTTCT	CCCAATCCCCAATCCCCAAAACCCAAAACC
	GATTGGGGATTGGGGATTGGGGATTGGGGG
	GGGATTGGGGAGTGGGGATTGGGGATTGGG
	GGGGATTGGGGATTGGGGATTGGGGATTGG
	TAAATTTAAAGAAGTTAAAGAAGAACAATT
	TCCCCAATCCCCAATCCCCAATCCCCATTA
	TCTTATACACATCTCGAGCCCACGAGACGA
	TGGGGATTGGGGATTGGGGAGTGGGGATTG
	TTGGGGATTGGGGAGTGGGGATTGGGGATT
	TTGGGGATTGGGGATTGGGGATTGGGGCCA

taken prior to the initiation of the trial, −10 indicates a 10% reduction, and 20 indicates a 20% increase in emissions.

- CH_4 **Change Control**: This column shows the mean percentage change in methane emissions for cows that did not receive the feed additive, serving as the control group.
- **Normalized Efficacy** η_{A,f_i}: Reflecting the normalized mean percentage change in methane emissions, accounting for variations in control and treatment groups (see formal definition in Section 6.3.1).
- **Effect Size**: This is a statistical measure that quantifies the size of the difference between the two groups, widely used to assess medical and nutritional treatment efficacy [39].
- **Cohen's D**: A Statistical effect size measure, indicating the standardized difference between the means in units of standard deviation [77].

In evaluating the effectiveness of various feed additives for reducing methane emissions across multiple farms, we observe a distinct variability in

Table 6.5: Top and Bottom k-mers markers for feed additive Kexxtone. See more details in Section 6.3.4

Top K-mers	**Bottom K-mers**
AAACACCATATATATTGAGAAAGAGAGGTG	AAACGCCTCAGGAGGCTTGACTCCCTTGAG
AAACATGGGCAGGCCTATGAAACCCACCGC	AAGAAGAAGAAGAAGAAGAAGAAGAAGAAG
AAATTAATGTTTATATATGTTAAATTAATG	AGGTACGACGGCGAGGTCAGTGAGCCTCTC
AAATTTAAAGAAGTTAAAGAAGAACAATTA	AGTGCATGATAGCCACGTGATCAGTGCATG
AAATTTATCTAATTTTAATACTAATAATGT	ATCAGTGCATGATAGCCACGTGATCAGTGC
AACTTCTTTAAATTTATCTAATTTTAATAC	ATCATGCACTGATCACGTGACTGATCATGC
AAGAAGAAGAAGAAGAAGAAGTTGAACATG	ATCTCGCGACCTCTCTCCAAACGCCTCAGG
AAGAAGAAGAAGAAGAAGTTGAACATGAAG	ATGCACTGATCACGTGGCTGATCATGCACT
AATTTTAATACTAATAATGTTAATAATATG	CACACACACACACACACACACACACACACA
AATTTTAATACTAATAATGTTACTGATATG	CAGGAGGCTTGACTCCCTTGAGTCCACCCA
ACACTAAACACCATATATATTGAGAAAGAG	CATGATAGCCACGTGATCAGTGCATGATCA
ACCATATATATTGAGAAAGAGAGGTGAGAA	CCTCAGGAGGCTTGACTCCCTTGAGTCCAC
AGATAAATTTAAAGAAGTTAAAGAAGAACA	CCTCTCTCCAAACGCCTCAGGAGGCTTGAC
AGGCCTATGAAACCCACCGCAGTCAAGAAG	CGGCTCGGCTCGGCTCGGCTCGGCTCGGCT
ATAAATGGGGATTGGGGATTGGGGATTGGG	CGTGATCAGTGCATGATAGCCACGTGATCA
ATCTAATTTTAATACTAATAATGTTAATAA	CTCCAAACGCCTCAGGAGGCTTGACTCCCT
ATGCGTCATGCGTCATGCGTCATGCGTCAT	CTGATCATGCACTGATCACGTGGCTATCAT
ATGGGGATTGGGGATTGGGGATTGGGGATT	GCGACCTCTCTCCAAACGCCTCAGGAGGCT
ATTATTATTATTATTATTATTATTATTATT	GTCCACCCAGTGAGCTCCAAGAGATACCCG
CCAATCCCCAATCCCCAATCCCCAATCCCC	TCTTATACACATCTCGAGCCCACGAGACTC
CGTCATGCGTCATGCGTCATGCGTCATGCG	
CTCTGCTCTGCTCTGCTCTGCTCTGCTCTG	
CTCTTATACACATCTCGAGCCCACGAGACG	
CTTATACACATCTCGAGCCCACGAGACAAC	
CTTATACACATCTCGAGCCCACGAGACTGT	
CTTTAAATTTATCTAATTTTAATACTAATA	
CTTTAACTTCTTTAAATTTATCTAATTTTA	
GATTGGGGATTGGGGATTGGGGATTGGGGA	
TTTATCTAATTTTAATACTAATAATGTTAA	

efficacy. As illustrated in an efficacy matrix (see Figure 6.10), each additive performs differently depending on the farm where it is applied. Notably, for every additive, there are at least a few farms where it either fails to reduce emissions or even exacerbates them. Similarly, the effectiveness of additives varies within individual farms, underscoring the complexity of methane reduction strategies and suggesting that a 'one-size-fits-all' approach may not be viable. This variability also highlights the economic and business challenges associated with the adoption of additives. Negative or non-existent efficacy, even if relatively rare, may discourage farmers from incorporating additives into their practices.

Figure 6.11 illustrates the variability in the efficacy of different additives across multiple farms. While a majority of the additives generally demonstrate positive efficacy—reducing methane emissions by at least 5% – the data also reveals cases where the additives either have a negligible impact or paradoxically even increase emissions. This high volatility in efficacy at the

Table 6.6: Top and Bottom k-mers markers for feed additive Relyon. See more details in Section 6.3.4

Top K-mers	**Bottom K-mers**
AATCATGCTGCTCAGCTGGCAATAATCAAG	AATACCCAAAACCCAAAACCCAAAACCCAA
AATCTTCCATTCGAGTTGCGAAGGAAAGCT	AATCCCCAATCCCCAAAACCCAAAACCCCA
ACACACACACACACACACACACACACACAC	AATTTTAATACTAATAATGTAACTAATATG
ACCTGCCCGCTCATCTGCCTGATACTCGCC	AGAGCAGAGCAGAGCAGAGCAGAGCAGAGC
ACGTGATCAGTGCATGATCAGTCACGTGAT	ATAATAATAATAATAATAATAATAATAATA
ACTGACCTCGCCGTCGTACCTCGTGAGAAA	ATATTAGTTACATTATTAGTATTAAAATTA
AGGTGTCGCGCGGCTCAGCTGGCGAGTATC	ATTATTATTATTATTATTATTATTATTATT
AGTATCAGGCAGATGAGCGGGCAGGTGTCG	ATTGGGCCCAATCCCCAATCCCCAAACCCC
AGTGCATGATAGCCACGTGATCAGTGCATG	ATTGGGGATTGGGGATTGGGGAGTGGGGAT
ATAGCCACGTGATCAGTGCATGATCAGTCA	CAATCCCCAAAACCCAAAACCCCAAACCCC
ATCACGTGACTGATCATGCACTGATCACGT	CACTGACTGCAGTGATAACACTGACTGCAG
ATCAGGCAGATGAGCGGGCAGGTGTCGCGC	CCAATCCCCAATCCCCAAACCCCAATCCCC
ATCAGTGCATGATAGCCACGTGATCAGTGC	CCAATCCCCAATCCCCAAACCCCCAAACCC
ATCATGCACTGATCACGTGACTGATCATGC	CCAATCCCCAATCCCCAATACCCAAAACCC
ATCATGCACTGATCACGTGGCTATCATGCA	CCCAATCCCCAATCCCCAATCCCCAATACC
ATCATGCACTGATCACGTGGCTGATCATAC	CCCCAATCCCCAAAACCCAAAACCCCAAAC
ATGATAGCCACGTGATCAGTGCATGATCAG	CCCCAATCCCCAATCCCCAAAACCCAAAAC
ATGATCAGTCACGTGATCAGTGCATGATCA	CTTATACACATCTCGAGCCCACGAGACACT
ATGCACTGATCACGTGGCTATCATGCACTG	CTTATACACATCTCGAGCCCACGAGACCTA
ATGCACTGATCACGTGGCTGATCATACACT	CTTATACACATCTCGAGCCCACGAGACGCT
CAGCTGGCGAGTATCAGGCAGATGAGCGGG	GACTGCAGTGATAACACTGACTGCAGTGAT
CATGATCAGTCACGTGATCTGTGCATGATC	GATAACACTGACTGCAGTGATAACACTGAC
CGCGGCTCAGCTGGCGAGTATCAGGCAGAT	GATTGGGGATTGGGGAGTGGGGATTGGGGA
CGTGATCAGTGCATGATAGCCACGTGATCA	GGGATTGGGGATTGGGGATTGGGGAGTGGG
CTCATCTGCCTGATACTCGCCAGCTGAGCC	GGGGATTGGGGATTGGGGAGTGGGGATTGG
CTGATCATGCACTGATCACGTGGCTATCAT	TATTGGGGATTGGGGATTGGGGATTGGGGA
CTTCCATTCGAGTTGCGAAGGAAAGCTGGG	TGCTTGCTTGCTTGCTTGCTTGCTTGCTTG
GCATGATCAGCCACGTGATCAGTGCATGAT	TTGGGGAGTGGGGATTGGGGATTGGGGATT
GCGAGTATCAGGCAGATGAGCGGGCAGGTG	
GCTCAGCTGGCGAGTATCAGGCAGATGAGC	
GGCAGGTGTCGCGCGGCTCAGCTGGCGAGT	
GTGCATGATCAGTCACGTGATCTGTGCATG	
GTGTGTGTGTGTGTGTGTGTGTGTGTGTGT	
TCATGCACTGATCACGTGGCTGATCATGCA	
TGCATGATCAGTCACGTGATCAGTGCATGA	
TGTCGCGCGGCTCAGCTGGCGAGTATCAGG	

farm level suggests that farmers who lack a rigorous selection methodology for additives are at a greater risk of experiencing poor outcomes. This variability can deter farmers from adopting additives, as a small number of poor matches can significantly undermine overall performance and satisfaction.

6.4.2 Optimized Additive Deployment

The following figures present the improvements in feed additive efficacy achieved using our proposed microbiome-based, AI-assisted predictive model. These improvements have significant potential economic implications by

Table 6.7: Summary of methane emission changes across farms for the additive Agolin. See a detailed explanation above for a full interpretation of the columns.

Farm	CH_4 Change (Treatment)	CH_4 Change (Control)	Normalized Efficacy η_{A,f_i}	Effect Size	Cohen's D
AB1	–46.8%	–41.2%	–9.5%	–8.3	–0.10
BT1	4.5%	20.7%	–13.4%	–20.6	-0.18
FG1	–56.6%	–49.1%	–14.8%	–11.6	–0.23
GR1	121.2%	155.5%	–13.4%	–47.5	–0.11
LV1	18.9%	22.3%	–2.8%	-9.5	–0.04
MP1	–58.6%	–48.3%	–19.9%	–20.5	–0.23
RZ1	–27.4%	–28.2%	1.1%	0.8	0.14
SI1	–38.7%	–25.5%	-17.7%	–13.7	–0.30
YE1	–26.0%	–24.3%	-2.3%	–2.0	–0.03
JN2	102.9%	100.7%	1.1%	–2.3	–0.01
ST2	59.6%	73.0%	–7.7%	–15.8	–0.08
TS2	32.3%	42.1%	–6.9%	–12.8	–0.08
YK2	–19.5%	–19.4%	–0.1%	–0.5	–0.00

Table 6.8: Summary of methane emission changes across farms for the additive Allimax. See a detailed explanation above for a full interpretation of the columns.

Farm	CH_4 Change (Treatment)	CH_4 Change (Control)	Normalized Efficacy η_{A,f_i}	Effect Size	Cohen's D
LV1	–27.9%	22.3%	–41.1%	–224.4	–0.76
YE1	–41.6%	–24.3%	–22.9%	–21.2	–0.28
BT2	–24.4%	–14.7%	–11.3%	–12.2	–0.16
FG2	78.7%	155.6%	–30.1%	–84.4	–0.33
SI2	161.0%	163.0%	–0.8%	–0.7	–0.01
AB3	–52.1%	–36.3%	–24.7%	–24.9	–0.32
JN3	–10.4%	16.5%	–23.1%	–27.5	–0.75
KS3	0.3%	34.6%	–25.5%	–67.3	–0.26
LV3	2.2%	47.8%	–30.9%	–60.9	–0.52
MP3	–61.0%	–63.2%	6.0%	2.4	0.11
VL3	–26.8%	19.6%	–38.8%	–71.0	–0.55

enabling more targeted and efficient use of additives. Such a targeted approach not only maximizes methane emission reduction but also optimizes resource allocation, thereby offering a compelling value proposition that could accelerate the widespread adoption of sustainable farming practices.

Figure 6.12 presents the efficacy analysis of the four additives examined in this study, as derived from the data in Tables 6.7, 6.8, 6.9, and 6.10. The efficacy is initially represented as a Normal distribution under naive deployment conditions, without farm selection (denoted as "Naive Deployment"). This is contrasted with an optimized deployment strategy where each additive is applied to only 50% of the farms, specifically selected based on our microbiome-based predictive model (denoted as "Optimized Deployment"). The comparison reveals that the optimized approach substantially improves

Table 6.9: Summary of methane emission changes across farms for the additive Kexxtone. See a detailed explanation above for a full interpretation of the columns.

Farm	CH_4 Change (Treatment)	CH_4 Change (Control)	Normalized Efficacy η_{A,f_i}	Effect Size	Cohen's D
AB2	–46.2%	–11.0%	–39.6%	–45.4	–0.62
JN2	30.1%	100.7%	–35.2%	–73.0	–0.46
KS2	106.3%	94.8%	5.9%	21.2	0.05
LV2	21.2%	66.6%	–27.2%	–47.0	–0.66
MP2	–50.6%	–39.7%	–18.2%	–21.0	–0.18
RZ2	–62.2%	–56.2%	–13.8%	–6.9	–0.14
ST2	24.9%	73.0%	–27.8%	–57.7	–0.29
TS2	3.7%	42.1%	–27.0%	–48.4	–0.30
BT3	–20.7%	–7.1%	–14.7%	–19.5	–0.15
FG3	–44.4%	–43.7%	–1.3%	–0.9	-0.04
SI3	–43.9%	–3.3%	–42.0%	–90.5	-0.58
SR3	–47.7%	–48.1%	0.7%	0.2	0.01
VL3	–5.1%	19.6%	–20.6%	–34.7	–0.33
YE3	–10.2%	2.9%	–12.7%	–14.2	–0.27

Table 6.10: Summary of methane emission changes across farms for the additive Relyon. See a detailed explanation above for a full interpretation of the columns.

Farm	CH_4 Change (Treatment)	CH_4 Change (Control)	Normalized Efficacy η_{A,f_i} η_{A,f_i}	Effect Size	Cohen's D
AB1	–47.3%	–41.2%	–10.4%	–8.5	–0.11
GR1	108.7%	155.5%	–18.3%	–64.9	–0.17
JN1	–38.3%	–36.5%	–2.8%	–2.6	–0.04
KS1	86.8%	120.1%	–15.1%	–45.9	–0.11
LV1	17.4%	22.3%	–4.0%	–12.6	–0.04
MP1	–67.8%	–48.3%	–37.8%	–40.2	–0.56
YE1	–25.1%	–24.3%	–1.0%	–0.9	–0.01
KS2	71.8%	94.8%	–11.8%	–45.6	–0.14
AB3	–41.5%	–36.3%	–8.0%	–7.6	–0.05
JN3	8.7%	16.5%	–6.7%	–8.7	–0.25
LV3	12.0%	47.8%	–24.2%	–46.3	–0.43
YE3	–13.6%	2.9%	–16.1%	–17.6	–0.35

additive efficacy by targeting farms where the highest impact is expected. This leads to an approximate 60% increase in the effectiveness of the additives in reducing methane emissions.

Figure 6.13 provides a complementary analysis to Figure 6.11, incorporating the optimization phase based on our microbiome-based predictive model. Observing this Figure it can be seen that not only does the optimized approach enhance the average additive performance by approximately 60%, but it also fundamentally alters the experience for farmers by shifting from a pattern of mixed successes and failures to a consistently positive performance profile. In other words, the targeted deployment avoids instances where additives could

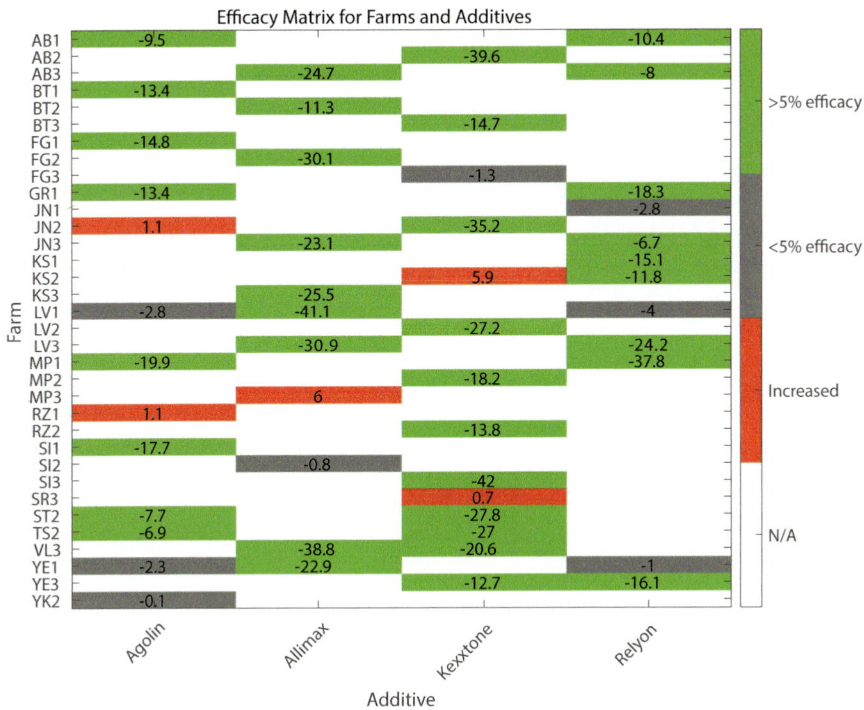

Figure 6.10: Efficacy matrix of various feed additives across multiple farms, illustrating how different feed additives affect methane emissions in a variety of farms. Each cell in the matrix represents the mean change in methane emissions, in percentage terms, for a particular farm-additive combination. The numbers are compared to the emissions level measured before the trial began, and are normalized by the respected change of the control group of the same farm. For instance, a value of '0' indicates no change in emmissions (or the same change that the control cows underwent), '–10' indicates a 10% normalized reduction, and '20' signifies a 20% increase. The color coding further aids interpretation: green cells indicate significant reductions in emissions (below –5%), red cells highlight increases, and gray cells show a negligible change (between –5% and 0). Notably, each feed additive has a varying level of efficacy across different farms. There are instances where a single additive either fails to lower emissions or even increases them in a subset of farms. This variation underscores the necessity for a nuanced approach in methane reduction strategies, as a one-size-fits-all solution may be ineffective or counterproductive.

yield poor or even detrimental outcomes. This transformative impact is likely to be a significant driver in increasing farmers' willingness to adopt feed additives, as it removes the unpredictability that has been a barrier to widespread adoption.

Figure 6.14 showcases the proficiency of our predictive model in accurately identifying the farms that are most likely to benefit from each specific additive.

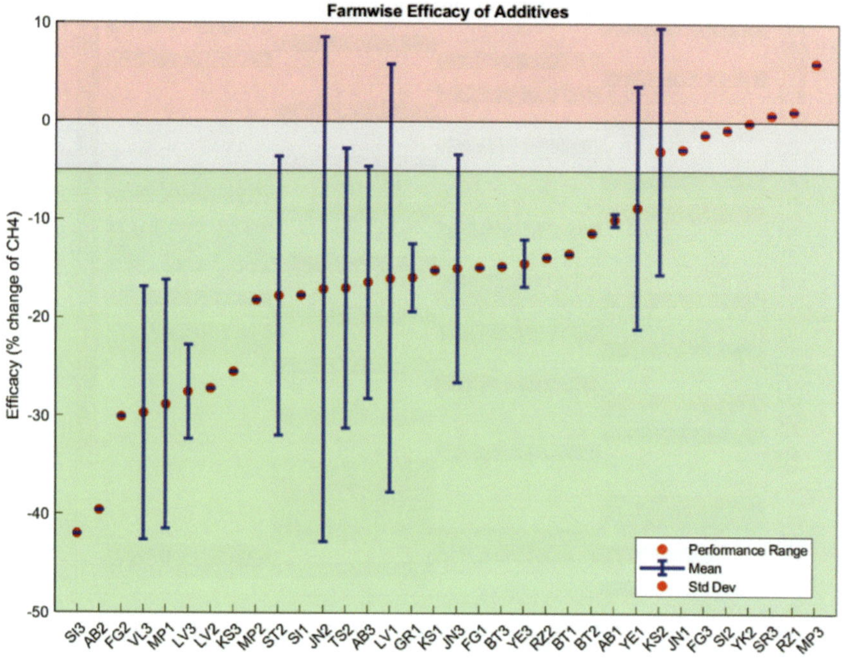

Figure 6.11: Efficacy bar chart of the farms across multiple feed additives. Each "stick" represents a single farm, extending from its lowest to highest additive performance. The mean efficacy is marked by a red dot, and the range of standard deviations is depicted by blue error bars. The chart is color-coded to facilitate interpretation: the green zone (below –5%) represents significant reductions in methane emissions, the gray zone (between –5% and 0%) indicates negligible changes, and the red zone (above 0%) highlights increases in emissions. While a cursory look at the data might suggest that most of the additive-farm combinations result in positive efficacy (as indicated by the prevalence of data in the green zone), a closer examination reveals a complex picture. Although the bulk of the data points indicate reduced methane emissions, the presence of occasional inefficacies (or even increases in emissions) significantly skews the overall performance. This is particularly evident when considering the upper range of efficacy (upper end of the error bars), which often falls within the gray or even red zones for many farms. This volatile efficacy performance underscores the risks associated with a non-strategic or arbitrary choice of additives, potentially explaining the hesitancy among farmers to adopt them.

The primary objective is to rank farms based on the anticipated efficacy of these additives, as estimated by the prediction model. For each additive, the scatter plot displays farms sorted by their predicted efficacy (x-axis) against their actual, post-factum measured efficacy (y-axis). Ideally, an accurate model would yield a scatter plot that approximates a monotonically decreasing line, since negative values indicate a reduction in methane emissions. Additionally,

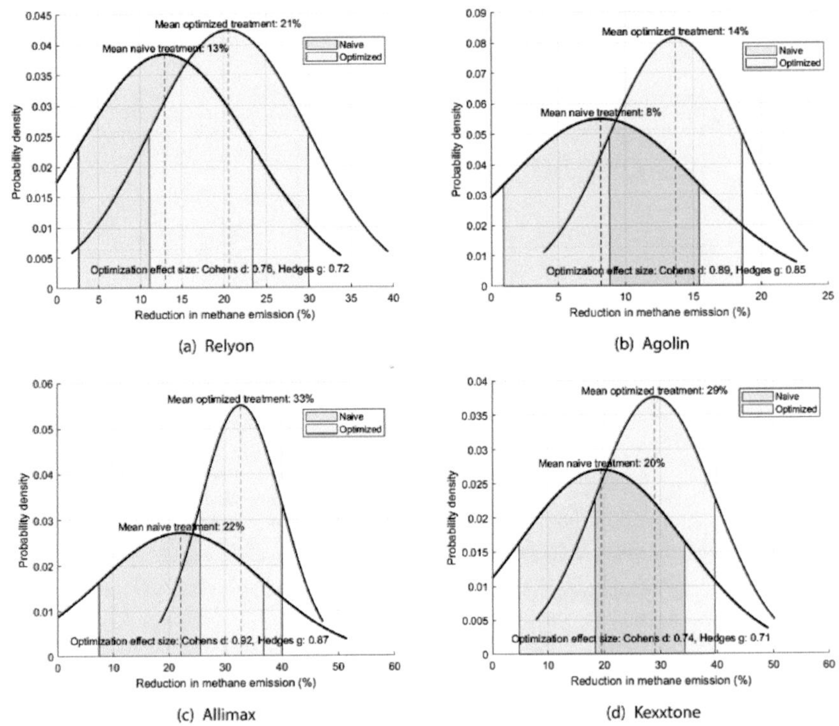

Figure 6.12: An illustration of the normalized efficacy distribution for the four feed additives measured in this study: Relyon, Agolin, Allimax, and Kexxtone. The data is based on Tables 6.7, 6.8, 6.9, and 6.10 and has been regressed to fit a Normal distribution. Each subfigure presents two distributions: one depicting the raw data from all farms (Dark Blue, denoted as 'Naive'), and the other (Green, denoted as 'Optimized') showing efficacy across the top 50% of farms as predicted by our microbiome-based model. Included in each chart are the Cohen's d [39] and Hedge's g [38] metrics, indicating strong statistical significance of the observed effects under optimized conditions.

each subplot provides two statistical measures: Spearman's ρ and Kendall's τ. Spearman's ρ quantifies the strength and direction of the association between the predicted and actual efficacies. Kendall's τ serves as a non-parametric measure to evaluate the strength of the correlation, focusing on the similarity in the ordering of data when both sets of quantities are ranked.

In Table 6.11 we present an extensive analysis of individual farm performances when applying our microbiome-based, AI-assisted predictive model for additive selection. Each farm is evaluated based on the average efficacy of feed additives for which the farm ranks in the top 33% or top 50% in terms of predicted efficacy, among the overall participating farms. These percentages represent the fraction of farms for which the model anticipates the

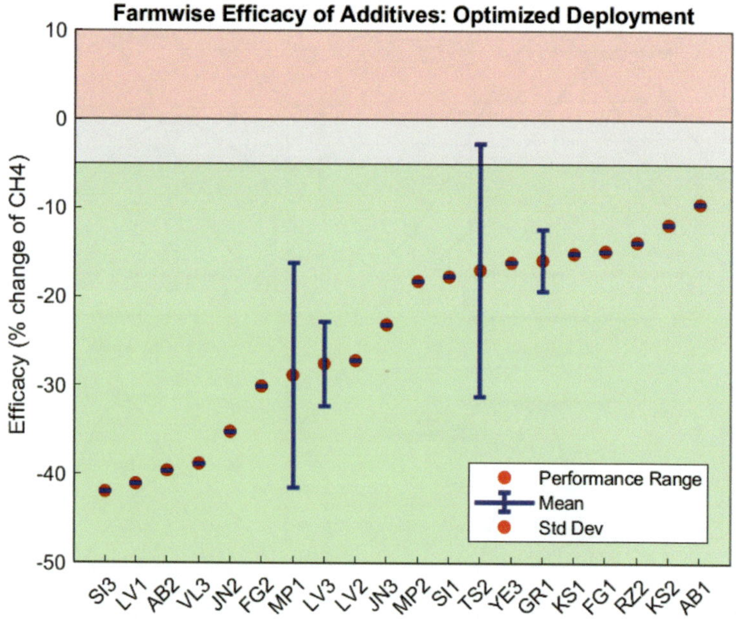

Figure 6.13: Efficacy bar chart of the farms across multiple feed additives, emphasizing the advantages of our microbiome-based, AI-assisted efficacy prediction. Each 'stick' represents a farm's performance range with different additives, extending from the lowest to the highest efficacy. The mean efficacy is highlighted by a red dot, and the standard deviation range is shown as blue error bars. The background color-coding aids in interpretation: the green zone (below –5%) signifies substantial reductions in methane emissions, the gray zone (–5% to 0%) suggests negligible impact, and the red zone (above 0%) indicates increases in emissions. Contrasted with the findings in Figure 6.11, the advantages of targeted optimization are clear. With the exception of a single farm, all data points are predominantly located within the green zone, indicating that farmers who employ this optimized strategy are likely to experience consistently positive results.

highest potential for methane emission reduction through the use of a particular additive. In other words, we choose to deploy additives in farms only for these farms that are predicted to benefit them the most, and if a farm is predicted to benefit from more than a single additive, we arbitrarily choose between them (taking the mean efficacy). A value of 'N/A' for a given farm implies that the farm does not fall within the top portion of predicted efficacy for any of the additives examined, and hence would not be administered any additive according to this targeted approach.

It is crucial to understand that although our strategy may leave some farms without additives, the optimization is primarily geared towards enhancing the overall reduction of methane emissions and increasing yield. These are key

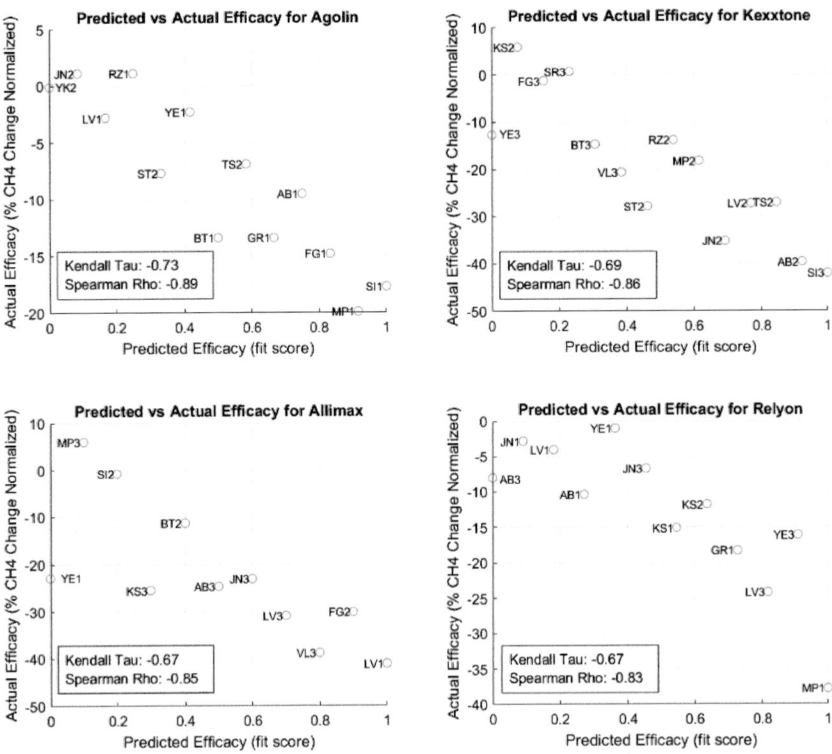

Figure 6.14: Evaluation of predictive model accuracy across four additives. Each subplot corresponds to a specific feed additive (Agolin, Kexxtone, Allimax, and Relyon) and presents a scatter plot of farms, ranked by their predicted efficacy (x-axis) against their actual, measured efficacy (y-axis). A more accurate model would manifest as a monotonically decreasing line, given that negative values indicate reduced methane emissions. Spearman's ρ and Kendall's τ are also displayed in each subplot, serving as statistical measures of correlation between the predicted and actual efficacies. These measures provide insights into the model's ability to correctly rank farms based on the anticipated benefits of each additive.

metrics not only for environmental regulators but also from a return on investment standpoint. This selective model is designed to optimize the use of resources dedicated to methane mitigation, thereby maximizing both environmental impact and profitability for farmers. The results of implementing this strategy are compelling: adopting the top 50% strategy results in additive deployment at 62% of farms and achieves an average emissions reduction efficacy of approximately 24%. Conversely, the more rigorous top 33% strategy is applicable to 44% of the farms but delivers a higher efficacy, exceeding 27% in emission reduction. Importantly, this performance surpasses the individual efficacy of each additive and closely aligns with the ambitious 30% reduction target set by major dairy stakeholders. Moreover, this tailored approach is likely to be more cost-effective than a naive deployment of the best—and potentially most expensive—additives, as it matches each farm with the most suitable, and often more economical, additive options.

Additionally, the scalability of our proposed model lends itself to easy integration with new additives. As we expand our catalog of additives, we anticipate improvements in two key areas: firstly, our ability to cater to a larger proportion of farms, and secondly, an increase in the overall average efficacy of the treatments.

6.4.3 Additional Benefits of Additive Optimization

Although this paper primarily targets methane emission reduction, we have also observed a consequential increase in yield, which took place when additives' efficacy was at its peak. This yield enhancement is not merely a fortuitous result, but it is inextricably linked to our methane reduction efforts. This relationship can be attributed to the metabolic energy redirection within the organism. As less energy is channeled towards methane production, more becomes accessible for other essential biological processes, such as milk production or body mass increase in cattle.

The correlation between methane emissions and yield has been well-documented in the literature. Studies like [16, 60], provide robust evidence substantiating this association. Our prediction model, initially designed to effectively facilitate methane emission reduction, also shows substantial promise in the sphere of yield maximization. This exciting potential demonstrates the dual environmental and economic benefits of our approach.

Although we will not delve into a comprehensive exploration of yield maximization in this paper, it's worth highlighting its relevancy and the utility of our predictive model in this context. We plan to detail the role of our model in maximizing yield alongside minimizing emissions in a forthcoming paper, thereby contributing to the ongoing effort for sustainable farming practices.

6.5 Microbial Data Analytics

6.5.1 Motivation and Overview

This research is predicated on the analysis of numerous microbiome samples collected from bovine subjects across diverse farm settings (ranging from different geographical locations, environmental conditions, herd sizes, and management practices). A subset of these subjects has been administered a feed additive, and subsequent methane emissions were measured, creating an

Table 6.11: Individual farm efficacy based on targeted additive allocation. Each row represents a farm and shows the average efficacy of feed additives for which the farm is ranked in the top 33% or 50% in terms of predicted additive efficacy. Values are expressed in percentages. The label 'N/A' indicates that the farm does not rank in the top 33% or 50% for any of the examined additives. This targeted approach is designed to optimize the aggregate reduction of methane emissions while maximizing economic returns. While some farms do not receive any additive under this model, the overall methane reduction efficacy is notably increased, aligning with both environmental conservation and economic objectives. In terms of scope and efficacy, following the top 50% strategy results in additive deployment at 62% of the farms, achieving an average efficacy of approximately 24% in emissions reduction. On the other hand, the more stringent top 33% strategy covers 44% of the farms but results in a higher average efficacy, exceeding 27% in emissions reduction.

Farm	Deployment for Top 33%	Deployment for Top 50%
AB1	–9.50%	–9.50%
AB2	–39.60%	–39.60%
AB3	N/A	N/A
BT1	N/A	–13.40%
BT2	N/A	N/A
BT3	N/A	N/A
FG1	–14.80%	–14.80%
FG2	–30.10%	–30.10%
FG3	N/A	N/A
GR1	–15.85%	–15.85%
JN1	N/A	N/A
JN2	–35.20%	–35.20%
JN3	N/A	–23.10%
KS1	N/A	–15.10%
KS2	N/A	–11.80%
KS3	N/A	N/A
LV1	–41.10%	–41.10%
LV2	–27.20%	–27.20%
LV3	–27.55%	–27.55%
MP1	–28.85%	–28.85%
MP2	N/A	–18.20%
MP3	N/A	N/A
RZ1	N/A	N/A
RZ2	N/A	–13.80%
SI1	–17.70%	–17.70%
SI2	N/A	N/A
SI3	–42.00%	–42.00%
SR3	N/A	N/A
ST2	N/A	N/A
TS2	–27.00%	–16.95%
VL3	–38.80%	–38.80%
YE1	N/A	N/A
YE3	–16.10%	–16.10%
YK2	N/A	N/A
Average efficacy	–27.42%	–23.65%
Farms treated	15 out of 34 (44%)	21 out of 34 (62%)

experimental group, while others remained a control group. Each sample encapsulates a plethora of 'reads', each representing sequences of 100 to 150 nucleotides. Our objective lies in the identification of significant microbial genetic patterns pertinent to the trait of interest, which, in this case, is the high efficacy of the feed additive.

A "k-mer" is a contiguous subsequence of length k derived from a longer string of nucleotides. In the context of genomics, a k-mer typically refers to a sequence of k nucleotides within a larger DNA or RNA sequence.

In a more formal mathematical context, if we denote the original longer sequence of nucleotides as the string S and its length as n, then a k-mer is a substring of S of length k. Given $S[i:j]$ represents the substring of S starting at position i and ending at position j (inclusive), a k-mer of S starting at position i would be represented as $S[i:i+k-1]$ (for $1 \leq i \leq n-k+1$). Consequently, the total number of distinct k-mers that can be extracted from a sequence S of length n is $n-k+1$.

Furthermore, considering the biological context where each position in the string can be one of four nucleotides (A, T, C, or G), the total number of possible k-mers of length k, without considering any specific longer sequence, is 4^k.

Our analytical approach is characterized by an unbiased exploration of large k-mers, specifically those with $k = 30$, though not confined to this value. Previous studies have illustrated the optimal expressivity of k-mers of length 30 or longer for predictive applications [20]. However, their use is typically constrained to cases of extreme data sampling or pre-set filtering criteria, both of which can introduce bias. Conversely, models that leverage k-mers as features in machine learning typically limit k to values of 6 or less, driven by concerns of data scarcity and potential model overfitting. Traditionally, an unbiased analysis of longer k-mers would be considered computationally impractical due to the vast number of possible combinations, approximately 2^{60}. Additionally, it would necessitate significant amounts of data to circumvent overfitting.

However, we leverage the understanding that the distribution of these 30-mers within DNA does not follow a uniform pattern but instead conforms to a power-law. This inherent property allows us to implement efficient analytic techniques and extract a significant number of k-mer groups automatically. Each of these groups is assuredly associated with a particular epigenetic trait. However, the relevancy of such traits to our current interest may vary.

The innovation presented within this study manifests in a dual capacity. Firstly, we extend our analysis beyond merely long k-mers, thereby enhancing their expressivity, to encompass groups of k-mers, which, in turn, fortifies their role as potent predictive features. Secondly, we address data paucity by utilizing a technique that capitalizes on our power-law distributed data, as opposed to a brute force examination of "all k-mers" or "all groups".

This technique facilitates the efficient detection of "correlated anomalies" – localized groups giving rise to network structures which do not naturally arise in power-law networks. Analytically, the presence of such groups is indicative of an underlying causality within the data, signifying an association with a specific property relevant to the group of genetic information.

Additionally, the nature of our approach, predicated on the holistic examination of microbiome samples, allows for each group of k-mers to potentially comprise DNA fragments derived from heterogeneous sources. This denotes that functionalities emanating from diverse microbes may concurrently contribute to the observed behavior of interest.

6.5.2 Architecture and Key Strengths

Outlined below is a systematic overview of our method to analyze microbial data. Each constituent step is elucidated in subsequent sections for deeper understanding:

1. **Representation of Sequenced Data:** Each sequenced microbial dataset is encapsulated as a network. While each network maintains a constant node count denoted by M, their edge configurations are susceptible to variability.
2. **Formation of Superposition Networks:** To achieve a more nuanced analysis, we formulate superposition networks. This is accomplished by layering individual networks based on specific criteria—be it samples originating from identical farms, those sourced from a particular geographic locale, share some biological attribute, or samples acquired under similar meteorological conditions.
3. **Unsupervised Analysis:** Both the original one-sample networks and the constructed superposition networks undergo an in-depth unsupervised exploration. The procedure for each network can be distilled into the following sub-steps:
   (a) For each unique degree d present in the network, nodes bearing the degree d are analyzed.
   (b) We then empirically assess the "internal connectivity" of this node cluster. This entails calculating the ratio of edges that both originate and culminate within this cluster to the edges that begin within this cluster but terminate externally.
   (c) Should this ratio surpass the benchmark delineated by Theorem 6.1, we infer analytically that the node cluster (e.g. k-mers) probably bears semantic relevance. The degree of this statistical confidence is symbolized by ε which we can calibrate as per our requirements.
4. **Storage of k-mer Clusters:** All discerned k-mer clusters are cataloged for subsequent utilization during the supervised phase. While every cluster

possesses its unique statistical confidence metric conducive for granular analysis, they collectively uphold a robustness standard that exceeds our initial ε setting.

5. **Retrospective Genetic Investigation:** The utility of superposition networks extends to retrospective analysis as well, offering invaluable insights when certain shared properties among cows are identified post-data collection. For instance, if disparate cows nationwide are later found to possess a common susceptibility to a specific disease, a dedicated superposition network can be easily crafted to encompass these particular cows. This allows for the extraction of unique genetic patterns potentially correlated with the identified trait. Consequently, superposition networks facilitate not only a dynamic reanalysis of the existing data but also enable researchers to unveil subtle, yet crucial, genetic markers tied to various biological characteristics discovered in hindsight. These markers then serve as pivotal reference points for future genetic investigations and strategies aimed at addressing the specific traits or susceptibilities uncovered.

The unsupervised learning phase, central to our approach, is initiated by utilizing all available microbiome samples. The essence of this phase lies in constructing networks derived from genetic material, necessitating the amalgamation of various sample combinations. Initially, individual samples stand autonomously, each constituting a unique source to form distinct networks. Following this, an aggregated network is derived by pooling samples from a complete farm. This composite structure, being inherently denser, serves as a robust representation of farm-level features.

Furthermore, our methodology permits the formation of several nuanced networks, shaped by criteria as diverse as geography, specific weather conditions, or a collation of all accessible data. One might assume that the creation of such intricate networks would be computationally arduous. However, by overlaying selected foundational networks, a superposition network is efficiently created, by executing a boolean OR operation on their edges. This design, which earmarks a dedicated network to each sample, streamlines the integration of new data or samples.

A pivotal aspect of this phase is its adaptability and inclusivity. Irrespective of the data source – whether being microbiome samples from cows, sheep, soil, or even humans – this phase seamlessly integrates diverse data. Moreover, its design ensures that the genetic markers identified are universally relevant across various label groups, all linked to a targeted biological condition or trait.

One of the salient features of our design is its capacity to harness the entirety of available raw data, including the long-tail genetic information. Long-tail data, denoting genetic information from microbes present in minimal quantities within a sample, often impedes mainstream metagenomic analytics, curtailing the diversity and precision of predictions obtainable from microbiome data. By leveraging our approach, we transcend this limitation, moving beyond traditional DNA read analysis and focusing on the intricate dynamics of networks instead.

Additionally, it is imperative to emphasize the unbiased nature of our approach. Contrary to being contingent on existing reference databases in scholarly literature, our method stands apart as inherently data-driven, refraining from zeroing in on selective information subsets.

6.5.3 Data Representation

As previously defined, from each farm $f_i \in F$ a random set of cows $C_{u,i}$ are selected, and their microbiome sampled and sequenced.

Each microbiome sample, denoted as X, contributes to the generation of a unique network, G_X, comprised of M nodes and E_X edges. The nodes' consistency across all networks originates from the initial data processing: during an automated preliminary analysis of the data, we exclude infrequently appearing k-mers (i.e., only k-mers that appear at least twice in at least two samples are retained[1]). We are thus left with a manageable quantity M that is conducive to our network analytics. This provision is facilitated by the power-law distribution of k-mer repetitions across the data, which guarantees that the majority of k-mers will indeed be unique and subsequently filtered out, while predicting that there will still be a substantial number of k-mers recurring multiple times.

Specifically, we work with 30-mers[2]. From a theoretical search space of 4^{30} possible 30-mers, we end up using approximately one million 30-mers, translating into networks of one million nodes, a scale that is computationally feasible to handle.

An enhanced model that repeats the flow described here for various values of k, each resulting in a different set of k-mers, can still be managed computationally efficiently. Even a model with k-mers for 60 different values of k would result in networks of around 100 million nodes. Although such networks may initially seem too large, due to the nature of our network analytics, they can still be analyzed with practical computational resources.

[1]This filtering criterion can, of course, be made stricter, further reducing the number of k-mers.

[2]The number $k = 30$ was arbitrarily selected for this study. It's large enough to allow for sufficient expressivity of the k-mers but low enough to allow different k-mers to appear in the same read, subsequently manifested as an edge in the k-mers network. Note that changing the value of k trades the maximum number of possible k-mers for the maximum number of possible connections between them. We believe various values from $k = 20$ to $k = 80$ could be used with our proposed method, potentially producing DNA patterns that would further enhance the model's predictive capabilities.

6.5.4 New Statistical Dimensions of Microbial Genetic Data

The realm of microbiome data has rapidly emerged as a thrilling frontier in the biological sciences. Rich in its complexity, it holds the potential to unlock myriad mysteries about agriculture efficiency, environmental challenges, as well as human health and disease. Estimates for the total number of microbial species on Earth range widely, from millions to a trillion [56]. While precise estimation is challenging, it is clear that only a small fraction has been characterized to date.

However, much of the current research in this space has been predominantly rooted in modeling. Modeling offers a structured way to simulate, predict, and infer mechanisms that underlie microbial dynamics. Yet, while it can be invaluable, modeling often tends to abstract away some of the intricate details and nuances that the raw data encapsulates. This orientation towards modeling, rather than a more exploratory, data-driven approach, might inadvertently overlook or overshadow key insights waiting to be discovered. Recent research reveals that the average human body hosts approximately 30 trillion human cells alongside 38 trillion bacteria [81].

A significant limitation of model-based analysis in microbiome research is its inherent dependency on reference databases. Such databases, though comprehensive, are not exhaustive. Relying heavily on them can introduce biases and potentially skew interpretations of microbial communities. Conversely, adopting a purely data-driven approach allows for an unbiased lens, grounding the analysis in the very essence of the data. A 2017 study using large-scale shotgun sequencing of circulating cell-free DNA, found that 99% of the genetic material identified was previously unknown to science, having little to no sequence homology in existing databases [52]. Furthermore, microbiota function as part of a larger community or microbiome, suggests that their characterization requires more than just listing individual species. Mapping their individual and collective functions within a network is essential to understand their roles in these complex systems.

In our study, we found that microbiome genetic data exhibits distinct power-law dynamics. This pattern is evident in the frequency of raw reads, the prevalence of k-mers within these reads, and the co-occurrence of these k-mers in the same samples. When we model k-mers as vertices in a network, with edges representing their co-occurrence in the same sample, the power-law dynamics are further underscored.

To the best of our knowledge, this observation stands as both unique and unprecedented in the realm of microbiome research. The recognition of such patterns may bear profound implications, potentially revolutionizing our understanding of microbial interactions and dynamics, serving as robust indicators or predictors of specific microbiome behaviors or states. Recognizing the potential of this discovery, in the following section, we delve into how these observed power-law dynamics can be harnessed for prediction purposes, potentially offering a novel toolset for researchers and practitioners in the field.

6.5.5 The Universal Footprint of Emergent Power Law Dynamics Across Varied Domains

The ubiquity of power law dynamics across a spectrum of systems and domains attests to its fundamental nature in characterizing complex phenomena. From the vast expanse of celestial bodies to the intricacies of minute cellular structures, power law distributions emerge as an underlying pattern that ties disparate realms into a coherent tapestry of universal behaviors.

Social Networks

The topology of social networks often follows a power law distribution. A minority of nodes (individuals) possess a disproportionately large number of connections, while the majority have relatively few. This phenomenon, often referred to as the "rich get richer" mechanism or preferential attachment, exemplifies the emergence of 'hubs' in networks, and it has profound implications for the spread of information, behaviors, and even epidemics [54].

Transportation

In transportation networks, whether examining the distribution of passenger flights among airports or the traffic flow across urban intersections, a power law dynamic becomes evident [33]. This pattern extends beyond traditional modes of transportation to include recent advancements in mobility services, such as ride-sharing systems [85]. In these systems, similar power law characteristics are observable, where a small number of hubs or nodes (major cities or busy urban centers, for instance) account for a disproportionately large volume of ride-sharing activities. This pattern reflects the concentration of demand and supply in these systems, further underscoring the ubiquity of power law dynamics across various transportation models, both conventional and modern [7].

Financial Activity

Stock market fluctuations, company sizes, and trading volumes frequently exhibit power law characteristics, a pattern that is emblematic of many economic systems [86]. This is particularly evident in the distribution of wealth, where a small fraction of entities control a disproportionately large share of resources, exemplifying classical power law dynamics in economics. Such distributions are not random but follow predictable patterns that can be quantitatively analyzed. Understanding these underlying patterns is not merely an academic exercise; it is instrumental in developing various predictive mechanisms [87]. These mechanisms are crucial for forecasting market trends, assessing economic risks, and devising strategies for financial management and policy-making. By leveraging the insights provided by power law dynamics, economists and financial analysts can better anticipate market behaviors, contributing to more robust and resilient economic systems [88].

Cyber Crime and Terrorism

Power law dynamics manifest significantly in the realms of cyber crime and counter-terrorism, providing a unique lens through which these complex and clandestine activities can be understood and addressed [8]. In the context of cyber crime, particularly with the rise of sophisticated botnets, power law distributions can often be observed in the behavior and spread of these networks [100]. Similarly, in the domain of blockchain and cryptocurrency, power law dynamics can be instrumental in uncovering illicit activities. This insight is particularly valuable for identifying hidden groups of illicit wallets which might be involved in money laundering or funding terrorist activities. By analyzing transaction patterns and identifying outliers that deviate from expected power law distributions, it becomes possible to flag suspicious activities and detect networks of wallets that may otherwise remain concealed within the vast expanse of blockchain transactions [69].

Large Language Models

The distribution of word popularity, as well as the context in which words appear, also exhibits a power law dynamic, a phenomenon that is particularly evident in the field of natural language processing (NLP). This observation is grounded in the principle that a small subset of words is used exceedingly frequently, while the vast majority are used much less often. Such a distribution can be observed in various languages and texts, ranging from everyday communication to specialized literature. This characteristic is also integral to the principles underpinning Large Language Models (LLMs) such as the Generative Pre-trained Transformer (GPT) series [98].

These aforementioned examples provide just a glimpse into the myriad of systems governed by power law dynamics. Recognizing and understanding these patterns is not just an academic exercise; it is crucial for optimizing and regulating these systems and predicting their behavior under various conditions. As this paper will later delve into, the prevalence of power law dynamics is not limited to these realms; it extends intriguingly into the domain of genetic data, offering profound insights into biological processes and evolutionary mechanics.

6.5.5.1 *The Uncharted Territory of Power Laws in Raw DNA Sequencing*

In our quest for deeper understanding, we direct our focus towards the statistical macroscopic patterns that naturally emerge from microbiome data. By analyzing the popularity and prevalence of various reads and k-mers, and studying the intricate web of interactions between them, we aim to capture the larger, holistic picture of microbial dynamics. A cornerstone of our investigation is the surprising observation of power-law properties manifesting within raw microbiome sequencing data. Power laws, characterized by phenomena where small occurrences are extremely common and large occurrences extremely rare, might hold the key to understanding overarching trends and rules governing microbial populations.

Initial observations regarding the emergence of "Network Motifs" in biological data were first documented as early as [1], where simple motifs across a variety of genetic and functional biological interactions were identified.

Most quantitative microbiome studies, such as those in [96] and [55], approach the microbiome as a system of discrete microbes interacting with each other. In [96], deep learning techniques identified key microbial species with significant environmental impacts relative to their abundance, although the likelihood of finding such keystone species in natural microbial communities is low. Another study [55] explored functional redundancy within microbial communities in the human body. It found that while metaproteomic data is crucial for detecting functional changes, factors like gut inflammation and chemical agents that reduce functional redundancy had minimal impact on the taxonomy of the human microbiome. This indicates that significant functional changes can occur with only minor shifts in the microbiome network's diversity.

Historical precedents in biological research have illuminated the emergence of power-law dynamics in diverse systems. For instance, interactions between proteins [44], as well as those between metabolites [43, 46] or cellular activities [11], have been shown to obey power-law distributions, hinting at universal principles that might be governing these interactions. However, when it comes to raw DNA – and especially in the context of microbiome DNA – such observations are virtually non-existent. Despite recent work that calls for the use of network theory in the study of microbiome data [53], and while power-law properties have been foundational in understanding various biological systems, their presence in sequenced data remains a largely unexplored domain. To the best of our knowledge, we tread new ground, aiming to uncover these elusive patterns and decipher what they signify for the world of microbiomes.

6.5.5.2 *Macroscopic Insights from Microbial DNA Sequencing*

Our initial analysis began with a singular microbiome sample from an arbitrary cow. We counted the occurrences of each unique sequence, 150 nucleotides in length. Given the sample's roughly 16 million reads, a randomly generated DNA sequence would predict an infrequent recurrence of unique reads. Contrary to this expectation, while over half of the unique segments appeared only once, hundreds repeated more than 10 times. A few segments even surfaced several dozen times. The distribution of these occurrences exhibited power law dynamics, with a R^2 value exceeding 0.9 when fit to a power law model. These results are detailed in Figure 6.15.

Expanding our scope, we amalgamated samples from five distinct cows, each from a different herd. The inherent microbiome diversity among herds amplified the number of unique reads. Nonetheless, this aggregated data mirrored the previously observed power law dynamics, with the R^2 of the regression fit nearing 0.95, as shown in Figure 6.16.

To mitigate potential biases from our rumen sampling or the sequencing protocol, we turned to the Human Microbiome Project [2, 3], curated by the National Institute of Health. This dataset, boasting over 1,300 analyzable microbiome samples, upheld our earlier observations. The power law dynamics persisted whether we treated the entire dataset as a singular entity or delved into individual samples (Figures 6.17 and 6.18).

In a subsequent phase, we construed networks from the samples taken from cows. Using k-mers as vertices, and establishing edges when two k-mers co-exist within a read, the resultant unweighted, undirected networks – after pruning low-occurrence k-mers to manage data volume – revealed the familiar power law dynamics. This held true for individual samples as well as amalgamated ones, as depicted in Figure 6.19. These findings further strengthen the argument that the observed dynamics stem from inherent microbial genetic properties rather than being external artifacts.

Lastly, for all instances above, the power law's validity received further affirmation using the methodology proposed by Clauset in [21].

6.5.6 Network Generation from k-mers

The generation of each network, denoted as G_X, involves a combination of genomic sequencing and preliminary data analysis. In the initial stage, the microbial DNA is fragmented into a set of k-mers, specifically, 30-mers, each of which serves as a potential node in the network we are going to produce in the next phase. This process results in a theoretical search space of 4^{30} k-mers.

To ensure the computational feasibility of our network analysis, we employ a preliminary filtering step. For each microbial sample, only those k-mers that appear at least twice in the sample are retained, significantly reducing the search space. The use of this filtering is made possible thanks to the fact that the repetition of k-mers across the data follows a power-law distribution, allowing us to significantly reduce the size of our search space (as most of the k-mers would indeed have a unique appearance throughout each sample), while still

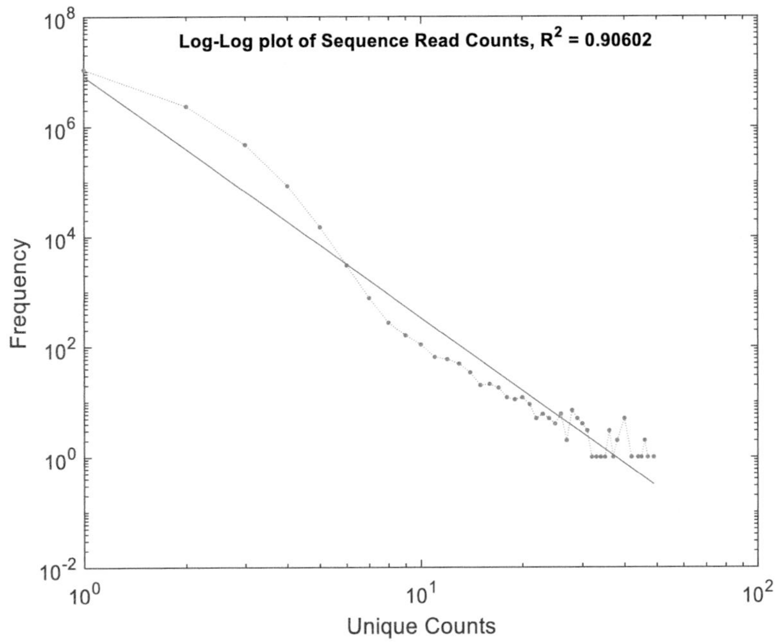

Figure 6.15: Log-log scale distribution of DNA read popularity from a sequenced microbiome sample of a representative cow. The Y-axis indicates the number of unique reads in the sample that appeared a given number of times, as specified by the X-axis. Notably, over half of the unique reads occur only once. Yet, approximately 0.01% of them surface 5 times or more. About 100 unique reads manifest 10 times, with a subset surfacing multiple dozens of times. The distribution's dynamics align closely with the power-law model, showcasing a compelling regression fit with $R^2 > 0.9$, and validated using the method suggested in [21].

statistically guaranteeing the existence of a large enough selection of k-mers that appear multiple times in each sample. This method results in approximately one million unique 30-mers that form the consistent nodes V used across all networks.

When new samples are received, they can be easily processed such that only k-mers from V are used for their analysis. From time to time, the nodes V can be refreshed by re-running this process, expanding the set of supported k-mers and updating the internal ordering of the networks, allowing for new k-mers originating from new data to be included in the overall analysis. This process, however, is not mandatory, ensuring flexibility in our analytical approach.

Once the initial construction of the set of supported k-mers is completed, each microbiome sample X can be analyzed and its corresponding network $G_X = G(V, E_X)$ can be constructed as follows:

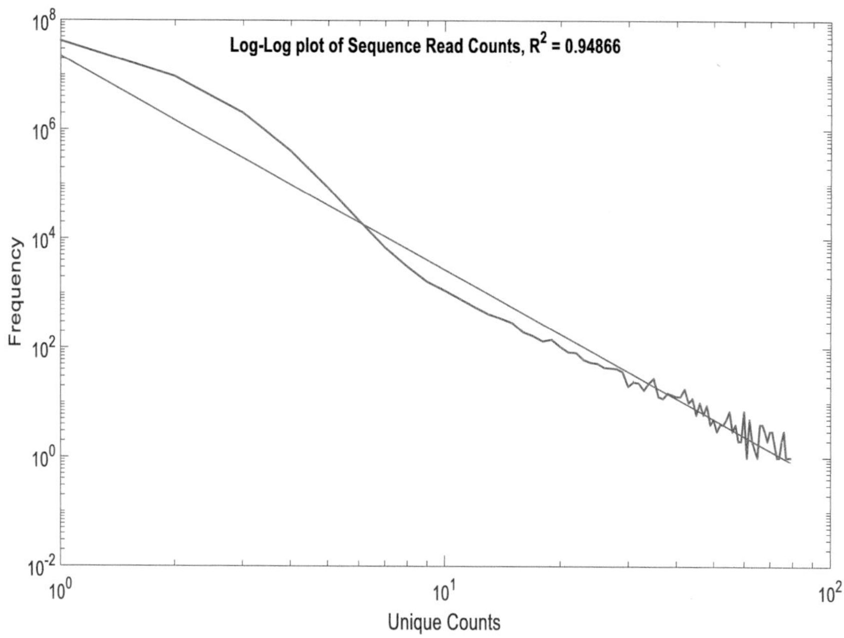

Figure 6.16: Log-log scale distribution of DNA read popularity from a combined microbiome sample, aggregating data from five cows, each from a distinct herd. The Y-axis represents the number of unique reads in the sample that appear a given number of times, denoted by the X-axis. This aggregated sample showcases dynamics strikingly similar to the individual cow sample depicted in Figure 6.15. The distribution adheres closely to the power-law model, evidenced by a robust regression fit with $R^2 > 0.94$, and validated using the method suggested in [21].

1. The sample X contains a large number of reads (usually between 1 and 20 million), each a string of nucleotides, 100 to 150 bases long.
2. For each read, extract all k-mers that are contained in the read and filter this list of k-mers as per the list of supported k-mers V.
3. For each pair v_1 and v_2 of supported k-mers that are contained in this read, add the edge (v_1, v_2) to E_X.

Note that whereas each network G_X has its unique set of edges E_X, the nodes in the network are the same group V, established during the initial filtering phase, using the multitude of available microbiome samples. In addition, we assume that it is enough that the adjacency matrix of the network G_X represents only the existence of connections without storing their actual strengths or quantities. In other words, a Boolean representation can be used for E_X, further reducing time and space complexities.

Once the networks G_X that are constructed from single microbial samples are available, further networks G_S can be constructed on demand, for various

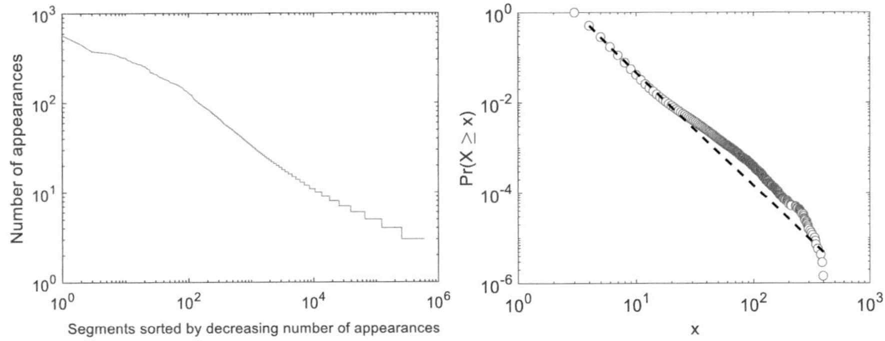

Figure 6.17: Log-log scale visualization of the popularity distribution for human microbiome data, sourced from a representative sample in the Human Microbiome Project [2]. The chart on the left illustrates the frequency of all unique segments, organized in descending order based on their popularity. The right-side chart depicts the likelihood of a segment from the sample to manifest at least x times, for varying values of x. A power-law fit is superimposed, and its validity is corroborated using the methodology proposed in [21].

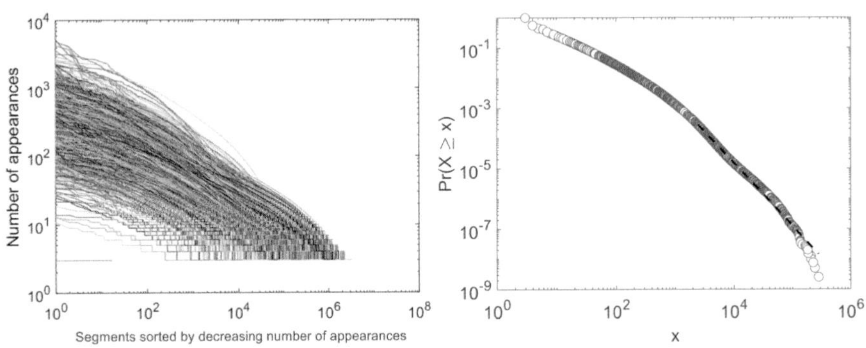

Figure 6.18: Log-log scale depiction of the popularity distribution derived from 1,300 microbiome samples in the Human Microbiome Project [2]. The chart on the left showcases each sample as an individual line, with a median regression fit to power-law standing at $R^2 = 0.88$. On the right, data from all samples is aggregated into a singular dataset, illustrating the probability of a segment occurring at least x times for various x values.

groups S, by easily applying a logical OR operator on the adjacency matrices corresponding to the networks G_X for every $X \in S$. Each of these networks can then be used as input for the network anomalies detection phase, allowing for the discovery of new data patterns, observable through the data-perspective provided by the semantics of the group S.

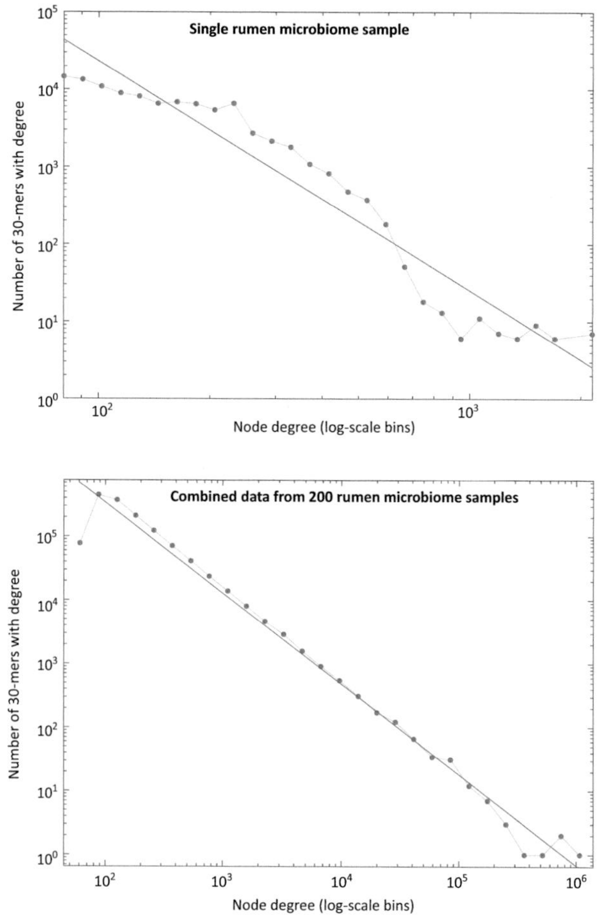

Figure 6.19: Degree distribution of networks constructed from 30-mer sequences: In this visualization, each node represents a unique 30-mer, and an edge is drawn between two 30-mers if they are found within the same read of a given sample. The left-hand chart displays the distribution from a singular cow's microbiome data, while the one on right amalgamates data from 200 distinct cows. Both charts exhibit a power-law dynamic in the node degree distribution, with a strong regression fit of $R^2 > 0.92$. Specifically, the Y-axis signifies the count of nodes with a specific degree x, as represented on the X-axis. The observed power-law trend, consistent across both scales, underscores the inherent structure and shared features in microbiome datasets.

A significant revelation from our exploration of the constructed networks G_X and G_S lies in the distribution of their node degrees. An intensive examination of these networks highlighted a consistent pattern: the degree distribution follows a power-law distribution, a finding that stands true for both individual microbial samples as well as composite groups. Furthermore, the consistency in observing this distribution across both individual and composite samples hints at an inherent and universal structure within microbial genetic data. A strong regression fit of $R^2 > 0.92$ further solidifies this observation. This is illustrated in Figure 6.19.

6.5.7 Detection of Local Network Anomalies in the k-mer Networks

Given a network $G_X(V, E_X)$ (or $G_S(V, E_S)$) we know that the degrees of its nodes closely follow a power-law distribution: $P(k) \sim k^{-\alpha}$. Literature shows that in many real-world networks that follow a power-law degree distribution this pattern applies only to degrees larger than some minimal degree $d_{\min}$.

Therefore, for some normalization constant c the probability of a node v to have degree $d \geq d_{\min}$ equals:

$$\forall v \in V \ , \ \forall d \geq d_{\min} \ , \ P[deg(v) = d] = c \cdot d^{-\alpha}$$

First, we will compute the expected value of the normalization constant c, utilizing values readily accessible from the data

Definition 6.1 Let $V_\star$ denote all nodes of degree at least $d_{\min}$ and let $E_\star$ denote the edges that have at least one side in $V_\star$:

$$V_\star \triangleq \{(v \in V) | (deg(v) \geq d_{\min})\}$$

$$E_\star \triangleq \{((v,u) \in E) | (v \in V_\star) \vee (u \in V_\star)\}$$

Lemma 6.1

If:

$$\forall v \in V, \ \forall d \geq d_{\min}, \ P[deg(v) = d] = c \cdot d^{-\alpha}$$

then:

$$c = \frac{2|E_\star|}{|V_\star| \cdot \sum_{d=d_{\min}}^{d_{max}} d^{1-\alpha}}$$

Proof 6.1 The sum of a graph's $G(V_\star, E_\star)$ degrees equals twice the number of its edges:

$$\sum_{v \in V_\star} deg(v) = 2|E_\star|$$

Taking into account the power-law distribution of the graph's nodes, and summing the expected number of nodes with degree d for every degree d from $d_{\min}$ to the maximal degree d_{max} we can see that:

$$2|E_\star| = \sum_{v \in V_\star} deg(v)$$

$$= E\left[\sum_{d=d_{\min}}^{d_{\max}} d \cdot |(v \in V_\star) \wedge (deg(v) = d)|\right]$$

$$= \sum_{d=d_{\min}}^{d_{\max}} d \cdot |V_\star| \cdot P(v \in V_\star, deg(v) = d)$$

$$= \sum_{d=d_{\min}}^{d_{\max}} d \cdot |V_\star| \cdot c \cdot d^{-\alpha} = 2|E_\star|$$

Implying:

$$c = \frac{2|E_\star|}{|V_\star| \cdot \sum_{d=d_{\min}}^{d_{\max}} d^{1-\alpha}}$$

□

From this we can easily calculate the expected number of nodes with some degree $\hat{d}$ in a network G with a power-law degree distribution:

Definition 6.2 Let $V_{\hat{d}} \subseteq V_\star$ denote all nodes of degree $\hat{d} \geq d_{\min}$:

$$V_{\hat{d}} \triangleq \left|(v \in V_\star) \wedge (deg(v) = \hat{d})\right|$$

Lemma 6.2
If:

$$\forall v \in V \;,\; \forall d \geq d_{\min} \;,\; P[deg(v) = d] = c \cdot d^{-\alpha}$$

then, for every degree $d_{\min} \leq \hat{d} \leq d_{max}$*:*

$$\left|V_{\hat{d}}\right| \approx \frac{2|E_\star| \cdot \hat{d}^{-\alpha}}{\frac{1}{2-\alpha}(d_{max}^{2-\alpha} - d_{min}^{2-\alpha})}$$

Proof 6.2 The number of nodes with degree $\hat{d}$ is :

$$\left|V_{\hat{d}}\right| = c \cdot |V_\star| \cdot \hat{d}^{-\alpha}$$

which using Lemma 6.1 equals:

$$\left|V_{\hat{d}}\right| = |V_\star| \cdot \hat{d}^{-\alpha} \cdot \frac{2|E_\star|}{|V_\star| \cdot \sum_{d=d_{\min}}^{d_{max}} d^{1-\alpha}}$$

Namely:

$$\left|V_{\hat{d}}\right| = \frac{2|E_\star| \cdot \hat{d}^{-\alpha}}{\sum_{d=d_{\min}}^{d_{max}} d^{1-\alpha}}$$

The denominator is a sum that goes from $d_{\min}$ to $d_{\max}$. Since $d_{\max}$ is large, we can approximate this sum by the integral:

$$\sum_{d=d_{\min}}^{d_{\max}} d^{1-\alpha} \approx \int_{d_{\min}}^{d_{\max}} x^{1-\alpha}\,dx$$

The integral can be evaluated by using the power rule for integration:

$$\int_{d_{\min}}^{d_{\max}} x^{1-\alpha}\,dx = \left.\frac{x^{2-\alpha}}{2-\alpha}\right|_{d_{\min}}^{d_{\max}} = \frac{d_{\max}^{2-\alpha}-1}{2-\alpha} - \frac{d_{\min}^{2-\alpha}-1}{2-\alpha} = \frac{d_{\max}^{2-\alpha}-d_{\min}^{2-\alpha}}{2-\alpha}$$

Substituting this back into the expression for $\left|V_{\hat{d}}\right|$ we obtain:

$$\left|V_{\hat{d}}\right| \approx \frac{2|E_\star|\cdot\hat{d}^{-\alpha}}{\frac{1}{2-\alpha}(d_{\max}^{2-\alpha}-d_{\min}^{2-\alpha})}$$

This approximation holds for $\alpha > 2$, where the integral is convergent.

□

Definition 6.3 Let λ denote the power-law normalizing constant for the graph $G(V_\star, E_\star)$:

$$\lambda = \frac{2(2-\alpha)|E_\star|}{(d_{\max}^{2-\alpha}-d_{\min}^{2-\alpha})}$$

such that Lemma 6.2 can be written as:

$$|V_{\hat{d}}| \approx \lambda\hat{d}^{-\alpha}$$

Definition 6.4 We define d_{critical} as the minimum degree threshold for nodes in network G, beyond which the approximations essential for Theorem 6.1 hold validity. This threshold is mathematically articulated as:

$$d_{\text{critical}} \triangleq \left(\frac{2(\alpha-2)\sqrt{|E_\star|}}{d_{\min}^{2-\alpha}-d_{\max}^{2-\alpha}}\right)^{\frac{1}{\alpha-1}}$$

Further explanation regarding the necessity of this property, including its implications and applications within the context of Theorem 6.1, is detailed within the Theorem itself.

Definition 6.5 Let $E_{\hat{d}} \subseteq E$ denote all the edges that touch at least one node in $V_{\hat{d}}$. We divide these edges to two complementing and mutually exclusive groups "internal edges" $E_{\hat{d}}^{in}$ (edges starting and ending in $V_{\hat{d}}$) and "external edges" $E_{\hat{d}}^{out}$ (edges that have one side in $V_{\hat{d}}$ and one side in $V \setminus V_{\hat{d}}$):

$$E_{\hat{d}} \triangleq \{(v_1,v_2)|(v_1 \in V_{\hat{d}}) \vee (v_2 \in V_{\hat{d}})\}$$

$$E_{\hat{d}}^{in} \triangleq \{(v_1,v_2)|(v_1 \in V_{\hat{d}}) \wedge (v_2 \in V_{\hat{d}})\}$$

$$E_{\hat{d}}^{out} \triangleq E_{\hat{d}} \setminus E_{\hat{d}}^{in}$$

Definition 6.6 We define $\xi_{\hat{d}}(\varepsilon)$ as the minimum value of $|E_{\hat{d}}^{in}|$ within network G that is necessary for the approximations crucial to Theorem 6.1 to hold validity, given a specified arbitrarily small confidence level ε. This threshold is mathematically expressed as:

$$\xi_{\hat{d}}(\varepsilon) \triangleq \min\left\{k \;\middle|\; \sum_{i=0}^{k} \frac{\mu_{\hat{d}}^{i} e^{-\mu_{\hat{d}}}}{i!} \geq 1 - \frac{\varepsilon}{2}\right\}$$

where :

$$\mu_{\hat{d}} \triangleq \frac{\hat{d}^2 \times |V_{\hat{d}}|^2}{|E_{\star}|}$$

The rationale behind this definition and its relevance are discussed in the proof of Theorem 6.1, providing the necessary context and justification for its use.

Note that for every value of $\hat{d}$ the values of $|E_{\hat{d}}^{in}|$, $|E_{\hat{d}}^{out}|$ and $\xi_{\hat{d}}(\varepsilon)$ can trivially be acquired from the data. We can therefore calculate the ratio between the number of "internal edges" and the number of overall edges for $\hat{d}$, which we denote as the internal connectivity for degree $\hat{d}$

Definition 6.7 Let $\beta_{\hat{d}}$ denote the internal connectivity of the network G for nodes of degree $\hat{d}$:

$$\beta_{\hat{d}} \triangleq \frac{\left|E_{\hat{d}}^{in}\right|}{\left|E_{\hat{d}}\right|}$$

Since the number of edges in $E_{\hat{d}}$ equals the sum of degrees of the nodes in $V_{\hat{d}}$ minus the edges that have their two sides in $V_{\hat{d}}$ (as these are counted twice), and since the degrees of the nodes in $V_{\hat{d}}$ is exactly $\hat{d}$ we get:

$$\beta_{\hat{d}} = \frac{\left|E_{\hat{d}}^{in}\right|}{|V_{\hat{d}}| \cdot \hat{d} - \left|E_{\hat{d}}^{in}\right|}.$$

Note that the value of $\beta_{\hat{d}}$ can be easily calculated from the data for every value of $\hat{d}$. Some groups of nodes of the same degree may have high internal connectivity and some may have a lower one. It is therefore interesting to ask: what is the expected value of the internal connectivity of various degrees, and above which value a given internal connectivity would be considered "too strong" (i.e. representing a local network structure whose probability to spontaneously emerge in a network with a power-law distribution of the degree is extremely low).

Theorem 6.1 provides a criterion that for every degree $\hat{d}$ greater than $d_{\min}$ can positively guarantee that the set of nodes of degree $\hat{d}$ are "too anomalous" with probability $(1-\varepsilon)$ for every small threshold ε. This criterion applies for the internal connectivity of this group of nodes (that is, the ratio between the number of edges connecting nodes of the same degree, and edges connecting nodes of different degrees), and depends only on the degree $\hat{d}$ and the network structural properties α, $|E_\star|$, $d_{\min}$ and $d_{\max}$.

Theorem 6.1

If for all vertices v in V, and for all degrees $d \geq d_{min}$, the probability $P[deg(v) = d]$ is given by $c \cdot d^{-\alpha}$, then for any arbitrarily small threshold ε and any degree $\hat{d} >$ $\max(d_{min}, d_{critical})$, a group of nodes with degree $\hat{d}$ possessing internal connectivity $\beta_{\hat{d}}$, and provided that $|E^{in}_{\hat{d}}| \geq \xi_{\hat{d}}(\varepsilon)$,is deemed "unlikely to spontaneously emerge" if $\beta_{\hat{d}}$ satisfies the condition below. Such a group is thereby inferred to possess statistical significance beyond mere chance, given the stipulated threshold ε:

$$\beta_{\hat{d}} \geq$$

$$\max\left\{ \frac{B_1\left(\sqrt{B_2^2 - 4B_3} - B_2\right)}{\lambda \hat{d}^{1-\alpha} - B_1\left(\sqrt{B_2^2 - 4B_3} - B_2\right)}, B_4 \right\}$$

where:

$$B_1 = \frac{1}{2 + \frac{\lambda \hat{d}^{1-\alpha}}{|E_\star|}}$$

$$B_2 = 1 - \frac{\lambda^2 \hat{d}^{2-2\alpha}}{|E_\star|} + \left(\Phi^{-1}(1-\varepsilon)\right)^2 \frac{\lambda \hat{d}^{1-\alpha}}{2|E_\star|}$$

$$B_3 = \left(0.5 - \frac{\lambda^2 \hat{d}^{2-2\alpha}}{2|E_\star|}\right)^2 - \left(\Phi^{-1}(1-\varepsilon)\right)^2 \frac{\lambda^2 \hat{d}^{2-2\alpha}}{2|E_\star|}\left(1 - \frac{\lambda \hat{d}^{1-\alpha}}{2|E_\star|}\right)$$

$$B_4 = \frac{\lambda^2 \hat{d}^{2-2\alpha}|E_\star|^{-1} - 1}{2\lambda \hat{d}^{1-\alpha} + 1}$$

Proof 6.3 For $\hat{d} \geq d_{\min}$ and for an edge $e \in E_{\hat{d}}$ we know by definition that at least one of its sides is in $V_{\hat{d}}$. The probability that the second side of this edge also ends in $V_{\hat{d}}$ as well is affected by two factors: first, the more nodes there are in $V_{\hat{d}}$ the greater the chance that they would be selected for the edge e. Second, as the network has a power-law degree distribution we assume it also follows the "preferential attachment" principle (meaning that nodes with higher degree are more likely to acquire new connections, leading to a skew in the distribution). Under this principle, the probability of a node being chosen for a new connection is proportional to its degree, and the probability for the edge e to have both sides in $V_{\hat{d}}$ is therefore a Bernoulli trial with success probability $p = \frac{\hat{d} \cdot |V_{\hat{d}}|}{\sum_{d=d_{\min}}^{d_{\max}} (d \cdot |V_d|)} = \frac{\hat{d} \cdot |V_{\hat{d}}|}{2|E_\star|}$:

$$E_{\hat{d}}^{in} \sim Binomial\left(\left|E_{\hat{d}}\right|, \frac{\hat{d} \cdot |V_{\hat{d}}|}{2|E_\star|}\right)$$

A value of $\beta_{\hat{d}}$ that is "too high" (i.e. unlikely to occur with a probability less than some threshold ε) means that the number of successes from $E_{\hat{d}}$ trials is greater than some threshold x_{thres}.

To calculate an upper threshold for the number of successes in our Bernoulli trials that is considered statistically significant at level ε, and since our number of trials N is large, and our success probability p is not too close to 0 or 1, the Binomial distribution can be approximated by a Normal distribution. If $X \sim Binomial(N, p)$, it can be approximated as $X \approx Y$ where $Y \sim Normal(\mu, \sigma^2)$ with $\mu = Np$ and $\sigma^2 = Np(1-p)$:

$$X \sim Normal\,(Np, Np(1-p))$$

This approximation allows us to express the cumulative distribution function (CDF) of X in terms of the CDF of the Normal distribution Φ as follows:

$$P(X \leq x) \approx \Phi\left(\frac{x + 0.5 - Np}{\sqrt{Np(1-p)}}\right)$$

Having $\Phi(x)$ formally defined as:

$$\Phi(x) = \frac{1}{\sqrt{2\pi}} \int_{-\infty}^{x} e^{-\frac{1}{2}t^2}\, dt$$

To find a threshold x_{thres} such that the probability of observing more than x_{thres} successes is less than some small ε, we can use the Normal approximation to write:

$$1 - \Phi\left(\frac{x_{\text{thres}} + 0.5 - Np}{\sqrt{Np(1-p)}}\right) < \varepsilon$$

Or equivalently,

$$\Phi^{-1}(1-\varepsilon) = \frac{x_{\text{thres}} + 0.5 - Np}{\sqrt{Np(1-p)}}$$

Solving for x_{thres} gives:

$$x_{\text{thres}} = \Phi^{-1}(1-\varepsilon)\sqrt{Np(1-p)} + Np - 0.5$$

This provides the threshold value of the number of successes for which the probability of achieving that number or more is less than ε.

Using the original expressions for N and p this means that the number of internal edges above which a group of nodes with a given degree is considered "unlikely" is:

$$x_{\text{thres}} = \Phi^{-1}(1-\varepsilon)\sqrt{\left|E_{\hat{d}}\right| \frac{\hat{d}|V_{\hat{d}}|}{2|E_{\star}|}\left(1-\frac{\hat{d}|V_{\hat{d}}|}{2|E_{\star}|}\right)} + \left|E_{\hat{d}}\right| \frac{\hat{d}|V_{\hat{d}}|}{2|E_{\star}|} - 0.5$$

Since we already know that:

$$\left|E_{\hat{d}}\right| = |V_{\hat{d}}| \cdot \hat{d} - \left|E_{\hat{d}}^{in}\right|$$

This means that:

$$x_{\text{thres}} = \Phi^{-1}(1-\varepsilon) \times \sqrt{\left(|V_{\hat{d}}|\hat{d} - x_{\text{thres}}\right) \frac{\hat{d}|V_{\hat{d}}|}{2|E_{\star}|}\left(1-\frac{\hat{d}|V_{\hat{d}}|}{2|E_{\star}|}\right)} + \left(|V_{\hat{d}}|\hat{d} - x_{\text{thres}}\right) \frac{\hat{d}|V_{\hat{d}}|}{2|E_{\star}|} - 0.5$$

Rearranging:

$$x_{\text{thres}}\left(1+\frac{\hat{d}|V_{\hat{d}}|}{2|E_{\star}|}\right) + 0.5 - \frac{\hat{d}^2|V_{\hat{d}}^2|}{2|E_{\star}|} = \Phi^{-1}(1-\varepsilon)\sqrt{\left(|V_{\hat{d}}|\hat{d} - x_{\text{thres}}\right) \frac{\hat{d}|V_{\hat{d}}|}{2|E_{\star}|}\left(1-\frac{\hat{d}|V_{\hat{d}}|}{2|E_{\star}|}\right)} \tag{6.1}$$

Squaring both sides:

$$A_1 x_{\text{thres}}^2 + A_2 x_{\text{thres}} + A_3 = 0$$

where:

$$A_1 = \left(1+\frac{\hat{d}|V_{\hat{d}}|}{2|E_{\star}|}\right)^2$$

$$A_2 = \left(1+\frac{\hat{d}|V_{\hat{d}}|}{2|E_{\star}|}\right) \times \left(1-\frac{\hat{d}^2|V_{\hat{d}}|^2}{|E_{\star}|} + \left(\Phi^{-1}(1-\varepsilon)\right)^2 \frac{\hat{d}|V_{\hat{d}}|}{2|E_{\star}|}\right)$$

$$A_3 = \left(0.5-\frac{\hat{d}^2|V_{\hat{d}}|^2}{2|E_{\star}|}\right)^2 - \left(\Phi^{-1}(1-\varepsilon)\right)^2 \frac{\hat{d}^2|V_{\hat{d}}|^2}{2|E_{\star}|}\left(1-\frac{\hat{d}|V_{\hat{d}}|}{2|E_{\star}|}\right)$$

Solving this quadratic equation, and noting that $A_1 > 0$, yields:

$$x_{\text{thres}} = B_1 \left(\sqrt{B_2^2 - 4B_3} - B_2 \right)$$
$$B_1 = \frac{1}{2 + \frac{\hat{d}|V_{\hat{d}}|}{|E_\star|}}$$
$$B_2 = 1 - \frac{\hat{d}^2|V_{\hat{d}}|^2}{|E_\star|} + \left(\Phi^{-1}(1-\varepsilon)\right)^2 \frac{\hat{d}|V_{\hat{d}}|}{2|E_\star|}$$
$$B_3 = \left(0.5 - \frac{\hat{d}^2|V_{\hat{d}}|^2}{2|E_\star|}\right)^2 - \left(\Phi^{-1}(1-\varepsilon)\right)^2 \frac{\hat{d}^2|V_{\hat{d}}|^2}{2|E_\star|}\left(1 - \frac{\hat{d}|V_{\hat{d}}|}{2|E_\star|}\right)$$

Using Lemma 6.2 this can be written as:

$$x_{\text{thres}} = B_1 \left(\sqrt{B_2^2 - 4B_3} - B_2 \right)$$
$$B_1 = \frac{1}{2 + \frac{\lambda \hat{d}^{1-\alpha}}{|E_\star|}}$$
$$B_2 = 1 - \frac{\lambda^2 \hat{d}^{2-2\alpha}}{|E_\star|} + \left(\Phi^{-1}(1-\varepsilon)\right)^2 \frac{\lambda \hat{d}^{1-\alpha}}{2|E_\star|}$$
$$B_3 = \left(0.5 - \frac{\lambda^2 \hat{d}^{2-2\alpha}}{2|E_\star|}\right)^2 - \left(\Phi^{-1}(1-\varepsilon)\right)^2 \frac{\lambda^2 \hat{d}^{2-2\alpha}}{2|E_\star|}\left(1 - \frac{\lambda \hat{d}^{1-\alpha}}{2|E_\star|}\right)$$

Recalling Definition 6.7 this means that for a given degree $\hat{d}$ a group of nodes would be considered "unlikely" for an internal connectivity greater than:

$$\beta_{\hat{d}} \geq \frac{x_{\text{thres}}}{|V_{\hat{d}}| \cdot \hat{d} - x_{\text{thres}}}$$

In addition, recalling Equation 6.1 and noting that both $\Phi^{-1}(1-\varepsilon)$ and the square root are always positive, we can see that :

$$x_{\text{thres}} \left(1 + \frac{\hat{d}|V_{\hat{d}}|}{2|E_\star|}\right) + 0.5 - \frac{\hat{d}^2|V_{\hat{d}}^2|}{2|E_\star|} \geq 0$$

meaning that :

$$x_{\text{thres}} \geq \frac{\hat{d}^2|V_{\hat{d}}^2| - |E_\star|}{\hat{d}|V_{\hat{d}}| + 2|E_\star|}$$

and using Lemma 6.2 written as:

$$x_{\text{thres}} \geq \frac{\lambda^2 \hat{d}^{2-2\alpha} - |E_\star|}{\lambda \hat{d}^{1-\alpha} + 2|E_\star|}$$

Recalling Definition 6.7 we can see that :

$$\beta_{\hat{d}} \geq \frac{\frac{\lambda^2 \hat{d}^{2-2\alpha} - |E_\star|}{\lambda \hat{d}^{1-\alpha} + 2|E_\star|}}{\lambda \hat{d}^{1-\alpha} - \frac{\lambda^2 \hat{d}^{2-2\alpha} - |E_\star|}{\lambda \hat{d}^{1-\alpha} + 2|E_\star|}} \geq \frac{\lambda^2 \hat{d}^{2-2\alpha}|E_\star|^{-1} - 1}{2\lambda \hat{d}^{1-\alpha} + 1}$$

producing B_4.

□

Given that the term x_{thres} denotes a count of graph edges its approximated value must be non-negative when applied in the calculation of $\beta_{\hat{d}}$. This necessitates:

$$\frac{\lambda^2 \hat{d}^{2-2\alpha} - |E_\star|}{\lambda \hat{d}^{1-\alpha} + 2|E_\star|} \geq 0$$

Given that $|E_\star| > 0$, and similarly, λ and $\hat{d}$ are positive, it follows that the denominator is also positive. Consequently, we must ensure the positivity of the numerator, which necessitates the condition:

$$\lambda^2 \hat{d}^{2-2\alpha} \geq |E_\star|$$

Applying λ's Definition 6.3 then yields :

$$\frac{4(2-\alpha)^2 |E_\star|}{(d_{\max}^{2-\alpha} - d_{\min}^{2-\alpha})^2} \hat{d}^{2-2\alpha} \geq 1$$

which must be satisfied by $\hat{d}$ in order for the $\beta_{\hat{d}}$ approximation to be valid. As $\alpha > 2$ and as for every $d_{\min} < d_{\max}$ it holds that $d_{\max}^{2-\alpha} \leq d_{\min}^{2-\alpha}$, incorporating Definition 6.4, this can be written as :

$$\hat{d} \geq \left(\frac{2(\alpha - 2)\sqrt{|E_\star|}}{d_{\min}^{2-\alpha} - d_{\max}^{2-\alpha}} \right)^{\frac{1}{\alpha - 1}} = d_{\text{critical}}$$

Lastly, we note that while the Normal approximation for the Binomial distribution holds for most values of $\hat{d}$, there are instances where the probability p may be too small, leading to inaccuracies in this approximation. Consequently, it is essential to develop a method to address these discrepancies when they arise.

When the number of Bernoulli trials N is very large and the success probability of a single trial p is very small, but the product $N \times p$ remains moderately kept between 1 and 10, the distribution of the number of successes X in N Bernoulli trials tends towards a Poisson distribution rather than a Binomial distribution.

Let μ denote the rate parameter for the Poisson distribution, defined as $\mu = N \times p$. Under these conditions, X can be better approximated by :

$$P(X = k) = \frac{\mu^k e^{-\mu}}{k!}$$

Given a desired threshold ε we can approximate the critical value k_ε as :

$$k_\varepsilon = \min\left\{k \middle| P(X \leq k) \geq 1 - \frac{\varepsilon}{2}\right\}$$

using the cumulative distribution function of the Poisson distribution :

$$P(X \leq k) = \sum_{i=0}^{k} \frac{\mu^i e^{-\mu}}{i!}$$

As $E_{\hat{d}}^{in}$ denotes the set of edges that originate from a node with degree $\hat{d}$ and terminate at a node with degree $\hat{d}$ we can define N as the product of the degree $\hat{d}$ and the cardinality of the set of vertices $V_{\hat{d}}$ with degree $\hat{d}$, expressed as $N = \hat{d} \times |V_{\hat{d}}|$. Similarly, the probability p that an edge connects two nodes of degree $\hat{d}$ is given by the ratio of N to the total number of edges $|E_\star|$, which can be formulated as $p = \frac{\hat{d} \times |V_{\hat{d}}|}{|E_\star|}$. The Poisson rate parameter μ would therefore be defined as :

$$\mu = \frac{\hat{d}^2 \times |V_{\hat{d}}|^2}{|E_\star|}$$

yielding $\xi_{\hat{d}}(\varepsilon)$ (see Definition 6.6).

6.5.8 Practical Applications and Insights of Theorem 6.1

Given the central role that Theorem 6.1 plays within the scope of this paper, it warrants an in-depth examination. Accordingly, this section is dedicated to a comprehensive exploration of the Theorem, illustrating its application through both simulated and real-world data. We aim to showcase not only the practical execution of Theorem 6.1 but also to delve into its appropriate usage and the far-reaching implications it holds. Through this thorough analysis, we aspire to provide readers with a deeper understanding of the theorem's significance and its potential impact on the field.

6.5.8.1 Computational Efficiency

Theorem 6.1 establishes an advanced computational framework, characterized by its minimal time and space complexity, tailored for analyzing large datasets. This theorem is particularly noted for its proficient use of macroscopic structural properties, facilitating an effective exploration of complex networks.

A prominent aspect of Theorem 6.1 is its capability to efficiently process large k-mers, allowing for the comprehensive analysis of extended sequences without a proportional increase in computational demand. This efficiency in leveraging macroscopic properties enables the theorem to extract significant patterns from vast datasets, optimizing computational resources.

Additionally, Theorem 6.1 is designed with adaptability in mind, accommodating the integration of new data into existing analyses without requiring extensive modifications to the framework. This adaptability is crucial for research environments where datasets are in a state of continual evolution. The efficiency and adaptability of Theorem 6.1, combined with the strategic use of long k-mers, significantly enhance prediction accuracy, as demonstrated in previous sections of this paper. The expressivity of these large k-mers, capturing the functional complexity of data, further augments the analysis's robustness and precision.

6.5.8.2 Finding d_{min}

Finding d_{min} is a pivotal step in analyzing and understanding the structure of complex networks. This parameter helps to delineate the range over which the power-law behavior is observable, distinguishing between the part of the network that adheres to a scale-free topology and the portion that might not. For calculating d_{min} we use the methodology suggested by Clauset et. al [21] that utilizes maximum likelihood estimation (MLE) to fit a power-law model to the network's degree distribution, seeking the value of d_{min} that maximizes the fit's goodness. The process iteratively evaluates the goodness-of-fit of the power-law model across different potential values of d_{min}, employing the Kolmogorov-Smirnov statistic [58] to measure the discrepancy between the empirical data and the model.

Upon determining the appropriate value of d_{min}, it is imperative to validate that it is not excessively small, as such a scenario could undermine the credibility of our theoretical framework. The selection of d_{min} significantly influences the graph segment under analysis, notably impacting $|E_\star|$, the number of edges, and the parameter α. These elements are integral to the graph's structure, particularly concerning nodes with a degree greater than d_{min}. Such interdependence highlights the need for careful consideration of d_{min}'s value.

Upon determining an initial d_{min}, a crucial subsequent step involves verifying that the clusters identified by Theorem 6.1 comply with an additional condition defined by d_{critical}. This d_{critical} serves as a lower boundary for the degrees of nodes, ensuring that the clusters under consideration truly reflect significant structural features of the network and are not merely artifacts of the analysis process. Ensuring each cluster's compatibility with d_{critical} is essential for upholding the methodological rigor of our analysis and affirming the theoretical validity of our findings.

6.5.8.3 *Illustration using Random Power Law Networks Using Simulated Data*

Theorem 6.1 is illustrated in Figures 6.20, 6.21, and 6.22. These figures contrast the thresholds predicted by the theorem for various confidence levels, ε, with graphs of differing sizes that conform to the power-law model. For each degree d that has at least 5 nodes, the internal connectivity of these nodes is contrasted with the "unlikelihood thresholds" outlined by Theorem 6.1. This highlights that while power-law graphs may inherently display certain patterns, arbitrary graphs – without specific embedded signals – are rarely expected to exhibit such local anomalies.

6.5.8.4 *Demonstration on Real-World Data*

In this section, we demonstrate the application of Theorem 6.1 using real-world data derived from a microbiome sample taken from an arbitrarily selected cow. This empirical application contrasts sharply with results obtained from simulated datasets designed to mimic random power-law distributions, as shown in Figures 6.20, 6.21, and 6.22. In these simulations, only a small fraction of the degrees exceed the critical threshold, and those that do surpass it only marginally, and only under less stringent values of ε.

Conversely, the analysis of the real-world microbiome data reveals a strikingly different pattern. As can be seen in Figure 6.23 there is a notable abundance of degrees for which the ratio of internal connectivity significantly exceeds the theoretical thresholds, even under the most rigorous conditions set by ε. This suggests the presence of genuine and statistically significant clusters of genetic k-mers. Specifically, the data indicates the existence of 151 clusters where $\varepsilon = 0.05$, 110 clusters for $\varepsilon = 0.01$, and 76 clusters for $\varepsilon = 0.001$. These results not only validate the robustness of Theorem 6.1 when applied to real biological data but also underscore the distinct characteristics of microbial communities that are not readily apparent in synthetic datasets. This clear distinction in the behavior of real and simulated data highlights the importance of employing rigorous mathematical frameworks like Theorem 6.1 in the analysis of complex biological systems.

6.5.8.5 *Practical Implications and Use Cases*

Drawing from Theorem 6.1, we possess a powerful tool to systematically probe k-mer networks, enabling us to swiftly identify and "extract" groups of nodes, representing specific k-mers. These k-mers are analytically established to be linked to distinct phenotypic attributes. The intriguing aspect of this linkage is that it might be directly or indirectly correlated with the biological state or condition we aim to elucidate. Several noteworthy implications emerge from this:

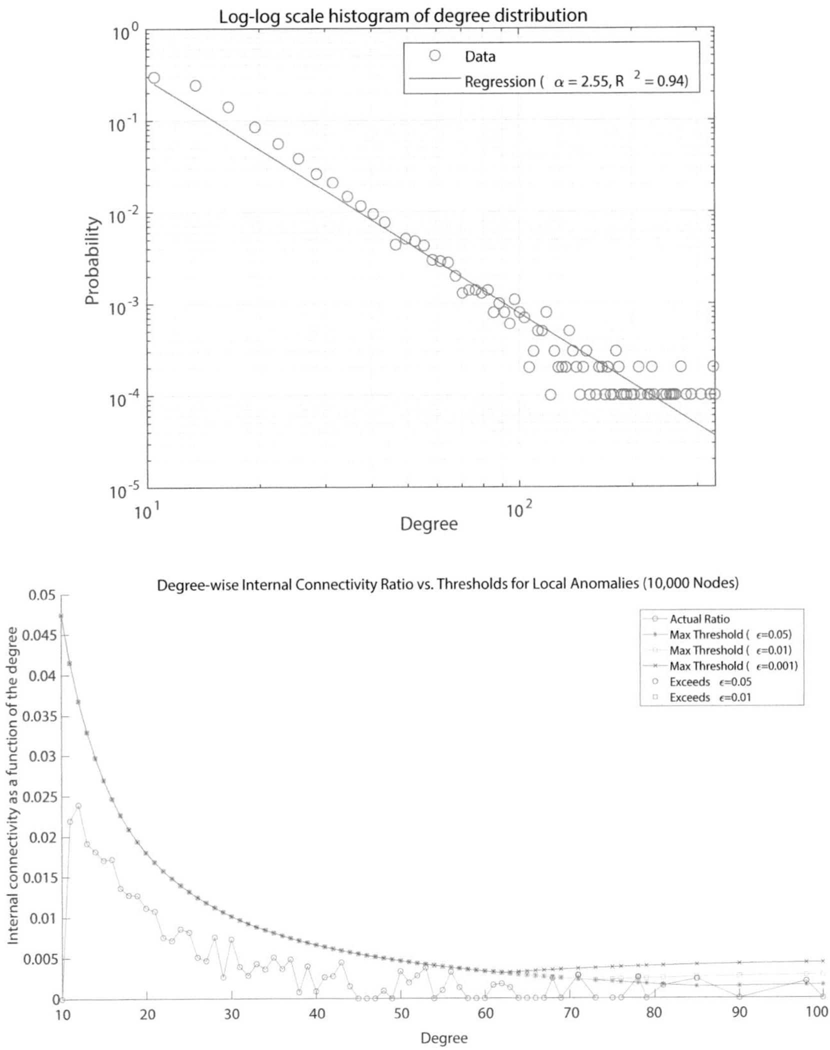

Figure 6.20: Illustration of Theorem 6.1 contrasting the probabilistic limit on the emergence of clusters of nodes with the same degree that exhibit high internal connectivity, against the internal connectivity observed in a randomly generated graph G using the Preferential Attachment method [66] for 10,000 nodes. The upper chart shows the graph's degree distribution, affirming its adherence to power-law dynamics. The chart below illustrates the theoretical constraints on the internal connectivity of nodes of degree d across all observed degrees in graph G, for varying statistical confidence levels $\varepsilon = 0.05, 0.01, 0.001$. As anticipated, a more stringent ε results in fewer degrees that can form node groups with internal connectivity exceeding the predicted bounds.

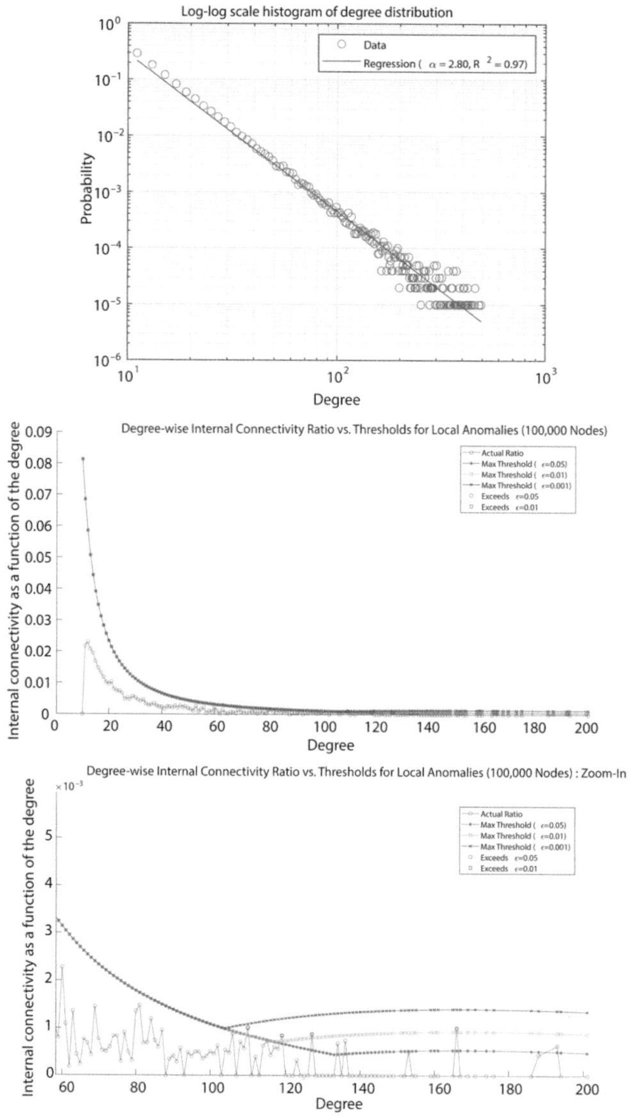

Figure 6.21: An illustration of Theorem 6.1 contrasting the probabilistic boundary regarding the appearance of node clusters of the same degree with elevated internal connectivity, with the observed connectivity in a graph G of 100,000 nodes, generated via the Preferential Attachment method [66]. The top chart displays the graph's degree distribution, reinforcing its alignment with power-law behavior. The middle chart delineates the theoretical limits on the internal connectivity for nodes of degree d throughout all present degrees in $(G$, set against distinct statistical confidence levels $\varepsilon = 0.05, 0.01, 0.001$. As expected, a tighter ε narrows the scope of degrees where node clusters exceed the anticipated connectivity values. The bottom chart offers a magnified perspective on the data from the middle chart.

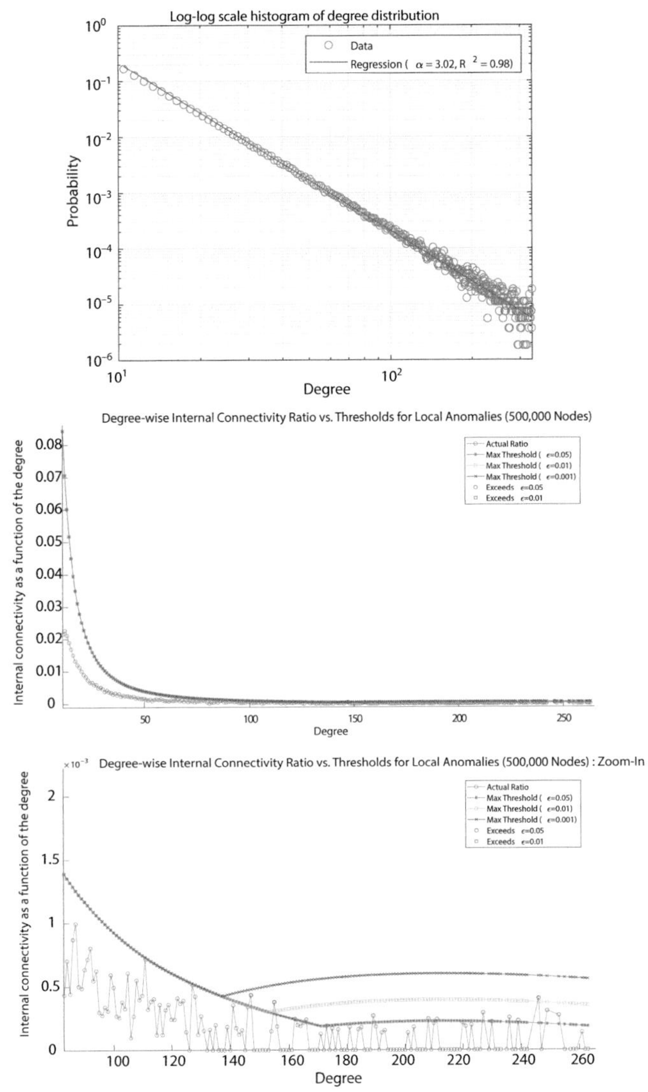

Figure 6.22: An illustration of Theorem 6.1 contrasting the probabilistic boundary regarding the appearance of node clusters of the same degree with elevated internal connectivity, with the observed connectivity in a graph G of 500,000 nodes, generated via the Preferential Attachment method [66]. The top chart displays the graph's degree distribution, reinforcing its alignment with power-law behavior. The middle chart delineates the theoretical limits on the internal connectivity for nodes of degree d throughout all present degrees in G, set against distinct statistical confidence levels $\varepsilon = 0.05, 0.01, 0.001$. As expected, a tighter ε narrows the scope of degrees where node clusters exceed the anticipated connectivity values. The bottom chart offers a magnified perspective on the data from the middle chart.

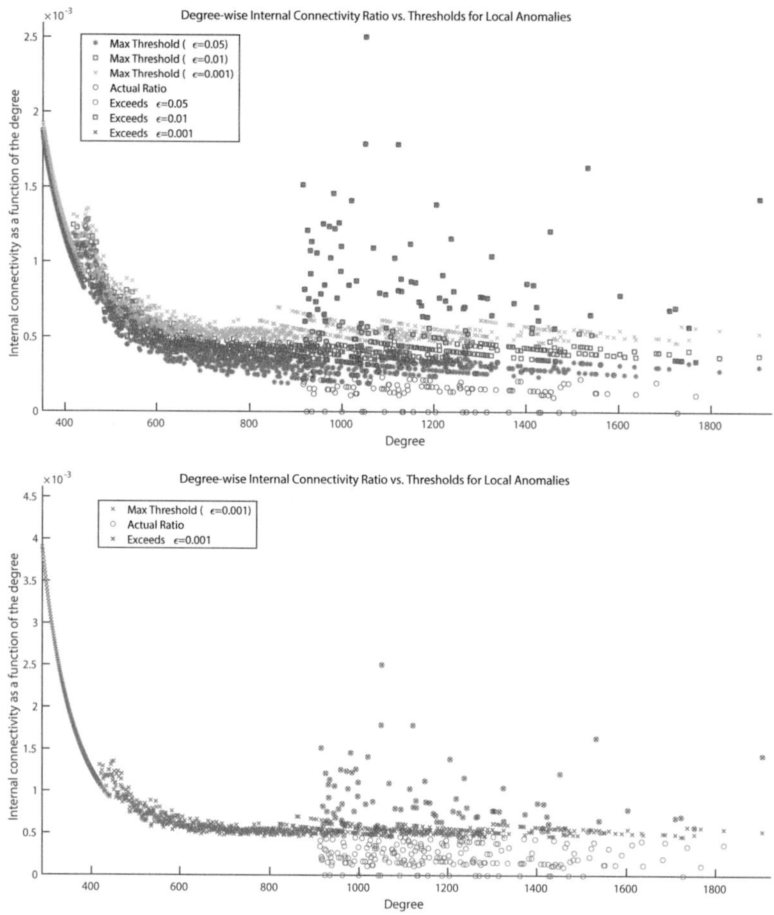

Figure 6.23: Demonstrating Theorem 6.1 using real-world data. This figure contrasts the theoretical probabilistic boundaries for the emergence of node clusters with similar degrees exhibiting heightened internal connectivity against actual connectivity observed in a graph G, which represents k-mers (with $k = 30$) co-occurring in the same reads from a microbiome sample of an arbitrarily selected cow. The top chart illustrates the theoretical bounds for varying levels of statistical significance, with ε values set at 0.05, 0.01, and 0.001, plotted alongside the connectivity ratio across different degrees. The bottom chart provides a focused view on the most stringent scenario, where $\varepsilon = 0.001$. Key structural parameters derived from the data include: $d_{\text{min}} = 203$, $d_{\text{max}} = 3559$, $|E_\star| = 118,862,706$, $\alpha = 4.8$, and $d_{\text{critical}} = 912$. The data reveals a substantial number of degrees where the nodes' connectivity significantly surpasses the theoretical threshold, indicative of statistically significant clusters of k-mers. Specifically, there are 151 clusters at $\varepsilon = 0.05$ (significantly higher than the approximately 20 expected in a comparable random dataset), 110 clusters at $\varepsilon = 0.01$ (far exceeding the roughly 4 anticipated in random data), and 76 clusters at $\varepsilon = 0.001$ (compared to none expected in random data). This robust demonstration underscores the utility of Theorem 6.1 in discerning meaningful biological structures within complex genetic data.

- **Relevancy Determination:** It is rather straightforward to assess the significance of a k-mer cluster concerning a particular phenotypic trait. This can be achieved by juxtaposing the frequency of its manifestation across various samples against the intensity or prevalence of the attribute in question.
- **Conditional Dynamics:** The pertinence of k-mer clusters is not static. For instance, a cluster deemed irrelevant to a specific property during certain periods (like summer) might emerge as highly pertinent in different settings (say, during winter, when the biological functionality associated with it becomes more relevant).
- **Evolution of the Repository:** As we continuously enrich our database with k-mer clusters, each endorsed by empirical evidence to bear some functional significance, the value of this repository amplifies. This augmentation manifests in two dimensions: first, the broad spectrum of biological attributes it can shed light on, and second, the precision with which it can make such predictions.

In essence, as we cultivate this burgeoning collection of functionally significant k-mer clusters, we enhance our capacity to decipher an array of biological phenomena with increased accuracy and breadth.

Illustrated in Figures 6.24 and 6.25 are two distinct k-mer clusters extracted from microbiome samples, using $\varepsilon = 0.001$. These clusters present anomalies within the microbiome k-mer network, indicating a high likelihood of their association with a particular biological trait. While one cluster does not show a significant statistical relationship with the performance of the feed additive Agolin, the other emerges as a compelling correlate. This strong association enables us to employ its presence in microbiome samples as a reliable marker for anticipating the enhanced effectiveness of the additive.

It's worth noting that the cluster depicted in Figure 6.24, which lacks correlation with Agolin's efficacy (as do many other clusters we extracted from the microbiome data), might still prove crucial in relation to other feed additives. Alternatively, this cluster could be relevant to other properties, unrelated to methane, that we might be keen on predicting in the future.

For the creation of the markers used in this study we have produced the following networks:

- A dedicated network for each cow.
- A dedicated network for each farm, created as the superposition of the networks of the cows of this farm.
- A network of the entire data, created as the superposition of the dedicated cow networks.

For each of the constructed networks, we employed Theorem 6.1 to identify and extract all anomalous 30-mer clusters at a statistical confidence level of $\varepsilon =$

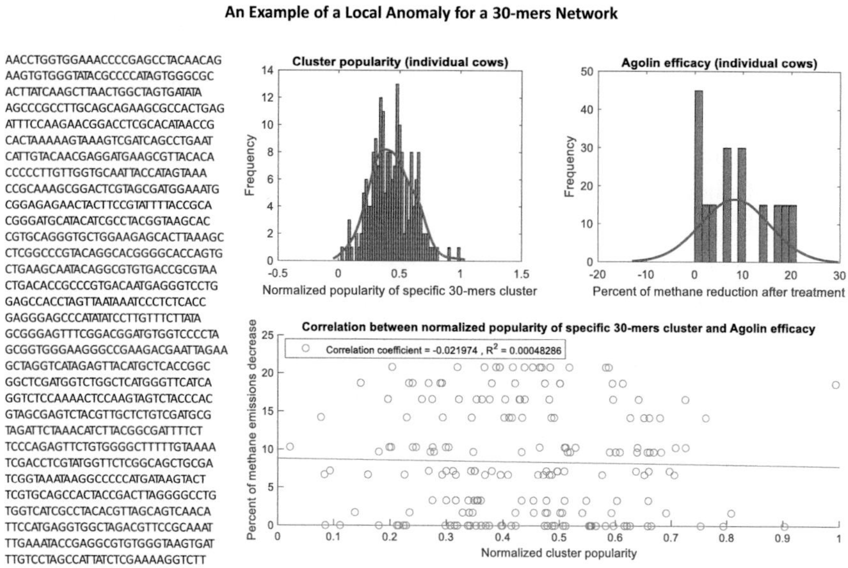

Figure 6.24: This illustration showcases a cluster of 32 k-mers (with $k = 30$) derived from combined microbiome samples of all collected samples, integrated into a single network. In this network, built using edges that represent the co-occurrence of k-mers in shared reads, all these k-mers exhibit the same degree. Their internal connectivity surpasses the threshold stipulated by Theorem 6.1, which suggests that such a pattern is improbable to arise spontaneously in a power-law network. The chart on the top-left visualizes the cluster's prevalence among the Microbiome Train cows, assessed by the frequency of k-mer appearances in their samples, normalized against the average frequency. In contrast, the top-right chart represents the efficacy of the Agolin additive, determined for the Methane Train cows that originate from the same farms as the Microbiome Train cows. Here, each Microbiome Train cow's efficacy is linked to its farm's average efficacy, calculated from the corresponding Methane Train cows. The bottom chart juxtaposes these two datasets in a scatter plot, seeking a potential linear relationship. Notably, it reveals an absence of linear correlation between the abundance of this k-mer cluster and Agolin's effectiveness.

0.001. This comprehensive extraction yielded a substantial collection of over 20,000 k-mer clusters that exhibited notable statistical significance.

To pinpoint clusters relevant to each feed additive, we undertook a rigorous filtering process. We sought out clusters that exhibited a strong linear correlation with the additive's efficacy, demanding both a correlation coefficient having an absolute value of at least 0.75 and an R^2 value surpassing 0.7. Through this meticulous filtering, we discerned between 17 to 31 distinct clusters that met our criteria for each specific additive.

We synthesized our results into two markers for each additive. Clusters that demonstrated a positive linear relationship with the additive's efficacy were

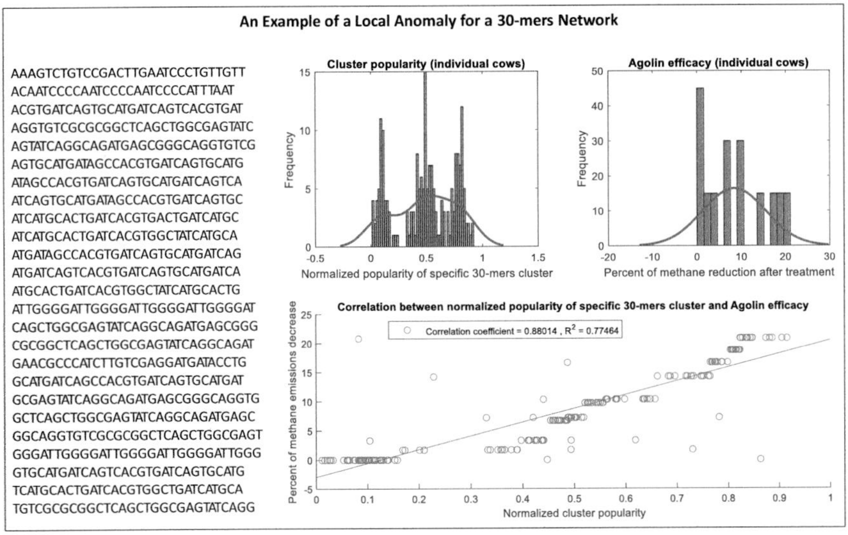

Figure 6.25: This illustration showcases a cluster of 25 k-mers (with $k = 30$) derived from combined microbiome samples of all collected samples, integrated into a single network. In this network, built using edges that represent the co-occurrence of k-mers in shared reads, all these k-mers exhibit the same degree. Their internal connectivity surpasses the threshold stipulated by Theorem 6.1, which suggests that such a pattern is improbable to arise spontaneously in a power-law network. The chart on the top-left visualizes the cluster's prevalence among the Microbiome Train cows, assessed by the frequency of k-mer appearances in their samples, normalized against the average frequency. In contrast, the top-right chart represents the efficacy of the Agolin additive, determined for the Methane Train cows that originate from the same farms as the Microbiome Train cows. Here, each Microbiome Train cow's efficacy is linked to its farm's average efficacy, calculated from the corresponding Methane Train cows. The bottom chart contrasts these two datasets in a scatter plot, revealing a discernible statistical association. Specifically, it illustrates that the efficacy of Agolin in farms exhibits a linear rise–with significant statistical backing–as the prevalence of this cluster's k-mers increases among the cows of these farms.

combined to form what we term the "top marker". Conversely, clusters that exhibited a negative linear association were amalgamated into the "bottom marker". During this merging process, we prioritized k-mers that were consistently recurrent, retaining only those that appeared in a minimum of two clusters of the same typology.

6.6 Discussion

The study implemented a novel methodology that utilizes the cow microbiome as a basis for predicting the efficacy of methane-reducing additives. The structure of the trial followed a two-step approach: initially, we collected microbiome samples from cattle across an array of farms. These farms were chosen to represent a broad spectrum of conditions, including geographical location, environmental context, herd size, and differing management methods. The genetic data obtained from these samples was then analyzed via unsupervised learning to discern meaningful microbial genetic patterns. In the subsequent stage, a selected subset of cows from these farms was subjected to methane emission measurements under both control and experimental conditions (the latter involving the feed additive). This stage utilized supervised learning, building upon the established microbial baseline from the first step.

The design of our study, combined with the unsupervised genetic material detection methodology, can be extended as a predictive framework for various other properties, given appropriate labeling for the supervised learning stage. This work highlights the positive network effect of this method: as more data are collected and incorporated into the unsupervised learning stage, the accuracy of the predictions improves. When new use-cases are introduced, they contribute additional microbiome samples, further improving the predictive capability across all use-cases.

The division of cows into different groups (e.g. "control methane cows", "train methane cows", and "test methane cows") was a critical part of our study design. It was done randomly, but with meticulous attention to mitigate any potential biases. Several key factors were taken into account, including the age of cows, their days in lactation, and average milk yield. By using these factors to inform our random selection, we were able to create groups that were representative of the general population of the herd. This ensured that the microbiome samples and methane measurements taken from each group were not skewed by any particular attribute, enhancing the generalizability and applicability of our findings.

Macroscopic Microbial Data Properties

Our analysis revealed a striking and noteworthy pattern in the distribution of the repetition of the reads and the k-mers. Specifically, we observed that the popularity of these repetitions follows a power-law distribution. This observation was consistent and sustained, evident in the data extracted from multiple cows, and across both read and k-mer repetitions.

It is crucial to underscore that the emergence of this power-law distribution is data-agnostic. This means that it is not an artifact of our data acquisition process or specific to the type of data being studied. Indeed, the power-law distribution we observed suggests the existence of underlying principles governing these distributions, which extend beyond the specifics of our study.

One might suggest that the 'shotgun' metagenomic sequencing method, which we employed to sequence the DNA samples, could somehow skew the data towards this power-law distribution. However, this is highly unlikely. While the sequencing method can indeed introduce certain biases into the data, there is no reason to suspect that it would give rise to a power-law distribution. This is because sequencing, by its nature, is a pseudo random process and, as such, is not predisposed towards creating any specific pattern, much less one as distinctive as a power-law distribution.

Moreover, when we pooled the data from several cows together and then analyzed the distribution of popularity, we consistently observed the power-law distribution. This further strengthens our assertion that the power-law distribution is not an artifact of data acquisition. If it were, the distribution would be different for each cow, reflecting the individual variances in data collection. The consistent manifestation of the power-law distribution across different cows suggests a more profound, universal principle at work.

Interestingly, we observed the power-law distribution in the repetition of k-mers, as well as whole reads. This consistency across different levels of granularity in the data adds another layer of validation to our findings. The emergence of a power-law distribution at different scales strongly suggests the existence of scale-free, fractal-like patterns in the data, which is a key characteristic of power-law distributions.

Taken together, our findings strongly suggest that the observed power-law distribution is not a mere artifact of our methodology, but rather points to fundamental biological phenomena. This could have profound implications for our understanding of the cow microbiome and the factors influencing the efficacy of methane-reducing additives. Further research is required to fully understand the implications of these findings, but they present promising avenues for future investigations.

Limitations

Despite the promising results of our study, we must also recognize its limitations. Firstly, while our methodology attempted to mitigate the potential for bias, with the division of cows into "control methane cows", "train methane cows", and "test methane cows" carefully executed to take into account factors such as the age of cows, their days in lactation, and average milk yield, some bias could still remain. Individual differences in cow physiology, diet, and environmental factors could potentially affect both the composition of the microbiome and the methane production. Future research should consider these factors in more depth, perhaps by introducing stratification or covariate balancing.

Moreover, while our study design can be extended as a predictive framework for various other properties, it is also reliant on the availability of accurate labeling data. Inadequate or inaccurately labeled data during the supervised learning stage could affect the model's predictive capability. The accuracy of our predictions is likely to improve with more data, highlighting the need for continued and broad-scale data collection efforts.

Lastly, our analysis primarily focused on the methane reduction potential of additives. It would be worthwhile to consider other potential effects of these additives on cow health, productivity, and the quality of dairy products.

Recognizing these limitations is vital to guide future research. While our study provides a significant step forward in the use of the cow microbiome to predict the efficacy of methane-reducing additives, there is still much work to be done to refine our understanding and improve our predictive capabilities.

Impact

Our results were promising, demonstrating a correlation between certain microbiome patterns and the efficacy of methane-reducing additives. This allows us to predict the impact of additives based on a cow's microbiome, enhancing the average effect by using the additive only when it is expected to be effective.

The potential impact of these findings is substantial. Given the ongoing need to reduce methane emissions for environmental reasons, the ability to predict the effectiveness of methane-reducing additives at an individual cow level could lead to significant improvements in emission reduction strategies.

Moreover, it is important to emphasize a few additional points. Even though methane measurements were only conducted at some farms, the broad-scale collection of microbiome data enriched the initial unsupervised learning phase, enhancing the detection of significant microbial genetic patterns. There was no overlap between the cow groups from which microbiomes were sampled and those from which methane was measured, affirming that a small microbiome sample is representative of the entire herd's microbiome.

Our approach rests on the assumption of strong microbiome homogeneity across the cows within a single herd. This means that while our microbiome samples were collected from one group of cows, we consider these samples as indicative of the microbiome composition of the entire herd. The validity of this assumption is confirmed when we successfully use these microbiome samples to predict the efficacy of the methane-reducing additives in a distinct group of cows within the same herd. This suggests that despite no overlap in the cows from which the microbiome and methane measurements were taken, our microbiome sample effectively represents the herd's overall microbiome. It's this feature that makes our method both robust and scalable, as it allows for a comprehensive prediction model without requiring invasive and extensive sampling from every cow.

Finally, while the initial implementation of our methodology was conducted in Israel, its design and underlying principles suggest a wide-ranging applicability. This makes it a valuable tool for dairy farming operations worldwide, particularly in the context of sustainable agriculture and environmental management.

Future Work

Our study opens up several intriguing avenues for future research. While we established a novel methodology for predicting the efficacy of methane-reducing additives, the potential of using the cow microbiome as a predictive tool extends beyond this singular application. In the future, our methodology could be employed to predict other biologically significant outcomes based on microbiome composition, such as disease susceptibility or productivity measures like milk yield and growth rate. Such predictive models could aid in early intervention strategies and improve the overall health and productivity of livestock.

Further, our research indicated that a small microbiome sample is representative of the entire herd's microbiome. This paves the way for non-invasive, rapid and cost-effective microbiome sampling techniques that could be used at scale. Developing these techniques and validating their efficacy would be a valuable next step.

Finally, while our study focused primarily on the methane reduction potential of additives, it would be worthwhile to investigate other potential effects of these additives. Are there any unintended consequences of their use, such as changes to cow health or milk quality? Understanding the holistic impact of these additives will provide a more comprehensive view and aid in making informed decisions regarding their widespread use.

Overall, this research has laid the groundwork for an exciting new direction in livestock management. Through continued refinement and expansion of this work, we envision a future where personalized additive regimens of the farm level, informed by each herd's dynamic microbiome, become a standard practice, optimizing both environmental impact and productivity outcomes.

References

[1] R. Milo, S. Shen-Orr, S. Itzkovitz, N. Kashtan, D. Chklovskii, and U. Alon. Network motifs: simple building blocks of complex networks. *Science*, 298(5594): 824–827, 2002.

[2] B.A. Methe, K.E. Nelson, M. Pop, H.H. Creasy, M.G. Giglio, C. Huttenhower, et al. A framework for human microbiome research. *Nature*, 486(7402): 215–221, 2012.

[3] C. Huttenhower, D. Gevers, R. Knight, S. Abubucker, J.H. Badger, A. Chinwalla, et al. Structure, function and diversity of the healthy human microbiome. *Nature*, 486(7402):207–214, 2012.

[4] Milk production. *National Agricultural Statistics Service*, 2023.

[5] T. Adesogan, W. Yang, C. Lee, P.J. Gerber, B. Henderson, et al. Mitigation of methane and nitrous oxide emissions from animal—special topics. *Journal of Animal Sciences*, 91: 5045–5069, 2013.

[6] J.S. Ahn, J.S. Shin, G.H. Son, S.S. Jang, B.K. Park, et al. Effect of allicin and illite supplementation on the methane production and growth performance of the beef cattle. *Indian Journal of Animal Research*, 1: 6, 2022.

[7] T. Altshuler, Y. Altshuler, R. Katoshevski and Y. Shiftan. Modeling and prediction of ride-sharing utilization dynamics. *Journal of Advanced Transportation*, 2019.

[8] Y. Altshuler, N. Aharony, A. Pentland, Y. Elovici, M. Cebrian, et al. Stealing reality: When criminals become data scientists (or vice versa). *Intelligent Systems, IEEE*, 26(6): 22–30, Nov.–Dec. 2011.

[9] Y. Altshuler, T. Chebach and S. Cohen. From microbes to methane: Ai-based predictive modeling of feed additive efficacy in dairy cows. *Arxiv preprint*, 2023.

[10] J.A.D. Ranga Niroshan Appuhamy, A.B. Strathe, S. Jayasundara, C. Wagner-Riddle, J. Dijkstra, et al. Anti-methanogenic effects of monensin in dairy and beef cattle: A meta-analysis. *Journal of Dairy Science*, 96(8): 5161–5173, 2013.

[11] A.-L. Barabasi and Z.N. Oltvai. Network biology: understanding the cell's functional organization. *Nature Reviews Genetics*, 5(2): 101–113, 2004.

[12] K.A. Beauchemin, M. Kreuzer, F. O'mara and T.A. McAllister. Nutritional management for enteric methane abatement: A review. *Australian Journal of Experimental Agriculture*, 48(2): 21–27, 2008.

[13] R. Bhatta and O. Enishi. Measurement of methane production from ruminants. *Asian-australasian Journal of Animal Sciences*, 20(8): 1305–1318, 2007.

[14] G.J. Bishop-Hurley, D. Paull, P. Valencia, L. Overs, K. Kalantar-zadeh, et al. Intra-ruminal gas-sensing in real time: a proof-of-concept. *Animal Production Science*, 56(3): 204–212, 2016.

[15] J.M. Bland and D.G. Altman. Statistical methods for assessing agreement between two methods of clinical measurement. *The lancet*, 327(8476): 307–310, 1986.

[16] T.M. Boland, K.M. Pierce, A.K. Kelly, D.A. Kenny, M.B. Lynch, et al. Feed intake, methane emissions, milk production and rumen methanogen populations of grazing dairy cows supplemented with various c 18 fatty acid sources. *Animals*, 10(12): 2380, 2020.

[17] J. Borlinghaus, F. Albrecht, M.C.H. Gruhlke, I.D. Nwachukwu, A.J. Slusarenko, et al. Allicin: chemistry and biological properties. *Molecules*, 19(8): 12591–12618, 2014.

[18] R. Brambila and J. Noricumbo-Saenz. Evaluation of agolin® ruminant, an essential oil blend, as a feed additive for cows at two levels of production. *Open Journal of Animal Sciences*, 12(3): 380–389, 2022.

[19] A.V. Carrazco, C.B. Peterson, Y. Zhao, Y. Pan, J.J. McGlone, et al. The impact of essential oil feed supplementation on enteric gas emissions and production parameters from dairy cattle. *Sustainability*, 12(24): 10347, 2020.

[20] R. Chikhi and P. Medvedev. Informed and automated k-mer size selection for genome assembly. *Bioinformatics*, 30(1): 31–37, 2014.

[21] A. Clauset, C.R. Shalizi and M.E.J. Newman. Power-law distributions in empirical data. *SIAM Review*, 51(4): 661–703, 2009.

[22] R. Danielsson, J. Dicksved, L. Sun, H. Gonda, B. Müller, et al. Methane production in dairy cows correlates with rumen methanogenic and bacterial community structure. *Frontiers in Microbiology*, 8: 226, 2017.

[23] C.F. de Assis Lage, S.E. Räisänen, A. Melgar, K. Nedelkov, X. Chen, et al. Comparison of two sampling techniques for evaluating ruminal fermentation and microbiota in the planktonic phase of rumen digesta in dairy cows. *Frontiers in Microbiology*, 11: 618032, 2020.

[24] S. Dengel, P.E. Levy, J. Grace, S.K. Jones, U.M. Skiba, et al. Methane emissions from sheep pasture, measured with an open-path eddy covariance system. *Global Change Biology*, 17(12): 3524–3533, 2011.

[25] G.F. Difford, D.R. Plichta, P. Løvendahl, J. Lassen, S.J. Noel, et al. Host genetics and the rumen microbiome jointly associate with methane emissions in dairy cows. *PLoS Genetics*, 14(10): e1007580, 2018.

[26] T. Duffield, J.C. Plaizier, A. Fairfield, R. Bagg, G. Vessie, et al. Comparison of techniques for measurement of rumen ph in lactating dairy cows. *Journal of Dairy Science*, 87(1): 59–66, 2004.

[27] H.S. Eggleston, L. Buendia, K. Miwa, T. Ngara, K. Tanabe, et al. 2006 IPCC Guidelines for National Greenhouse Gas Inventories. 2006.

[28] European Medicine Agency. Kexxtone, 2013.

[29] J.Q. Fouts, M.C. Honan, B.M. Roque, J.M. Tricarico, E. Kebreab, et al. Enteric methane mitigation interventions. *Translational Animal Science*, 6(2): txac041, 2022.

[30] P.C. Garnsworthy, G.F. Difford, M.J. Bell, A.R. Bayat, P. Huhtanen, et al. Comparison of methods to measure methane for use in genetic evaluation of dairy cattle. *Animals*, 9(10): 837, 2019.

[31] T. Geishauser and A. Gitzel. A comparison of rumen fluid sampled by oro-ruminal probe versus rumen fistula. *Small Ruminant Research*, 21(1): 63–69, 1996.

[32] Davide Giavarina. Understanding bland altman analysis. *Biochemia Medica*, 25(2): 141–151, 2015.

[33] M.C. Gonzalez, C.A. Hidalgo and A.-L. Barabasi. Understanding individual human mobility patterns. *Nature*, 453(7196): 779–782, 06 2008.

[34] R.D. Goodrich, J.E. Garrett, D.R. Gast, M.A. Kirick, D.A. Larson, et al. Influence of monensin on the performance of cattle. *Journal of Animal Science*, 58(6): 1484–1498, 1984.

[35] J.V. Hagey, M. Laabs, E.A. Maga and E.J. DePeters. Rumen sampling methods bias bacterial communities observed. *PLoS One*, 17(5): e0258176, 2022.

[36] K.J. Hammond, D.J. Humphries, L.A. Crompton, C. Green, C.K. Reynolds, et al. Methane emissions from cattle: Estimates from short-term measurements using a greenfeed system compared with measurements obtained using respiration chambers or sulphur hexafluoride tracer. *Animal Feed Science and Technology*, 203: 41–52, 2015.

[37] K.J. Hammond, G.C. Waghorn and R.S. Hegarty. The greenfeed system for measurement of enteric methane emission from cattle. *Animal Production Science*, 56(3): 181–189, 2016.

[38] L.V. Hedges. Distribution theory for glass's estimator of effect size and related estimators. *Journal of Educational Statistics*, 6(2): 107–128, 1981.

[39] Melinda R Hess and Jeffrey D Kromrey. Robust confidence intervals for effect sizes: A comparative study of cohen'sd and cliff's delta under non-normality and heterogeneous variances. In: *Annual Meeting of The American Educational Research Association*, Volume 1. Citeseer, 2004.

[40] M. Honan, X. Feng, J.M. Tricarico and E. Kebreab. Feed additives as a strategic approach to reduce enteric methane production in cattle: Modes of action, effectiveness and safety. *Animal Production Science*, 2021.

[41] A.N. Hristov, J. Oh, F. Giallongo, T. Frederick, H. Weeks, et al. The use of an automated system (greenfeed) to monitor enteric methane and carbon dioxide emissions from ruminant animals. *JoVE (Journal of Visualized Experiments)*, (103): e52904, 2015.

[42] P. Huhtanen, M. Ramin and A.N. Hristov. Enteric methane emission can be reliably measured by the greenfeed monitoring unit. *Livestock Science*, 222: 31–40, 2019.

[43] M. Huss and P. Holme. Currency and commodity metabolites: their identification and relation to the modularity of metabolic networks. *IET Systems Biology*, 1(5): 280–285, 2007.

[44] T. Ito, K. Tashiro, S. Muta, R. Ozawa, T. Chiba, et al. Toward a protein–protein interaction map of the budding yeast: a comprehensive system to examine two-hybrid interactions in all possible combinations between the yeast proteins. *Proceedings of the National Academy of Sciences*, 97(3): 1143–1147, 2000.

[45] E. Jami, A. Israel, A. Kotser and I. Mizrahi. Exploring the bovine rumen bacterial community from birth to adulthood. *The ISME Journal*, 7(6): 1069–1079, 2013.

[46] H. Jeong, B. Tombor, R. Albert, Z.N. Oltvai, A.-L. Barabási, et al. The large-scale organization of metabolic networks. *Nature*, 407(6804): 651–654, 2000.

[47] C.-F. Johannesson, K.S. Larsen and J. Nordén. Forest soil and deadwood ch4 fluxes in response to climate change and forest management. In: *EGU General Assembly Conference Abstracts*, pages EGU22–7061, 2022.

[48] K.A. Johnson and D.E. Johnson. Methane emissions from cattle. *Journal of Animal Science*, 73(8): 2483–2492, 1995.

[49] E. Kebreab, K. Clark, C. Wagner-Riddle and J. France. Methane and nitrous oxide emissions from canadian animal agriculture: A review. *Canadian Journal of Animal Science*, 86(2): 135–157, 2006.

[50] M.R. Kekana, D. Luseba and M.C. Muyu. Effects of garlic supplementation on in vitro nutrient digestibility, rumen fermentation, and gas production. *South African Journal of Animal Science*, 51(2): 271–279, 2021.

[51] Piotr Korben. *Estimation of Methane Emissions and Investigation of Isotopic Composition of Methane from Selected Sources in Germany, Poland and Romania*. PhD Thesis, 2021.

[52] M. Kowarsky, J. Camunas-Soler, M. Kertesz, I.D. Vlaminck, W. Koh, et al. Numerous uncharacterized and highly divergent microbes which colonize humans are revealed by circulating cell-free dna. *Proceedings of the National Academy of Sciences*, 114(36): 9623–9628, 2017.

[53] M. Layeghifard, D.M. Hwang and D.S. Guttman. Disentangling interactions in the microbiome: A network perspective. *Trends in Microbiology*, 25(3): 217–228, 2017.

[54] D. Lazer, A. Pentland, L. Adamic, S. Aral, A.-L. Barabasi, et al. Social science: Computational social science. *Science*, 323(5915): 721–723, 2009.

[55] L. Li, T. Wang, Z. Ning, X. Zhang, J. Butcher, et al. Revealing proteome-level functional redundancy in the human gut microbiome using ultra-deep metaproteomics. *Nature Communications*, 14(1): 3428, 2023.

[56] K.J. Locey and J.T. Lennon. Scaling laws predict global microbial diversity. *Proceedings of the National Academy of Sciences*, 113(21): 5970–5975, 2016.

[57] T. Ma, D. Chen, Y. Tu, N. Zhang, B. Si, et al. Effect of supplementation of allicin on methanogenesis and ruminal microbial flora in dorper crossbred ewes. *Journal of Animal Science and Biotechnology*, 7(1): 1–7, 2016.

[58] J.L. Marumo, P.A. LaPierre and M.E. Van Amburgh. Enteric methane emissions prediction in dairy cattle and effects of monensin on methane emissions: A meta-analysis. *Animals*, 13(8): 1392, 2023.

[59] D. McDermitt, G. Burba, L. Xu, T. Anderson, A. Komissarov, et al. A new low-power, open-path instrument for measuring methane flux by eddy covariance. *Applied Physics B*, 102: 391–405, 2011.

[60] A. Melgar, C.F.A. Lage, K. Nedelkov, S.E. Räisänen, H. Stefenoni, et al. Enteric methane emission, milk production, and composition of dairy cows fed 3-nitrooxypropanol. *Journal of Dairy Science*, 104(1): 357–366, 2021.

[61] M.E. Mercadante, A.P. de Melo Caliman, R.C. Canesin, S.F.M. Bonilha, A. Berndt, et al. Relationship between residual feed intake and enteric methane emission in nellore cattle. *Revista Brasileira de Zootecnia*, 44: 255–262, 2015.

[62] I. Mizrahi, R.J. Wallace and S. Moraïs. The rumen microbiome: balancing food security and environmental impacts. *Nature Reviews Microbiology*, 19(9): 553–566, 2021.

[63] S. Moraïs and I. Mizrahi. Islands in the stream: from individual to communal fiber degradation in the rumen ecosystem. *FEMS Microbiology Reviews*, 43(4): 362–379, 04 2019.

[64] W. Muizelaar, P. Bani, B. Kuhla, M. Larsen, I. Tapio, et al. Rumen fluid sampling via oral stomach tubing method. 2020.

[65] N. Nachar. The mann-whitney u: A test for assessing whether two independent samples come from the same distribution. *Tutorials in Quantitative Methods for Psychology*, 4(1): 13–20, 2008.

[66] M.E.J. Newman. Clustering and preferential attachment in growing networks. *Physical Review E*, 64(2): 025102, 2001.

[67] J.E. Nocek. Bovine acidosis: Implications on laminitis. *Journal of Dairy Science*, 80(5): 1005–1028, 1997.

[68] M. Odai, I. Nonaka, W. Sumamal, R. Narmsilee, P. Pholsen, et al. Estimation of methane production by lactating and dry crossbred holstein cows in thailand. *Japan Agricultural Research Quarterly: JARQ*, 44(4): 429–434, 2010.

[69] A. Pentland and Y. Altshuler. *Social Physics and Cybercrime*, chapter New Solutions for Cybersecurity, pages 351–364. MIT Press, 2018.

[70] Phibro. Phibro data 2021. Available on request, 2021.

[71] S.E. Place, Y. Pan, Y. Zhao and F.M. Mitloehner. Construction and operation of a ventilated hood system for measuring greenhouse gas and volatile organic compound emissions from cattle. *Animals*, 1(4): 433–446, 2011.

[72] Z. Qi, Y. Yi, W. Qiaohua, W. Zhihao, L. Zhe, et al. Study on the online dissolved gas analysis monitor based on the photoacoustic spectroscopy. In: *2012 IEEE International Conference on Condition Monitoring and Diagnosis*, pages 433–436. IEEE, 2012.

[73] Y. Ramayo-Caldas, L. Zingaretti, M. Popova, J. Estellé, A. Bernard, et al. Identification of rumen microbial biomarkers linked to methane emission in holstein dairy cows. *Journal of Animal Breeding and Genetics*, 137(1): 49–59, 2020.

[74] E. Ramos-Morales, A. Arco-Pérez, A.I. Martín-García, D.R. Yáñez-Ruiz, P. Frutos, et al. Use of stomach tubing as an alternative to rumen cannulation to study ruminal fermentation and microbiota in sheep and goats. *Animal Feed Science and Technology*, 198: 57–66, 2014.

[75] J. Rey, R. Atxaerandio, R. Ruiz, E. Ugarte, O. González-Recio, et al. Comparison between non-invasive methane measurement techniques in cattle. *Animals*, 9(8): 563, 2019.

[76] A.A. Ribeiro, L.A. Pinedo, L.-D.C. Codognoto, J. Cavali, M.O. Porto, et al. Comparison of methods to measure enteric methane emissions from ruminants: an integrative review. *Research, Society and Development*, 9(11): e8259118143–e8259118143, 2020.

[77] R. Rosenthal, H. Cooper and L. Hedges. Parametric measures of effect size. *The Handbook of Research Synthesis*, 621(2): 231–244, 1994.

[78] C.T. Roulston. Emissions of fine particulate matter from tropical peat fires. 2018.

[79] S.A. Mootral. Vm0041 methodology for the reduction of enteric methane emissions from ruminants through the use of feed ingredients. *Verra*, 2021.

[80] G. Sasson, S.K. Ben-Shabat, E. Seroussi, A. Doron-Faigenboim, N. Shterzer, et al. Heritable bovine rumen bacteria are phylogenetically related and correlated with the cow's capacity to harvest energy from its feed. *mBio*, 8(4): 10.1128/mbio.00703–17, 2017.

[81] R. Sender, S. Fuchs and R. Milo. Revised estimates for the number of human and bacteria cells in the body. *PLoS Biology*, 14(8): e1002533, 2016.

[82] Y. Shaani, T. Zehavi, S. Eyal, J. Miron, I. Mizrahi, et al. Microbiome niche modification drives diurnal rumen community assembly, overpowering individual variability and diet effects. *The ISME Journal*, 12(10): 2446–2457, 2018.

[83] A. Shah, O. Laurent, L. Lienhardt, G. Broquet, R.R. Martinez, et al. Characterising the methane gas and environmental response of the figaro taguchi gas sensor (tgs) 2611-e00. *Atmospheric Measurement Techniques*, 16(13): 3391–3419, 2023.

[84] J.S. Shen, Z. Chai, L.J. Song, J.X. Liu, Y.M. Wu, et al. Insertion depth of oral stomach tubes may affect the fermentation parameters of ruminal fluid collected in dairy cows. *Journal of Dairy Science*, 95(10): 5978–5984, 2012.

[85] E. Shmueli, I. Mazeh, L. Radaelli, A.S. Pentland, Y. Altshuler, et al. Ride sharing: A network perspective. In: *Social Computing, Behavioral-Cultural Modeling, and Prediction*, pages 434–439. Springer, 2015.

[86] S. Somin, Y. Altshuler, G. Gordon, A. Pentland, E. Shmueli, et al. network dynamics of a financial ecosystem. *Scientific Reports*, 10(1): 1–10, 2020.

[87] S. Somin, Y. Altshuler, A.S. Pentland and E. Shmueli. Beyond preferential attachment: falling of stars and survival of superstars. *Royal Society Open Science*, 9(8): 220899, 2022.

[88] S. Somin, Y. Altshuler and E. Shmueli. Remaining popular: power-law regularities in network dynamics. *EPJ Data Science*, 11(1): 61, 2022.

[89] I.M.L.D. Storm, A.L.F. Hellwing, N.I. Nielsen and J. Madsen. Methods for measuring and estimating methane emission from ruminants. *Animals*, 2(2): 160–183, 2012.

[90] T. Takano and M. Ueyama. Spatial variations in daytime methane and carbon dioxide emissions in two urban landscapes, sakai, Japan. *Urban Climate*, 36: 100798, 2021.

[91] Z. Tao, C. Chen, Q. Yang, Z.Zhong, Y. Wan, S. Chen, et al. Understanding the impact of allicin for organic matter release and microorganism community in anaerobic co-digestion of food waste and waste activated sludge. *Science of The Total Environment*, 776: 145598, 2021.

[92] C. J. Van Nevel and D.I. Demeyer. Effect of monensin on rumen metabolism in vitro. *Applied and Environmental Microbiology*, 34(3): 251–257, 1977.

[93] H. Vranken, M. Suenkel, P.R. Hargreaves, L. Chew, E. Towers, et al. Reduction of enteric methane emission in a commercial dairy farm by a novel feed supplement. *Open Journal of Animal Sciences*, 9: 286–296, 2019.

[94] R.J. Wallace, J.A. Rooke, N. McKain, C.-A. Duthie, J.J. Hyslop, et al. The rumen microbial metagenome associated with high methane production in cattle. *BMC Genomics*, 16(1): 1–14, 2015.

[95] M. Wang, R. Wang, P.H. Janssen, X.M. Zhang, X.Z. Sun, et al. Sampling procedure for the measurement of dissolved hydrogen and volatile fatty acids in the rumen of dairy cows. *Journal of Animal Science*, 94(3): 1159–1169, 2016.

[96] X.-W. Wang, Z. Sun, H. Jia, S. Michel-Mata, M.T. Angulo, et al. Identifying keystone species in microbial communities using deep learning. *Biorxiv*, pages 2023–03, 2023.

[97] J.I. Weller, E. Ezra, E. Seroussi and M. Gershoni. Genetic and genomic analysis of cow mortality in the israeli holstein population. *Genes*, 14(3): 588, 2023.

[98] S. Wolfram. *What Is ChatGPT Doing... and Why Does It Work?* Stephen Wolfram, 2023.

[99] L. Xu and R. Vath. Chamber-based soil methane flux measurement system. In: *AGU Fall Meeting Abstracts*, Volume 2018, pages B41H–2824, 2018.

[100] M. Zwang, S. Somin, A.S. Pentland and Y. Altshuler. Detecting bot activity in the ethereum blockchain network, 2018.

Index